Weibull Models

WILEY SERIES IN PROBABILITY AND STATISTICS

Established by WALTER A. SHEWHART and SAMUEL S. WILKS

A complete list of the titles in this series appears at the end of this volume.

Weibull Models

D. N. PRABHAKAR MURTHY

University of Queensland
Department of Mechanical Engineering
Brisbane, Australia

MIN XIE

National University of Singapore
Department of Industrial and Systems Engineering
Kent Ridge Crescent, Singapore

RENYAN JIANG

University of Toronto
Department of Mechanical and Industrial Engineering
Toronto, Ontario, Canada

WILEY-
INTERSCIENCE

A JOHN WILEY & SONS, INC., PUBLICATION

Published by John Wiley & Sons, Inc., Hoboken, New Jersey.
Published simultaneously in Canada.

Library of Congress Cataloging-in-Publication Data:

Murthy, D. N. P.
Weibull models / D.N. Prabhakar Murthy, Min Xie, Renyan Jiang.
p. cm. – (Wiley series in probability and statistics)
Includes bibliographical references and index.
ISBN 0-471-36092-9 (cloth)
1. Weibull distribution. I. Xie, M. (Min) II. Jiang, Renyan, 1956-
III. Title. IV. Series.

QA273.6.M87 2004
519.2′4–dc21 2003053450

Printed in the United States of America.

10 9 8 7 6 5 4 3 2 1

Dedicated to our Families

Contents

Preface xiii

PART A OVERVIEW 1

Chapter 1 Overview 3

1.1 Introduction, 3
1.2 Illustrative Problems, 5
1.3 Empirical Modeling Methodology, 7
1.4 Weibull Models, 9
1.5 Weibull Model Selection, 11
1.6 Applications of Weibull Models, 12
1.7 Outline of the Book, 15
1.8 Notes, 16
Exercises, 16

Chapter 2 Taxonomy for Weibull Models 18

2.1 Introduction, 18
2.2 Taxonomy for Weibull Models, 18
2.3 Type I Models: Transformation of Weibull Variable, 21
2.4 Type II Models: Modification/Generalization of Weibull Distribution, 23
2.5 Type III Models: Models Involving Two or More Distributions, 28
2.6 Type IV Models: Weibull Models with Varying Parameters, 30
2.7 Type V Models: Discrete Weibull Models, 33
2.8 Type VI Models: Multivariate Weibull Models, 34
2.9 Type VII Models: Stochastic Point Process Models, 37
Exercises, 39

PART B BASIC WEIBULL MODEL 43

Chapter 3 Model Analysis 45

3.1 Introduction, 45
3.2 Basic Concepts, 45
3.3 Standard Weibull Model, 50
3.4 Three-Parameter Weibull Model, 54
3.5 Notes, 55
Exercises, 56

Chapter 4 Parameter Estimation 58

4.1 Introduction, 58
4.2 Data Types, 58
4.3 Estimation: An Overview, 60
4.4 Estimation Methods and Estimators, 61
4.5 Two-Parameter Weibull Model: Graphical Methods, 65
4.6 Standard Weibull Model: Statistical Methods, 67
4.7 Three-Parameter Weibull Model, 74
Exercises, 82

Chapter 5 Model Selection and Validation 85

5.1 Introduction, 85
5.2 Graphical Methods, 86
5.3 Goodness-of-Fit Tests, 89
5.4 Model Discrimination, 93
5.5 Model Validation, 94
5.6 Two-Parameter Weibull Model, 95
5.7 Three-Parameter Weibull Model, 99
Exercises, 100

PART C TYPES I AND II MODELS 103

Chapter 6 Type I Weibull Models 105

6.1 Introduction, 105
6.2 Model I(a)-3: Reflected Weibull Distribution, 106
6.3 Model I(a)-4: Double Weibull Distribution, 108
6.4 Model I(b)-1: Power Law Transformation, 109

6.5 Model I(b)-2: Log Weibull Transformation, 111
6.6 Model I(b)-3: Inverse Weibull Distribution, 114
Exercises, 119

Chapter 7 Type II Weibull Models 121

7.1 Introduction, 121
7.2 Model II(a)-1: Pseudo-Weibull Distribution, 122
7.3 Model II(a)-2: Stacy–Mihram Model, 124
7.4 Model II(b)-1: Extended Weibull Distribution, 125
7.5 Model II(b)-2: Exponentiated Weibull Distribution, 127
7.6 Model II(b)-3: Modified Weibull Distribution, 134
7.7 Models II(b)4–6: Generalized Weibull Family, 138
7.8 Model II(b)-7: Three-Parameter Generalized Gamma, 140
7.9 Model II(b)-8: Extended Generalized Gamma, 143
7.10 Models II(b)9–10: Four- and Five-Parameter Weibulls, 145
7.11 Model II(b)-11: Truncated Weibull Distribution, 146
7.12 Model II(b)-12: Slymen–Lachenbruch Distributions, 148
7.13 Model II(b)-13: Weibull Extension, 151
Exercises, 154

PART D TYPE III MODELS 157

Chapter 8 Type III(a) Weibull Models 159

8.1 Introduction, 159
8.2 Model III(a)-1: Weibull Mixture Model, 160
8.3 Model III(a)-2: Inverse Weibull Mixture Model, 176
8.4 Model III(a)-3: Hybrid Weibull Mixture Models, 179
8.5 Notes, 179
Exercises, 180

Chapter 9 Type III(b) Weibull Models 182

9.1 Introduction, 182
9.2 Model III(b)-1: Weibull Competing Risk Model, 183
9.3 Model III(b)-2: Inverse Weibull Competing Risk Model, 190
9.4 Model III(b)-3: Hybrid Weibull Competing Risk Model, 191
9.5 Model III(b)-4: Generalized Competing Risk Model, 192
Exercises, 195

Chapter 10 Type III(c) Weibull Models **197**

10.1 Introduction, 197
10.2 Model III(c)-1: Multiplicative Weibull Model, 198
10.3 Model III(c)-2: Inverse Weibull Multiplicative Model, 203
Exercises, 206

Chapter 11 Type III(d) Weibull Models **208**

11.1 Introduction, 208
11.2 Analysis of Weibull Sectional Models, 210
11.3 Parameter Estimation, 216
11.4 Modeling Data Set, 219
11.5 Applications, 219
Exercises, 220

PART E TYPES IV TO VII MODELS **221**

Chapter 12 Type IV Weibull Models **223**

12.1 Introduction, 223
12.2 Type IV(a) Models, 224
12.3 Type IV(b) Models: Accelerated Failure Time (AFT) Models, 225
12.4 Type IV(c) Models: Proportional Hazard (PH) Models, 229
12.5 Model IV(d)-1, 231
12.6 Type IV(e) Models: Random Parameters, 232
12.7 Bayesian Approach to Parameter Estimation, 236
Exercises, 236

Chapter 13 Type V Weibull Models **238**

13.1 Introduction, 238
13.2 Concepts and Notation, 238
13.3 Model V-1, 239
13.4 Model V-2, 242
13.5 Model V-3, 243
13.6 Model V-4, 244
Exercises, 245

Chapter 14 Type VI Weibull Models (Multivariate Models) **247**

14.1 Introduction, 247

14.2 Some Preliminaries and Model Classification, 248
14.3 Bivariate Models, 250
14.4 Multivariate Models, 256
14.5 Other Models, 258
Exercises, 258

Chapter 15 Type VII Weibull Models **261**

15.1 Introduction, 261
15.2 Model Formulations, 261
15.3 Model VII(a)-1: Power Law Process, 265
15.4 Model VII(a)-2: Modulated Power Law Process, 272
15.5 Model VII(a)-3: Proportional Intensity Model, 273
15.6 Model VII(b)-1: Ordinary Weibull Renewal Process, 274
15.7 Model VII(b)-2: Delayed Renewal Process, 277
15.8 Model VII(b)-3: Alternating Renewal Process, 278
15.9 Model VII(c): Power Law–Weibull Renewal Process, 278
Exercises, 278

PART F WEIBULL MODELING OF DATA **281**

Chapter 16 Weibull Modeling of Data **283**

16.1 Introduction, 283
16.2 Data-Related Issues, 284
16.3 Preliminary Model Selection and Parameter Estimation, 285
16.4 Final Model Selection, Parameter Estimation, and Model Validation, 287
16.5 Case Studies, 290
16.6 Conclusions, 299
Exercises, 299

PART G APPLICATIONS IN RELIABILITY **301**

Chapter 17 Modeling Product Failures **303**

17.1 Introduction, 303
17.2 Some Basic Concepts, 304
17.3 Product Structure, 306
17.4 Modeling Failures, 306
17.5 Component-Level Modeling (Black-Box Approach), 306

17.6 Component-Level Modeling (White-Box Approach), 308
17.7 Component-Level Modeling (Gray-Box Approach), 312
17.8 System-Level Modeling (Black-Box Approach), 313
17.9 System-Level Modeling (White-Box Approach), 316

Chapter 18 Product Reliability and Weibull Models **324**

18.1 Introduction, 324
18.2 Premanufacturing Phase, 325
18.3 Manufacturing Phase, 332
18.4 Postsale Phase, 336
18.5 Decision Models Involving Weibull Failure Models, 341

References **348**
Index **377**

Preface

Mathematical models have been used in solving real-world problems from many different disciplines. This requires building a suitable mathematical model. Two different approaches to building mathematical models are as follows:

1. *Theory-Based Modeling* Here, the modeling is based on theories (from physical, biological, and social sciences) relevant to the problem. This kind of model is also called a *physics-based model* or *white-box model.*
2. *Empirical Modeling* Here the data available forms the basis for model building, and it does not require an understanding of the underlying mechanisms involved. This kind of model is also called as *data-dependent model* or *black-box model.*

In the black-box approach to modeling, one first carries out an analysis of the data, and then one determines the type of mathematical formulation appropriate to model the data.

Many data exhibit a high degree of variability or randomness. These kinds of data are often best modeled by a suitable probability model (such as a distribution function) so that the data can be viewed as observed outcomes (values) of random variables from the distribution.

Black-box modeling is a multistep process. It requires a good understanding of probability and of statistical inference. In probability, we use the model to make statements about the nature of the data that may result if the model is correct. This involves model analysis using analytical and simulation techniques. The principal objective of statistical inference is to use the available data to make statements about the probability model, either in terms of probability distribution itself or in terms of its parameters or some other characteristics. This involves topics such as model selection, estimation of model parameters, and model validation. As a result, probability and statistical inference may be thought of as inverses of one another as indicated below.

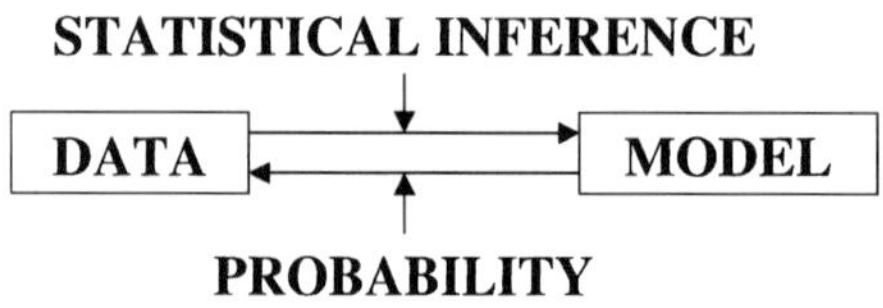

Procedures of statistical inference are the basic tools of *data analysis*. Most are based on quite specific assumptions regarding the nature of the probabilistic mechanism that gave rise to the data.

Many standard probability distribution functions (e.g., normal, exponential) have been used as models to model data exhibiting significant variability. More complex models are distributions derived from standard distributions (e.g., lognormal). One distribution of particular significance is the Weibull distribution. It is named after Professor Waloddi Weibull who was the first to promote the usefulness of this to model data sets of widely differing characteristics.

Over the last two decades several new models have been proposed that are either derived from, or in some way related to, the Weibull distributions. We use the term *Weibull models* to denote such models. They provide a richness that makes them appropriate to model complex data sets.

The literature on Weibull models is vast, disjointed, and scattered across many different journals. There are a couple of books devoted solely to the Weibull distribution, but these are oriented toward training and/or consulting purposes. There is no book that deals with the different Weibull models in an integrated manner. This book fills that gap.

The aims of this book are to:

1. Integrate the disjointed literature on Weibull models by developing a proper taxonomy for the classification of such models.
2. Review the literature dealing with the analysis and statistical inference (parameter estimation, goodness of fit) for the different Weibull models.
3. Discuss the usefulness of the Weibull probability paper (WPP) plot in the model selection to model a given data set.
4. Highlight the use of Weibull models in reliability theory.

The book would be of great interest to practitioners in reliability and other disciplines in the context of modeling data sets using univariate Weibull models. Some of the exercises at the end of each chapter define potential topics for future research. As such, the book would also be of great interest to researchers interested in Weibull models.

The book is organized into the following seven parts (Parts A to G).

Part A consists of two chapters. Chapter 1 gives an overview of the book. Chapter 2 deals with the taxonomy for Weibull models and gives the mathematical structure of the different models. The taxonomy involves seven different categories that we denote as Types I to VII.

Part B consists of three chapters. Chapter 3 deals with model analysis and discusses various model-related properties. Chapter 4 deals with parameter estimation and examines different data structures, estimation methods, and their properties. Chapter 5 deals with model selection and validation, where the focus is on deciding whether a specific model is appropriate to model a given data set or not. In these chapters many concepts and techniques are introduced, and these are used in later chapters. In these three chapters, the analysis, estimation, and validation are discussed for the standard Weibull model as well as the three-parameter Weibull model, as the two models are very similar.

Part C consists of two chapters. Chapter 6 deals with Type I models derived from nonlinear transformations of random variables from the standard Weibull model. Chapter 7 deals with Type II models, which are obtained by modifications of the standard Weibull model and in some cases involving one or more additional parameters.

Part D consists of four chapters and deals with Type III models. Chapter 8 deals with the mixture models, Chapter 9 with the competing risk models, Chapter 10 with the multiplicative models, and Chapter 11 with the sectional models.

Part E consists of four chapters. Chapter 12 deals with Type IV models, Chapter 13 with Type V models, Chapter 14 with Type VI models, and Chapter 15 with Type VII models.

In Parts C to E, for each model we review the available results (analysis, statistical inference, etc.) relating to the model.

Part F consists of a single chapter (Chapter 16) dealing with model selection to model a given data set.

Part G deals with the application of Weibull models in reliability theory and consists of two chapters. Chapter 17 deals with modeling failures. Chapter 18 discusses a variety of reliability-related decision problems in the different phases (premanufacturing, manufacturing, and postsale) of the product life cycle and reviews the literature relating to Weibull failure models.

Reliability engineers and applied statisticians involved with reliability and survival analysis should find this as a valuable reference book. It can be used as a textbook for a course on probabilistic modeling at the graduate (or advanced undergraduate) level in industrial engineering, operations research, and statistics.

We would like to thank Professor Wallace Blischke (University of Southern California) for his comments on several chapters of the book and Dr Michael Bulmer and Professor John Eccleston (University of Queensland) for their contributions to Chapter 16. Several reviewers of the book have given detailed and encouraging comments and we are grateful for their contributions. Special thanks to Steve Quigley, Heather Bergman, Susanne Steitz, and Christine Punzo at Wiley for their patience and support.

We would like to thank several funding agencies for their support over the last few years. In particular, the funding from the National University of Singapore for the first author's appointment as a distinguished visiting professor in 1998 needs a special mention as it lead to the initiation of this project.

Finally, the encouragement and support of our families are also greatly appreciated.

D. N. Prabhakar Murthy
Brisbane, Queensland, Australia

Min Xie
Kent Ridge, Singapore

Renyan Jiang
Toronto, Canada

PART A

Overview

CHAPTER 1

Overview

1.1 INTRODUCTION

In the real world, problems arise in many different contexts. Problem solving is an activity that has a history as old as the human race. Models have played an important role in problem solving and can be traced back to well beyond the recorded history of the human race. Many different kinds of models have been used. These include physical (full or scaled) models, pictorial models, analog models, descriptive models, symbolic models, and mathematical models. The use of mathematical models is relatively recent (roughly the last 500 years). Initially, mathematical models were used for solving problems from the physical sciences (e.g., predicting motion of planets, timing of high and low tides), but, over the last few hundred years, mathematical models have been used extensively in solving problems from biological and social sciences. There is hardly any discipline where mathematical models have not been used for solving problems.

Two different approaches to building mathematical models are as follows:

1. *Theory-Based Modeling*. Here, the modeling is based on the established theories (from physical, biological, and social sciences) relevant to the problem. This kind of model is also called *physics-based model* or *white-box model* as the underlying mechanisms form the starting point for the model building.
2. *Empirical Modeling*. Here, the data available forms the basis for the model building, and it does not require an understanding of the underlying mechanisms involved. As such, these models are used when there is insufficient understanding to use the earlier approach. This kind of model is also called *data-dependent model* or *black-box model*.

In empirical modeling, the type of mathematical formulations needed for modeling is dictated by a preliminary analysis of data available. If the analysis indicates

Weibull Models, by D.N.P. Murthy, Min Xie, and Renyan Jiang.
ISBN 0-471-36092-9

that there is a high degree of variability, then one needs to use models that can capture this variability. This requires probabilistic and stochastic models to model a given data set.

Effective empirical modeling requires good understanding of (i) the methodology needed for model building, (ii) properties of different models, and (iii) tools and techniques to determine if a particular model is appropriate to model a given data set.

A variety of such models have been developed and studied extensively. One such class of models is the *Weibull models*. These are a collection of probabilistic and stochastic models derived from the Weibull distribution. These can be divided into univariate and multivariate models and each, in turn, can be further subdivided into continuous and discrete. Weibull models have been used in many different applications to model complex data sets.

1.1.1 Aims of the Book

This book deals with Weibull models and their applications in reliability. The aims of the book are as follows:

1. Develop a taxonomy to integrate the different Weibull models.
2. Review the literature for each model to summarize model properties and other issues.
3. Discuss the use of Weibull probability paper (WPP) plots in model selection. It allows the model builder to determine whether one or more of the Weibull models are suitable for modeling a given data set.
4. Highlight issues that need further study.
5. Illustrate the application of Weibull models in reliability theory.

The book provides a good foundation for empirical model building involving Weibull models. As such, it should be of interest to practitioners from many different disciplines. The book should also be of interest to researchers as some topics for future research are defined as part of the exercises at the end of several chapters.

1.1.2 Outline of Chapter

The outline of the chapter is as follows. We start with a collection of real-world problems in Section 1.2 and discuss the data aspects and empirical models to obtain solutions to the problems. Section 1.3 deals with the modeling methodology, and we discuss the different issues involved. We highlight the role of statistics, probability theory, and stochastic processes in the context of the link between data and model. Section 1.4 starts a brief historical perspective and then introduces the standard Weibull model (involving the two-parameter Weibull distribution). Following this, a taxonomy to classify the different Weibull models is briefly discussed. Given a univariate continuous data set, a question of great interest to model builders is whether one of the Weibull models is suitable for modeling the given data set or not. This topic is discussed in Section 1.5. Section 1.6 deals with the applications of Weibull models where we start with a short list of applications to highlight the diverse range of applications of the Weibull models in different disciplines. However, in this book we focus on the application of Weibull models in the context of product reliability from a product life perspective. We discuss this briefly so as to set the scene for the discussion on Weibull model applications later in the book. We finally conclude with an outline of the book in Section 1.7.

1.2 ILLUSTRATIVE PROBLEMS

In this section we give a few illustrative problems and the types of data available to build models to obtain solutions to the problems.

Example 1: Tidal Heights
At a popular tourist beach the cyclone season precedes the tourist season. Very high tides during the cyclone season cause the erosion of sand on the beach. The erosion is related to the amplitude of the high tide, and it takes a long time for the beach to recover naturally from the effect of such erosion. Often, sand needs to be pumped to restore the loss and to ensure high tourist numbers. A problem of interest to the city council responsible for the beach is the probability that a high tide during the cyclone season exceeds some specified height resulting in the council incurring the sand pumping cost. The data available is the amplitude of high tides over several years.

Example 2: Efficacy of Treatment
In medical science, a problem of interest is in determining the efficacy of a new treatment to control the spread of a disease (e.g., cancer). In this case, clinical trials are carried out for a certain period. The data available are the number entering the program, the time instants, and the age at death for the patients who died during the trial period, ages of the patients who survived the test period, and so on. Similar data for a sample not given the new treatment might also be available. The problem is to determine if the new treatment increases the life expectancy of the patients.

Example 3: Strength of Components
Due to manufacturing variability, the strength of a component varies significantly. The component is used in an environment where it fails immediately when put into use if its strength is below some specified value. The problem is to determine the probability that a component manufactured will fail under a given environment. If this probability is high, changing the material, the process of manufacturing, or redesigning might be the alternatives that the manufacturer might need to explore. The data available is the laboratory test data. Here items are subjected to increasing levels of stress and the stress level at failure being recorded.

Example 4: Insurance Claims
Whenever there is a legitimate claim, a car insurance company has to pay out. The pay out indicates a high degree of variability (since it can vary from a small to a

very large amount). The insurance company has used the expected value as the basis for determining the annual premium it should charge its customers. It is planning to change the premium and is interested in assessing the probability of an individual claim exceeding five times the premium charged. The data available is the insurance claims over the last few years.

Example 5: Growth of Trees

Paper manufacturing requires wood chips. One way of producing wood chips is through plantations where trees are harvested when the trees reach a certain age. The height of the tree at the time of the harvesting is critical as the volume of wood chips obtained is related to this height. The heights of trees vary significantly. As a result, the output of a plantation can vary significantly, and this has an impact on the profitability of the operation. The operator of a plantation is faced with the problem of choosing between two different types of trees. The data available (from other plantations) are the heights of trees at the time of harvesting for both species.

Example 6: Maintenance of Street Lights

The life of electric bulbs used for street lighting is uncertain and is influenced by a variety of factors (variability in the material used and in the manufacturing process, fluctuations in the voltage, etc.). Replacement of an individual failed item is in general expensive. In this case the road authority might decide on some preventive maintenance action where the bulbs are replaced by new ones at set time instants $t = kT, k = 1, 2, \ldots$. The cost of replacing a bulb under such a replacement policy is much cheaper, but it involves discarding the remaining useful life of the bulb. Any failure in between results in the failed item being replaced by a new one at a much higher cost. The problem facing the authority is to determine the optimal T that minimizes the expected cost. The data available is the historical record of failures and preventive replacements in the past.

Example 7: Stress on Offshore Platform

An offshore platform must be designed to withstand the buffeting of waves. The impact of each wave on the structure is determined by the energy contained in the wave. The wave height is an indicator of the energy in a wave. The data available are the heights of successive waves over a certain time interval, and this exhibits a high degree of variability. The problem is to determine the risk of an offshore platform collapsing if designed to withstand waves up to a certain height.

Example 8: Wind Velocity

Windmills are structures that harness the energy in the wind and convert it into electrical or mechanical energy. The wind velocity fluctuates, and as a result the output of the windmill fluctuates. The economic viability of a windmill is dependent on it being capable of generating a certain minimum level of output for a specified fraction of the day. The problem is to determine the viability of windmills based on the data for wind speeds measured every 5 minutes over a week.

Example 9: Rock Blasting
Mining involves blasting ore formation using explosives. The effect of explosion is that it fragments the ore into different sizes. Ore smaller than the minimum acceptable size is of no value as it are unsuited for processing. Ore lumps bigger than the maximum acceptable size need to be broken down, which involves additional cost. The problem of interest to a mine operator is to determine the size distribution of ore under different blasting strategies so as to decide on the best blasting strategy. In this case, the data available are the size distribution of ore randomly sampled after a blast.

Example 10: Spare Part Planning
For commercial equipment (e.g., aircraft, locomotive) downtime implies a loss of revenue. Downtime occurs due to failure of one or more components of the equipment. Failure of a component is dependent on the reliability of the component. The downtime is dependent on whether a spare is available or not and the time to get a spare if one is not available. When the component is expensive, one must manage the inventory of spare parts properly. Carrying a large inventory implies too much capital being tied up. On the other hand, having a small inventory can lead to high downtimes. The problem is to determine the optimal spare part inventory for components. The data available are the failure times for the different components over a certain period of time.

1.3 EMPIRICAL MODELING METHODOLOGY

The empirical modeling process involves the following five steps:

Step 1: Collecting data
Step 2: Analysis of data
Step 3: Model selection
Step 4: Parameter estimation
Step 5: Model validation

In this section we briefly discuss each of these steps.

Step 1: Collecting Data
Data can be either laboratory data or field data. Laboratory data is often obtained under controlled environment and based on a properly planned experiment. In contrast, field data suffers from variability in the operation environment as well as other uncontrollable factors.

The form of data can vary. In the case of reliability data, it could be continuous valued (e.g., life of an individual item) or discrete valued (e.g., number of items failing in a specified interval). In the former case, it could represent failure times or censored times (the lives of nonfailed items when data collection was stopped) for items. We shall discuss this issue in greater detail in Chapter 4.

Finally, when the data needed for modeling is not available, one needs to collect data based on a proper experiment on expert judgment in some cases. The experiment, in general, is discipline specific. We will discuss this issue in the context of product reliability later in the book.

Step 2: Preliminary Analysis of Data

Given a data set, one starts with a preliminary analysis of the data. Suppose that the data set available is given by $(t_1, t_2, \ldots, t_n)$. In the first stage, one computes various sample statistics (such as max, min, mean, sample variance, median, and first and third quartiles) based on the data. If the range (= max – min) is small relative to the sample mean, one might ignore the variability in the data and model the data by the sample mean. However, when this is not the case, then the model needs to mimic this variability in the data. In the case of time-ordered data, preliminary analysis is used to determine properties such as trends (increasing or decreasing), correlation over time, and so forth.

The main aim of the analysis is to assist in determining whether a particular model is appropriate or not to model a given data set. Many different plots have been developed to assist in this. Some of these plots (e.g., histogram) are general and others (e.g., Weibull probability paper plot) were originally developed for a particular model but have since been used for a broader class of models.

Step 3: Model Selection

Suppose that the data set $(t_1, t_2, \ldots, t_n)$ exhibits significant variability. In this case the data set needs to be viewed as an observed value of a set of random variables $(T_1, T_2, \ldots, T_n)$. If the random variables are statistically independent, then each T can be modeled by a univariate probability distribution function:

$$F(t; \theta) = P(T \le t) \qquad -\infty < t < \infty \tag{1.1}$$

where θ denotes the set of parameters for the distribution. In some cases the range of t is constrained. For example, if T represents the lifetime of an item, then it is constrained to be nonnegative so that $F(t; \theta)$ is zero for $t < 0$.

Model selection involves choosing an appropriate model formulation (e.g., a distribution function) to model a given data set. In order to execute this step, one needs to have a good understanding of the properties of different model formulations suitable for modeling. Some basic concepts are discussed in Chapter 3. Probability theory deals with such study for a variety of model formulations. An important feature of modeling is that often there is more than one model formulation that will adequately model a given data set. In other words, one can have multiple models for a given data set.

The data source often provides a clue to the selection of an appropriate model. In the case of failure data, for example, lognormal or Weibull distributions have been used for modeling failures due to fatigue and exponential distributions for failure of electronic components. In order to use this knowledge, the model builder must be familiar with earlier models for failures of different items.

If the data are not independent, one needs to use models involving multivariate distribution functions. If time is a factor that needs to be included in the model explicitly, then the model becomes more complex. The building of such models requires concepts from stochastic processes.

Step 4: Parameter Estimation
Once a model is selected, one needs to estimate the model parameters. The estimates are obtained using the data available. A variety of techniques have been developed, and these can be broadly divided into two categories—graphical and analytical. The accuracy of the estimate is dependent on the size of the data and the method used. Graphical methods yield crude estimates while analytical methods yield better estimates and confidence limits for the estimates. The basic concepts are discussed in in Chapter 4 and in later chapters in the context of specific models.

Step 5: Model Validation
One can always fit a model to a given data set. However, the model might not be appropriate or adequate. An inappropriate model, in general, will not yield the desired solution to the problem. Hence, it is necessary to check the validity of the model selected. There are several methods for doing this. The basic concepts are discussed in Chapter 5 and in later chapters in the context of specific models.

Comments

1. Steps 2, 4, and 5 deal with *statistical inference*. In probability theory, one models the uncertainty (randomness) through a distribution function, and then makes statements, based on the model, about the nature (e.g., variability) of the data that may result if the model is correct. The principal objective of statistical inference is to use data to make statements about the model, either in terms of probability distribution itself or in terms of its parameters or some other characteristics. Thus, probability theory and statistical inference may be thought of as inverse of one another as indicated:

$$\begin{aligned} \text{Probability theory:}\quad & \text{Model} \rightarrow \text{Data} \\ \text{Statistics:}\quad & \text{Data} \rightarrow \text{Model} \end{aligned}$$

2. Statistical inference requires concepts, tools, and techniques from the *theory of statistics*. Understanding a model requires studying the properties of the model. This requires concepts, tools, and techniques from the *theory of probability* and the *theory of stochastic processes*.
3. In this book we discuss both model properties and statistical inference for Weibull models.

1.4 WEIBULL MODELS

1.4.1 Historical Perspective

The three-parameter Weibull distribution is given by the distribution function

$$F(t;\theta) = 1 - \exp\left[-\left(\frac{t-\tau}{\alpha}\right)^{\beta}\right] \qquad t \geq \tau \tag{1.2}$$

The parameters of the distribution are given by the set $\theta = \{\alpha, \beta, \tau\}$ with $\alpha > 0, \beta > 0$, and $\tau \geq 0$. The parameters α, β, and τ are the *scale, shape*, and *location parameters* of the distribution, respectively. The distribution is named after Waloddi Weibull who was the first to promote the usefulness of this to model data sets of widely differing character. The initial study by Weibull (Weibull, 1939) appeared in a Scandinavian journal and dealt with the strength of materials. A subsequent study in English (Weibull, 1951) was a landmark work in which he modeled data sets from many different disciplines and promoted the versatility of the model in terms of its applications in different disciplines.

A similar model was proposed earlier by Rosen and Rammler (1933) in the context of modeling the variability in the diameter of powder particles being greater than a specific size. The earliest known publication dealing with the Weibull distribution is a work by Fisher and Tippet (1928) where this distribution is obtained as the limiting distribution of the smallest extremes in a sample. Gumbel (1958) refers to the Weibull distribution as the third asymptotic distribution of the smallest extremes.

Although Weibull was not the first person to propose the distribution, he was instrumental in its promotion as a useful and versatile model with a wide range of applicability. A report by Weibull (Weibull, 1977) lists over 1000 references to the applications of the basic Weibull model, and a recent search of various databases indicate that this has increased by a factor of 3 to 4 over the last 30 years.

1.4.2 Taxonomy

The two-parameter Weibull distribution is a special case of (1.2) with $\tau = 0$ so that

$$F(t;\theta) = 1 - \exp\left[-\left(\frac{t}{\alpha}\right)^{\beta}\right] \qquad t \geq 0 \tag{1.3}$$

We shall refer to this as the *standard* Weibull model with $\alpha(> 0)$ and $\beta(> 0)$ being the scale and shape parameters respectively. The model can be written in alternate parametric forms as indicated below:

$$F(t;\theta) = 1 - \exp\left[-(\lambda t)^{\beta}\right] \tag{1.4}$$

with $\lambda = 1/\alpha$;

$$F(t;\theta) = 1 - \exp\left(-\frac{t^{\beta}}{\alpha'}\right) \tag{1.5}$$

with $\alpha' = \alpha^{\beta}$; and

$$F(t;\theta) = 1 - \exp(-\lambda' t^{\beta}) \tag{1.6}$$

with $\lambda' = (1/\alpha)^\beta$. Although they are all equivalent, depending on the context a particular parametric representation might be more appropriate. In the remainder of the book, the form for the standard Weibull model is (1.3) unless indicated otherwise.

A variety of models have evolved from this standard model. We propose a taxonomy for classifying these models, and it involves seven major categories labeled Types I to VII. In this section, we briefly discuss the basis for the taxonomy, and the different models in each category are discussed in Chapter 2.

Let T denote the random variable from the standard Weibull model. Let the distribution function for the derived model be $G(t;\theta)$, and let Z denote the random variable from this distribution. The links between the standard Weibull model and the seven different categories of Weibull models are as follows:

Type I Models Here Z and T are related by a transformation. The transformation can be either (i) linear or (ii) nonlinear.

Type II Models Here $G(t,\theta)$ is related to $F(t,\theta)$ through some functional relationship.

Type III Models These are univariate models derived from two or more distributions with one or more being a standard Weibull distribution. As a result, $G(t,\theta)$ is a univariate distribution function involving one or more standard Weibull distributions.

Type IV Models The parameters of the standard Weibull model are constant. For models belonging to this group, this is not the case. As a result, they are either a function of the variable t or some other variables (such as stress level) or are random variables.

Type V Models In the standard Weibull model, the variable t is continuous valued and can assume any value in the interval $[0,\infty)$. As a result, T is a continuous random variable. In contrast, for Type V models Z can only assume nonnegative integer values, and this defines the support for $G(t,\theta)$.

Type VI Models The standard Weibull model is a univariate model. Type VI models are multivariate extensions of the standard Weibull model. As a result, $G(\cdot)$ is a multivariate function of the form $G(t_1, t_2, \ldots, t_n)$ and related to the standard Weibull in some manner.

Type VII Models These are stochastic point process models with links to the standard Weibull model.

1.5 WEIBULL MODEL SELECTION

Model selection tends to be a trial-and-error process. For Types I to III models the Weibull probability paper plot provides a systematic procedure to determine whether one of these models is suitable for modeling a given data set or not. It is based on the Weibull transformations

$$y = \ln\{-\ln[1 - F(t)]\} \quad \text{and} \quad x = \ln(t) \tag{1.7}$$

A plot of y versus x is called the Weibull probability plot. In the early 1970s a special paper was developed for plotting the data under this transformation and was referred to as the Weibull probability paper (WPP) and the plot called the WPP plot. These days, most reliability software packages contain programs to produce these plots automatically given a data set. We use the term WPP plot to denote the plot using computer packages.

1.6 APPLICATIONS OF WEIBULL MODELS

Weibull models have been used in many different applications and for solving a variety of problems from many different disciplines. Table 1.1 gives a small sample of the application of Weibull models along with references where interested readers can find more details.

1.6.1 Reliability Applications

All man-made systems (ranging from simple products to complex systems) are unreliable in the sense that they degrade with time and/or usage and ultimately fail. The following material is from Blischke and Murthy (2000).

The *reliability* of a product (system) is the probability that the product (system) will perform its intended function for a specified time period when operating under normal (or stated) environmental conditions.

Product Life Cycle and Reliability

A product life cycle (for a consumer durable or an industrial product), from the point of view of the manufacturer, is the time from initial concept of the product to its withdrawal from the marketplace. It involves several stages as indicated in Figure 1.1.

The process begins with an idea to build a product to meet some customer requirements regarding performance (including reliability) targets. This is usually based on a study of the market and the potential demand for the product being planned. The next step is to carry out a feasibility study. This involves evaluating whether it is possible to achieve the targets within the specified cost limits. If this analysis indicates that the project is feasible, an initial product design is undertaken. A prototype is then developed and tested. It is not unusual at this stage to find that achieved performance level of the prototype product is below the target value. In this case, further product development is undertaken to overcome the problem. Once this is achieved, the next step is to carry out trials to determine performance of the product in the field and to start a preproduction run. This is required because

Table 1.1 Sample of Weibull Model Applications

Application	Reference
Yield strength of steel	Weibull (1951)
Size distribution of fly ash	Weibull (1951)
Fiber strength of Indian cotton	Weibull (1951)
Fatigue life of ST-37 steel	Weibull (1951)
Tensile strength of optical fibers	Phani (1987)
Fineness of coal	Rosen and Rammler (1933)
Size of droplets in sprays	Fraser and Eisenklam (1956)
Particle size	Fang et al. (1993)
Pitting corrosion in pipe	Sheikh et al. (1990)
Time to death following exposure to carcinogens	Pike (1966), Peto and Lee (1973)
Strength of fibers from coconut husks	Kulkarni et al. (1973)
Forecasting technology change	Sharif and Islam (1980)
Dielectric breakdown voltage	Nossier et al. (1980)
	Mu et al. (2000)
	Wang et al. (1997)
Fracture strength of glass	Keshvan et al. (1980)
Size of Antarctic icebergs	Neshyba (1980)
Inventory lead time	Tadikamalla (1978)
Wave heights in English Channel	Henderson and Weber (1978)
Size of rock fragments	Rad and Olson (1974)
Ball bearing failures	Lieblein and Zelen (1956)
Failure of carbon fiber composites	Durham and Padgett (1997),
	Padgett et al. (1995)
Machining center failures	Yazhou et al. (1995)
Traffic conflict in expressway merging	Chin et al. (1991)
Partial discharge phenomena	Cacciari et al. (1995)
Flight load variation in helicopter	Boorla and Rotenberger (1997)
Precipitation in Pacific Northwest	Duan et al. (1998)
Adhesive wear in metals	Quereshi and Sheikh (1997)
Fracture in concrete	Xu and Barr (1995)
Latent failures of electronic products	Yang et al. (1995)
Damage in laminated composites	Kwon and Berner (1994)
Software reliability growth	Yamada et al. (1993)
Brittle material	Fok et al. (2001)
Thermoluminescence glow	Pagonis et al. (2001)
Flood frequency	Heo et al. (2001)
Temperature fluctuations	Talkner et al. (2000)
Wind speed distribution	Seguro and Lambert (2000)
	Lun and Lam (2000)
Earthquakes	Huillet and Raynaud (1999)
Failure of coatings	Almeida (1999)
Rain drop size	Jiang et al. (1997)
Discharge inference	Contin et al. (1994)

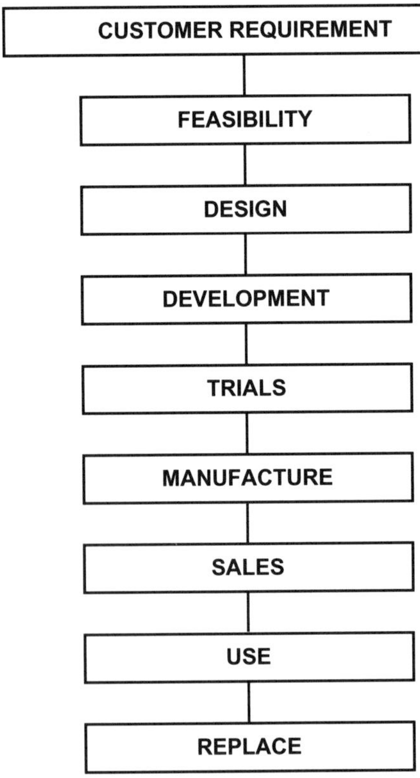

Figure 1.1 Different stages of product life cycle. (From Blischke and Murthy, 2000.)

the manufacturing process must be fine tuned and quality control procedures established to ensure that the items produced have the same performance characteristics as those of the final prototype. After this, the production and marketing efforts begin. The items are produced and sold. Production continues until the product is removed from the market because of obsolescence and/or the launch of a new product.

We focus our attention on the reliability of the product over its life cycle. Although this may vary considerably, a typical scenario is as shown in Figure 1.2. A feasibility study is carried out using the specified target value for product reliability. During the design stage, product reliability is assessed in terms of part and component reliabilities. Product reliability increases as the design is improved. However, this improvement has an upper limit. If the target value is below this limit, then the design using available parts and components achieves the desired target value. If not, then a development program to improve the reliability through test–fix–test cycles is necessary. Here the prototype is tested until a failure occurs and the causes of the failure are analyzed. Based on this, design and/or manufacturing changes are introduced to overcome the identified failure causes. This process is continued until the reliability target is achieved.

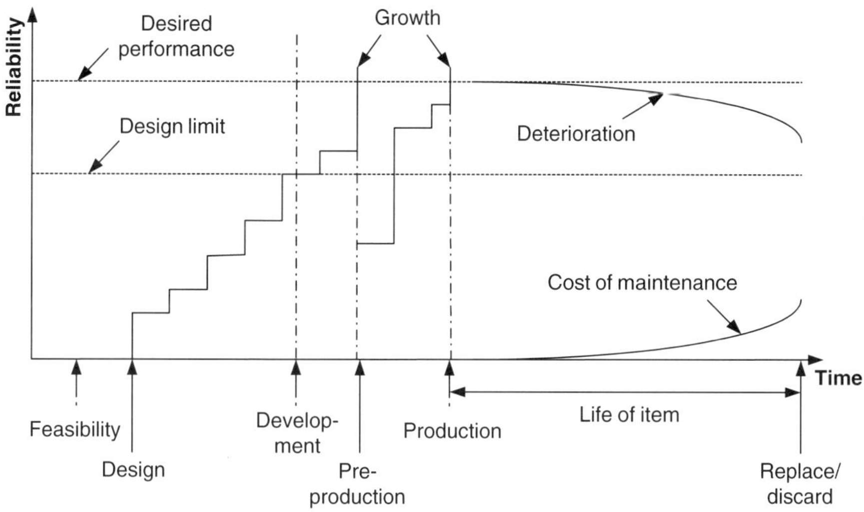

Figure 1.2 Reliability issues over product life cycle. (From Blischke and Murthy, 2000.)

The reliability of the items produced during the preproduction run is usually below that for the final prototype. This is caused by variations resulting from the manufacturing process. Through proper process and quality control, these variations are identified and reduced or eliminated, and the reliability of items produced is increased until it reaches the target value. Once this is achieved, full-scale production commences and the items are released for sale.

The reliability of an item in use deteriorates with age. This deterioration is affected by several factors, including environment, operating conditions, and maintenance. The rate of deterioration can be controlled through preventive maintenance as shown in Figure 1.2.

It is worth noting that if the reliability target values are too high, they might not be achievable with development. In this case, the manufacturer needs to revise the target value and start with a new feasibility study before proceeding further.

Weibull models have been used not only to model the reliability of the product but also to study other issues in the different stages of the product life cycle. We can group the different stages into two groups: (i) premanufacturing and (ii) postmanufacturing. The former deals with issues such as reliability assessment, accelerated testing, and reliability growth, and the latter with issues such as maintenance and warranties.

1.7 OUTLINE OF THE BOOK

The book is structured in seven parts (Parts A to G) with each part containing one or more chapters.

Part A consists of two chapters. Chapter 1 gives an overview of the book, and Chapter 2 deals with a detailed discussion of the taxonomy for Weibull models and the mathematical structure of these models.

Part B consists of three chapters. Chapter 3 deals with the analysis and presents various results dealing with model properties. Chapter 4 deals with parameter estimation and examines different data structures and estimation methods and their properties. Chapter 5 deals with model selection and validation, where the focus is on deciding whether a specific model is appropriate to model a given data set or not. In these chapters various concepts and techniques are introduced, and these are referred to in later chapters. In these three chapters, the analysis, estimation, and validation issues for the standard Weibull model as well as the three-parameter Weibull model are discussed as the two models are very similar.

Part C consists of two chapters. Chapter 6 deals with Type I models and Chapter 7 deals with Type II models. Part D consists of four chapters and deals with Type III models, and each chapter looks at a different family of models. Chapter 8 deals with mixture models, Chapter 9 with competing risk models, Chapter 10 with the multiplicative models, and Chapter 11 with sectional models. Part E consists of four chapters. Chapter 12 deals with Type IV models and looks at four different models. Chapter 13 deals with Type V models, Chapter 14 with Type VI models, and Chapter 15 with Type VII models. In Chapters 6 to 15, for each model, the model properties and the statistical inference issues are discussed.

Part F consists of a single chapter (Chapter 16) and deals with model selection. Here the focus is on deciding whether one or more Type I to III models are suitable for modeling a given univariate continuous valued data set.

Part G deals with the application of Weibull models in reliability theory. It consists of three chapters. Chapter 17 deals with modeling failures at system and component levels. Chapter 18 deals with a variety of reliability-related decision problems at the premanufacturing, manufacturing, and postsale service stages of the product life cycle. We briefly review the literature dealing with such problems with failures modeled by Weibull models.

1.8 NOTES

The notion of random variable and some basic concepts from probability theory can be found in most textbooks on the probability theory or models. See, for example, Ross (1983). The four volumes by Johnson and Kotz (1969a, 1969b, 1970a, 1970b) deal with discrete, univariate, and multivariate distributions. Basic concepts from the theory of stochastic processes can be found in most texts on the subject. See, for example, Ross (1980).

EXERCISES

1.1. Define a set of problems from each of the following areas that requires the use of probability distribution function to model one or more of the variables of importance to the problem:

a. Biological sciences

b. Social sciences

c. Health sciences

d. Physical sciences

e. Environmental science

1.2. Carry out a search of computer databases (such as Inspec, Compendex, etc.) to compile a list of studies dealing with the problems defined in Exercise 1.1.

1.3. A city council operates a fleet of buses. At the workshop level, for each bus the date of failure and the component that caused the failure are recorded. Discuss what other data should have been recorded for proper understanding of bus failures? The recorded data is converted into monthly failures for reporting to top management. Discuss if this is appropriate or not.

1.4. Discuss why the modeling of a data set is both an art as well as a science.

1.5. The lifetime of a human is a random variable and affected by several factors that can be grouped into genetic, ethnic, environmental, and lifestyle. One can model the lifetime by either ignoring these factors or by incorporating them. Discuss problems when it is appropriate to ignore them and when one needs to include them in the model building.

1.6. An automobile is a complex system comprised of several components. The failure of an automobile is due to the failure of one or more components. As such, one can model the failures either at the system or component level. Discuss when it is more appropriate to use the modeling at the system level as opposed to the component level.

1.7. Discuss the different kinds of data that is generated at each stage of the product life cycle that is relevant for determining the reliability of the product.

1.8. Weibull distribution is one of many probability distributions that have been proposed and studied. Make a list of distributions that you are familiar with and can be used to model nonnegative continuous random variable.

1.9. Repeat Exercise 1.8 for nonnegative discrete random variable.

CHAPTER 2

Taxonomy for Weibull Models

2.1 INTRODUCTION

In Chapter 1 we proposed a taxonomy to categorize the Weibull models into different categories. In this chapter we give the precise details of the model formulations and a concise overview of the relationships between the models. Specific details of the models are discussed in later chapters. We have retained the original terminology as far as possible while the notations have been modified to ensure consistency in the presentation.

The outline of the chapter is as follows. In Section 2.2 we indicate the basis for the taxonomy. The taxonomy involves seven categories (labeled Types I to VII). Sections 2.3 to 2.9 discuss the different models for each of the seven categories. For some of the models, the authors of the model have proposed a name. We indicate this in our discussion of the model. Because of the lack of a consistent terminology, in some cases the same name has been used for two different models and in some other cases the same model has been called by more than one name.

2.2 TAXONOMY FOR WEIBULL MODELS

The taxonomy for Weibull models involves seven categories and each category can be divided into several subcategories as shown in Figure 2.1. In this section, we discuss the basis for these subgroupings and give the structure of the different models in each subgrouping.

The starting point is the standard Weibull model given by (1.3). Let $G(t;\theta)$ denote the derived Weibull model; T is a random variable from $F(t;\theta)$ and Z is a random variable from $G(t;\theta)$.

2.2.1 Type I Models: Transformation of Weibull Variable

For these models Z and T are related by a transformation. The transformation can be either (a) linear or (b) nonlinear. Based on this we have the following two subgroups.

Weibull Models, by D.N.P. Murthy, Min Xie, and Renyan Jiang.
ISBN 0-471-36092-9

Type I(a): Linear Transformation This yields four models: Model I(a)-1 through Model I(a)-4.

Type I(b): Nonlinear Transformation This yields three models: Model I(b)-1 through Model I(b)-3.

2.2.2 Type II Models: Modification/Generalization of Weibull Distribution

For these models $G(t; \theta)$ is related to $F(t; \theta)$ by some relationship. One can subdivide these into two subgroups.

Type II(a) These involve no additional parameters and hence can be viewed as modified models. There are two models: Model II(a)-1 and II(a)-2.

Type II(b) These involve one or more additional parameters and hence can be viewed as generalized models as the standard Weibull model is obtained as special cases of these models. There are 13 models: Model II(b)-1 through II(b)-13.

2.2.3 Type III Models: Models Involving Two or More Distributions

These models are univariate models derived from two or more distributions with at least one being either the standard Weibull model or a distribution derived from it. One can subdivide these into four subgroups as indicated below.

Type III(a) Mixture models.

Type III(b) Competing risk models.

Type III(c) Multiplicative models.

Type III(d) Sectional models.

Each of these contains several models depending on the number of distributions involved, whether one or more of the distributions is non-Weibull and the forms of the non-Weibull distributions.

2.2.4 Type IV Models: Weibull Models with Varying Parameters

The parameters of the standard Weibull model are constant. For models belonging to this group, this is not the case. As a result, they are either a function of the independent variable (t) or some other variables (such as stress level), or are random variables. These lead to the following three subgroups.

Type IV(a) Parameters are functions of the variable t (time-varying parameters if t represents time).

Type IV(b–d) Parameters are functions of some other variables (regression models, proportional hazard models, etc.).

Type IV(e) Parameters are random variables (Bayesian models).

2.2.5 Type V Models: Discrete Weibull Models

In the standard Weibull model, the variable t is continuous and can assume any value in the interval $[0, \infty)$. As a result, T is a continuos random variable. In contrast, for Type V models T can only assume nonnegative integer values, and this defines the support for $G(t; \theta)$.

2.2.6 Type VI Models: Multivariate Weibull Distributions

The standard Weibull model is a univariate model. Type VI models are multivariate extensions of the standard Weibull model. As a result, $G(\cdot)$ is a multivariate function of the form $G(t_1, t_2, \ldots, t_n)$ and related to the basic Weibull in some manner. These can be divided into two categories:

Type VI(a) Bivariate Weibull models.

Type VI(b) Multivariate Weibull models.

Again, there are several models under each subcategory.

2.2.7 Type VII Models: Stochastic Point Process Models

These are stochastic point process models with links to the standard Weibull model. Three subgroups that have received some attention are the following:

Type VII(a) Weibull intensity function models.

Type VII(b) Weibull renewal process models.

Type VII(c) Power law–Weibull renewal model.

2.2.8 Notation

Often we will suppress the parameter for convenience and write $F(t)$ instead of $F(t; \theta)$ and similarly for $G(t; \theta)$. For a distribution function $F(t)$ we have the following:

- Density function: $f(t) = dF(t)/dt$ if the derivative exists.
- Survivor function: $\bar{F}(t) = 1 - F(t)$.

- Hazard function (failure rate): $h(t) = f(t)/\bar{F}(t)$.
- Cumulative hazard function: $H(t) = \int_0^t h(x)\,dx$.
- Quantile: $Q(u) : F(Q(u)) = u, u \in [0, 1]$.

2.3 TYPE I MODELS: TRANSFORMATION OF WEIBULL VARIABLE

2.3.1 Linear Transformation

Consider the linear transformation

$$Z = aT + b \tag{2.1}$$

Model I(a)-1: One-Parameter Weibull
The terms $a = 1/\alpha$ and $b = 0$ in (2.1) yield

$$G(t) = 1 - \exp(-t^{\beta}) \qquad t \geq 0 \tag{2.2}$$

This is the *one-parameter Weibull model*. The support for $G(t)$ is the same as that for $F(t)$. Note that this can also be obtained as a special case of the standard Weibull model with $\alpha = 1$.

Model I(a)-2
The terms $a = 1$ and $b = \tau$ in (2.1) yield

$$G(t) = 1 - \exp\left[-\left(\frac{t-\tau}{\alpha}\right)^{\beta}\right] \qquad t > \tau \tag{2.3}$$

This is the *three-parameter Weibull distribution* [see (1.2)]. The model is characterized by three parameters and the support for $G(t)$ is different from that for $F(t)$.

Some special cases of this model are as follows:

- $\beta = 1$: Two-parameter exponential distribution
- $\beta = 2$ and $\tau = 0$: Rayleigh distribution

Model I(a)-3
The terms $a = -1$ and $b = \tau$ in (2.1) yield

$$G(t) = \exp\left[-\left(\frac{\tau-t}{\alpha}\right)^{\beta}\right] \qquad -\infty < t < \tau \tag{2.4}$$

The supports for $G(t)$ and $F(t)$ are disjoint. The model can also be viewed as a reflection of the three-parameter Weibull model [given by (1.2)] with $1 - G(t)$

being the mirror image of $F(t)$. The model was proposed in Cohen (1973) and called the *reflected Weibull distribution*.

Model I(a)-4
This model is a combination of the one-parameter Weibull model [given by (2.2)] and its reflection about the vertical axis through the origin. As a result, the model is given by the density function

$$g(t) = \beta\left(\tfrac{1}{2}\right)|t|^{(\beta-1)}\exp(-|t|^{\beta}) \qquad -\infty < t < \infty \tag{2.5}$$

Note that the support for $G(t)$ is bigger than that for the basic model. This model was proposed in Balakrishnan and Kocherlakota (1985) and is called the *double Weibull distribution*.

2.3.2 Nonlinear Transformation

Model I(b)-1
The model is obtained using the transformation

$$\frac{Z-\tau}{\eta} = \left(\frac{T}{\alpha}\right)^{\beta} \tag{2.6}$$

and is given by

$$G(t) = 1 - \exp\left[-\left(\frac{t-\tau}{\eta}\right)\right] \qquad t \geq \tau \tag{2.7}$$

As a result, the support for $G(t)$ is smaller than that for $F(t)$. Note that this is the *two-parameter exponential distribution*. This can also be obtained as a special case of the three-parameter Weibull model with the shape parameter $\beta = 1$.

If $\tau = 0$ in (2.6), then (2.7) is the *standard (one-parameter) exponential distribution*. This can also be obtained as a special case of the standard Weibull model with the shape parameter $\beta = 1$.

The transformation in (2.6) is often referred to as the *power law transformation* that links the Weibull and exponential distributions. This transformation forms the basis for several multivariate Weibull models discussed in Section 2.8.

Model I(b)-2
The model is obtained using the transformation

$$\frac{Z-\tau}{\eta} = \beta \ln\left(\frac{T}{\alpha}\right) \tag{2.8}$$

and is given by

$$G(t) = 1 - \exp\left[-\exp\left(\frac{t-\tau}{\eta}\right)\right] \qquad -\infty < t < \infty \tag{2.9}$$

The support for $G(t)$ is greater than that for $F(t)$; $G(t)$ is a *type 1 extreme value distribution* while the basic Weibull distribution itself is related to Gumbel's Type II distribution (Gumbel, 1958, p. 157). This model is also commonly called the *log Weibull model.*

As with the earlier model, setting $\tau = 0$ and/or $\eta = 1$ leads to several special cases.

Model I(b)-3

The model follows from the transformation

$$Z = \frac{\alpha^2}{T} \tag{2.10}$$

and is given by

$$G(t) = \exp\left[-\left(\frac{t}{\alpha}\right)^{-\beta}\right] \qquad t \geq 0 \tag{2.11}$$

This is referred to as the *inverse* (or *reverse*) *Weibull model.* It is also a limiting distribution of the largest order statistics (Type II asymptotic distribution of the largest extreme). Drapella (1993) calls it the *complementary Weibull distribution.* Mudholkar and Kollia (1994) call it the *reciprocal Weibull distribution.*

2.4 TYPE II MODELS: MODIFICATION/GENERALIZATION OF WEIBULL DISTRIBUTION

For several models in this group, the link between the model and the standard Weibull model is better described in terms of the quantile function. For the standard Weibull model [given by (1.3)] the quantile function is given by

$$Q(u) = \alpha[-\ln(1-u)]^{1/\beta} \tag{2.12}$$

2.4.1 Modification of Weibull Distribution

Model II(a)-1

The density function is given by

$$g(t) = \frac{t}{\mu} f(t) \qquad t \geq 0 \tag{2.13}$$

where μ is the mean of the random variable from the standard Weibull model. As a result

$$g(t) = \frac{\beta t^{\beta}}{\alpha^{\beta+1}\Gamma(1+1/\beta)} \exp\left[-\left(\frac{t}{\alpha}\right)^{\beta}\right] \qquad t \geq 0 \tag{2.14}$$

Note that it is not possible to express $G(t)$ analytically. This model was proposed in Voda (1989) and called the *pseudo-Weibull distribution*.

Model II(a)-2

The distribution function is given by

$$G(t) = \begin{cases} 1 - e^{-(t/\alpha)^{\beta}} & \beta > 0 \\ e^{-(t/\alpha)^{\beta}} & \beta < 0 \end{cases} \tag{2.15}$$

where β can take either positive or negative values. When $\beta > 0$, this reduces to the standard Weibull model [given by (1.3)] and to the inverse Weibull model [given by (2.11)] when $\beta < 0$. This model is referred to in Fisher and Tippet (1928) and Gumbel (1958). It was made more explicit in Stacy and Mihram (1965) and called the *generalized Weibull distribution.*

2.4.2 Generalization of Weibull Distribution

Model II(b)-1

This model involves an additional parameter $\nu > 0$. The survivor function $\bar{G}(t)$ is given by

$$\bar{G}(t) = \frac{\nu \bar{F}(t)}{F(t) + \nu \bar{F}(t)} \qquad t \geq 0 \tag{2.16}$$

with $F(t)$ given by (1.4). As a result

$$G(t) = 1 - \frac{\nu e^{-(\lambda t)^{\beta}}}{1 - (1-\nu) e^{-(\lambda t)^{\beta}}} \tag{2.17}$$

This model was proposed in Marshall and Olkin (1997) and called the *extended Weibull distribution*. Note that the model reduces to the standard Weibull model when $\nu = 1$.

Model II(b)-2

This model involves one additional parameter $\nu(\geq 0)$ and is as follows:

$$G(t) = 1 - \exp(-\lambda t^{\beta} e^{\nu t}) \qquad t \geq 0 \tag{2.18}$$

with $\lambda = (1/\alpha)^{\beta}$. This model was proposed in Lai et al. (2003) and called the *modified Weibull distribution.*

Model II(b)-3

The model, described in terms of the quantile function, is given by

$$Q(u) = \alpha[-\ln(1-u)^{1/\nu}]^{1/\beta} \tag{2.19}$$

and involves one additional parameter $\nu > 0$. As a result, model is given by

$$G(t) = [F(t)]^{\nu} = \left\{1 - \exp\left[-\left(\frac{t}{\alpha}\right)^{\beta}\right]\right\}^{\nu} \qquad t \geq 0 \tag{2.20}$$

When $\nu = 1$, this reduces to the standard Weibull distribution given by (1.3). This model was proposed in Mudholkar and Srivasatava (1993) and called the *exponentiated Weibull distribution.*

Model II(b)-4

The quantile function for the model is

$$Q(u) = \alpha\left[\frac{1-(1-u)^{\nu}}{\nu}\right]^{1/\beta} \tag{2.21}$$

where the new parameter ν is unconstrained so that $-\infty < \nu < \infty$. This implies

$$G(t) = 1 - \left[1 - \nu\left(\frac{t}{\alpha}\right)^{\beta}\right]^{1/\nu} \qquad t \geq 0 \tag{2.22}$$

An interesting feature of the model in the support for $G(t)$ is $(0, \infty)$ for $\nu \leq 0$ and $(0, \alpha/\nu^{1/\beta})$ for $\nu > 0$. This model was proposed in Mudholkar et al. (1996) and called the *generalized Weibull family.* Note that the model reduces to the standard two-parameter Weibull model when $\nu \to 0$.

Model II(b)-5

A different version of (2.21) is

$$Q(u) = \beta\left[\frac{1-(1-u)^{\nu}}{\nu}\right]^{1/\beta} - \beta \tag{2.23}$$

with ν unconstrained so that $-\infty < \nu < \infty$. Here the parameter β can take negative values so that the inverse Weibull is a special case of this distribution. This yields

$$G(t) = 1 - \left[1 - \nu\left(\frac{\beta + t}{\beta}\right)^{\beta}\right]^{1/\nu} \tag{2.24}$$

The support for $G(t)$ depends on the model parameters and is as follows. For $\beta < 0$; if $\nu < 0$, $(-\infty, -\beta)$; and if $\nu > 0$, $(-\infty, (\beta/\nu^{1/\beta} - \beta))$; while for $\beta > 0$, it is, if $\nu < 0$, $(-\beta, \infty)$ and if $\nu > 0$, $(-\beta, \beta/\nu^{1/\beta} - \beta)$. This model was proposed in Mudholkar and Kolia (1994) and called a *more generalized Weibull family.*

Model II(b)-6

The model, defined in terms of the quantile function, is given by

$$Q(u) = \begin{cases} [(-\ln(1-u))^{\beta} - 1]/\beta & \text{for} \quad \beta \neq 0 \\ \ln(-\ln(1-u)) & \text{for} \quad \beta = 0 \end{cases} \tag{2.25}$$

As a result,

$$G(t) = \begin{cases} 1 - \exp[-(1+\beta t)^{1/\beta}] & \text{for} \quad \beta \neq 0 \\ 1 - \exp[-\exp(t)] & \text{for} \quad \beta = 0 \end{cases} \tag{2.26}$$

The support for $G(t)$ depends on the model parameters. It is $(-\infty, -1/\beta)$ for $\beta < 0$, $(-1/\beta, \infty)$ for $\beta > 0$, and $(-\infty, \infty)$ for $\beta = 0$. This model was proposed in Freimer et al. (1989) and called the *extended Weibull distribution.*

Model II(b)-7

The model, proposed by Stacy (1962), is best described in terms of the density function

$$g(t;\theta) = \frac{\alpha^{-\beta k}\beta t^{\beta k-1}}{\Gamma(k)} \exp\left[-\left(\frac{t}{\alpha}\right)^{\beta}\right] \qquad t \geq 0 \tag{2.27}$$

with $k > 0$. $\Gamma()$ is the gamma function [see Abramowitz and Stegun (1964)]. This is called the *generalized gamma model.* Note that when $k = 1$, this reduces to the standard Weibull model and when $\beta = 1$ it gets reduced to a gamma distribution.

An extension to include a location parameter was proposed by Harter (1967) and has the density function

$$g(t;\theta) = \frac{\alpha^{-\beta k}\beta(t-\tau)^{\beta k-1}}{\Gamma(k)} \exp\left[-\left(\frac{t-\tau}{\alpha}\right)^{\beta}\right] \qquad t \geq \tau \tag{2.28}$$

where $\theta = \{\alpha, \beta, \tau, k\}$ and is called the *four-parameter generalized gamma model.*

Model II(b)-8

Ghitany (1998) suggested an extension to (2.28) and the model has the density function

$$g(t;\theta) = \frac{\alpha^{(m-1)/\beta+1-\lambda}\beta t^{m-\beta-2}(n+t^{\beta})^{-\lambda}}{\Gamma_{\lambda}[(m-1)/\beta, \alpha n]} \exp\left[-\left(\frac{t}{\alpha}\right)^{\beta}\right] \tag{2.29}$$

where $\theta = \{\alpha, \beta, \lambda, m, n\}$ and $\Gamma_\lambda(,)$ is the incomplete gamma function [see Abramowitz and Stegun (1964)]. This can be viewed as the *extended generalized gamma model*.

Agarwal and Al-Saleh (2001) proposed a model given by

$$g(t;\theta) = \frac{\beta\lambda^\beta t^{\beta-1}}{\Gamma_\lambda(1,n)[(\lambda t)^\beta + n]^\delta}\exp[-(\lambda t)^\beta] \qquad t \geq 0 \tag{2.30}$$

where β, n, and $\lambda > 0$ and $\delta \in [-1, 0]$ and call it a *Weibull-type* distribution. Note that this reduces to the two-parameter Weibull distribution [given by (1.3)] when $\delta = 0$.

Model II(b)-9

This model was proposed by Kies (1958) and is given by

$$G(t) = 1 - \exp\left[-\lambda\left(\frac{t-a}{b-t}\right)^\beta\right] \qquad 0 \leq a \leq t \leq b < \infty \tag{2.31}$$

with $\lambda > 0$ and $\beta > 0$. This implies that the support is a finite interval. This is also called a *four-parameter Weibull distribution*. Smith and Hoeppner (1990) also deal with a similar four-parameter Weibull model.

Model II(b)-10

Phani (1987) extended the model due to Kies and the model is given by

$$G(t) = 1 - \exp\left\{-\lambda\frac{(t-a)^{\beta_1}}{(b-t)^{\beta_2}}\right\} \qquad 0 \leq a \leq t \leq b < \infty \tag{2.32}$$

with $\lambda > 0$, $\beta_1 > 0$, and $\beta_2 > 0$. This has been called a *modified Weibull distribution* and also as a *five-parameter Weibull distribution*.

Model II(b)-11

This model is obtained by a truncation of the standard Weibull model and is given by

$$G(t) = \frac{F(t) - F(a)}{F(b) - F(a)} \qquad 0 < a \leq t \leq b < \infty \tag{2.33}$$

with $F(t)$ given by (1.3). This is referred to as the *doubly truncated Weibull distribution*. Two special cases are as follows:

1. $a = 0$ and $b < \infty$: Right truncated Weibull distribution
2. $a > 0$ and $b \to \infty$: Left truncated Weibull distribution

For all three cases, the support for $G(t)$ is smaller than that for $F(t)$.

Model II(b)-12

The Slymen–Lachenbruce (1984) modified Weibull model is given by the following distribution function:

$$G(t) = 1 - \exp\left\{-\exp\left[\alpha + \frac{\beta(t^k - t^{-k})}{2k}\right]\right\} \tag{2.34}$$

The model is not related to the standard Weibull model but has a close connection with the Weibull transformation (see Section 1.5).

Model II(b)-13

Xie et al. (2002b) proposed a slight modification to a model first proposed by Chen (2000), and the model is given by

$$G(t) = 1 - \exp\left[\lambda\alpha\left(1 - e^{(t/\alpha)^{\beta}}\right)\right] \qquad t \geq 0 \tag{2.35}$$

with $\alpha, \beta, \lambda > 0$. They call it the *Weibull extension*. When α is large, we have $e^{(t/\alpha)} \approx 1 + (t/\alpha)$ and in this case, (2.35) reduces to a two-parameter Weibull distribution given by (1.3).

2.5 TYPE III MODELS: MODELS INVOLVING TWO OR MORE DISTRIBUTIONS

These models involve two or more distributions with one or more being Weibull distributions. The distributions involved are called the subpopulations.

2.5.1 Mixture Models

A general n-fold mixture model involves n subpopulations and is given by

$$G(t) = \sum_{i=1}^{n} p_i F_i(t) \quad p_i > 0 \qquad \sum_{i=1}^{n} p_i = 1 \tag{2.36}$$

Model III(a)-1

In this case the $F_i(t), i = 1, 2, \ldots,$ are either two- or three-parameter Weibull distributions. The model is called *finite Weibull mixture model*. In the literature, the Weibull mixture model has been referred to by many other names, such as, additive-mixed Weibull distribution, bimodal-mixed Weibull (for a two fold mixture), mixed-mode Weibull distribution, Weibull distribution of the mixed type, multimodal Weibull distribution, and so forth.

Model III(a)-2

This is similar to Model III(a)-1 except that the $F_i(t), i = 1, 2, \ldots,$ are inverse Weibull distributions [given by (2.11)].

Model III(a)-3

This is similar to Model III(a)-2 except that some of the subpopulations are Weibull distributions and the remaining are non-Weibull distributions.

2.5.2 Competing Risk Models

A general n-fold competing risk model involving n subpopulations is given by

$$G(t) = 1 - \prod_{i=1}^{n} [1 - F_i(t)] \tag{2.37}$$

The competing risk model is also called the compound model, series system model, multirisk model, and poly-Weibull model (if the subpopulations are Weibull distributions) in the reliability literature.

Model III(b)-1

In this case the $F_i(t), i = 1, 2, \ldots,$ are all two-parameter Weibull distributions. When $\beta_i = \beta_j, i \neq j$, then the two subpopulations can be merged into one subpopulation with shape parameter β_i, and as a result, the n-fold model gets reduced to a $(n - 1)$-fold model. Hence, we assume, without loss of generality, that $\beta_i < \beta_j$, $i < j$.

Model III(b)-2

This is similar to Model III(b)-1 except that $F_i(t), i = 1, 2, \ldots,$ are inverse Weibull distributions.

Model III(b)-3

This is similar to Model III(b)-2 except that some of the subpopulations are Weibull distributions and the remaining are non Weibull distributions.

Comments

1. Shooman (1968) proposed a model with the hazard function given by

$$h(t) = \sum_{i=0}^{n} k_i t^i \qquad t \geq 0 \tag{2.38}$$

When $k_i > 0$, the model is a special case of (2.37) with the shape parameters being the set of integers from 1 to $n + 1$.

2. The additive Weibull model proposed by Xie and Lai (1996) is also a special case of the competing risk model.

2.5.3 Multiplicative Models

A general n-fold multiplicative model is given by

$$G(t) = \prod_{i=1}^{n} F_i(t) \tag{2.39}$$

The general model has received very little attention. Basu and Klein (1982) deal with a general competing risk model and a so-called *complementary risk model*. The latter is identical to the multiplicative model defined above.

Model III(c)-1

In this case the $F_i(t), i = 1, 2, \ldots,$ are either two- or three-parameter Weibull distributions. When the $F_i(t)$ are the same for all i, the model is the same as the case of the exponentiated Weibull model.

When all the subpopulations are two-parameter Weibull distributions, then we can assume without loss of generality, $\beta_1 \leq \beta_2 \leq \cdots \leq \beta_n$ and when $\beta_i = \beta_j$, $i < j$, we assume $\alpha_i \geq \alpha_j$.

Model III(c)-2

This is similar to Model III(c)-1 except that the $F_i(t), i = 1, 2, \ldots,$ are inverse Weibull distributions.

Model III(c)-3

This is similar to Model III(c)-2 except that some of the subpopulations are Weibull distributions and the remaining are non-Weibull distributions.

2.5.4 Sectional Models

A general n-fold sectional model is given by

$$G(t) = \begin{cases} k_1 F_1(t) & 0 \leq t \leq t_1 \\ 1 - k_2 \bar{F}_2(t) & t_1 < t \leq t_2 \\ \cdots & \\ 1 - k_n \bar{F}_n(t) & t > t_{n-1} \end{cases} \tag{2.40}$$

where the subpopulations $F_i(t)$ are two- or three-parameter Weibull distributions and the t_i's (called partition points) are an increasing sequence. The distribution function, density function, and hazard function can be either continuous or discontinuous at the partition points. In the former case the k_i's are constrained.

The sectional model is also known as composite model, piecewise model, and step-function model.

2.6 TYPE IV MODELS: WEIBULL MODELS WITH VARYING PARAMETERS

The parameters of the models discussed so far are fixed constants. This section deals with models where some of the parameters are (i) functions of the variable t, (ii) functions of some supplementary variables (denoted by S), and (iii) random variables. The first category can be divided into two subgroups, and, as a result, we have four types of models and these are discussed in the remainder of the section.

2.6.1 Time-Varying Parameters

In these models the scale parameter (α) and/or the shape parameter (β) of the standard model given by (1.3) are functions of the variable t.

Model IV(a)-1

Here the scale parameter α is a function of the independent variable t. Shrivasatava (1974) considers the case $\alpha(t)$ is binary valued and changes in a periodic manner. Zacks (1984) deals with a model where the shape parameter $\beta = 1$ for $t < t_0$ and changes to $\beta > 1$ for $t > t_0$.

Model IV(a)-2

Here both α and β are functions of t. Zuo et al. (1999) consider the case where

$$\beta(t) = a\left(1 + \frac{1}{t}\right)^b e^{c/t}, \quad \alpha(t) = a' t^{b'} e^{c'/t} \tag{2.41}$$

Model IV(a)-3

The model is given by

$$G(t) = 1 - e^{-\Lambda(t)} \tag{2.42}$$

where $\Lambda(t)$ is a nondecreasing function with $\Lambda(0) = 0$ and $\Lambda(\infty) = \infty$. Srivastava (1989) calls this the *generalized Weibull distribution* and looks at the special case where

$$\Lambda(t) = \sum_{c=1}^{m} \lambda_i \phi_i(t) \qquad 1 \leq i \leq m \tag{2.43}$$

The special case

$$\Lambda(t) = \left(\frac{t}{\alpha}\right)^{\beta} \tag{2.44}$$

corresponds to the standard two-parameter Weibull model given by (1.3).

2.6.2 Weibull Accelerated Life Models

In these models the scale parameter α is a function of some supplementary variable S. In reliability applications, S represents the stress on the item and the life of the item [a random variable with distribution $G(\cdot)$] is a function of S. The shape parameter is unaffected by S.

Model IV(b)-1

The term $\alpha(S)$ has the form

$$\alpha(s) = \exp(\gamma_0 + \gamma_1 S) \tag{2.45}$$

so that $\ln[\alpha(S)]$ is linear in S. The model is called the *Arrhenius Weibull model*. For the log Weibull model [Model I(b)-2] this implies that scale parameter is linear in S.

Model IV(b)-2

The term $\alpha(S)$ has the form

$$\alpha(s) = \frac{e^{\gamma_0}}{S^{\gamma_1}} \tag{2.46}$$

and the model is called the *power Weibull model*.

Comment

These types of models have been used extensively in accelerated life testing [see Nelson (1990)] in reliability theory. As a result, they are referred to as the *accelerated failure models*. A more general formulation is one where the scale parameter has the form

$$\alpha(S) = \exp\left(b_0 + \sum_{i=1}^{k} b_i s_i\right) \tag{2.47}$$

where $S[S^T = (s_1, s_2, \ldots, s_k)]$ is a k-dimensional vector of supplementary variables.

2.6.3 Weibull Proportional Hazard Models

In Type IV(b) models, the scale parameter is a function of the supplementary variable S. An alternate approach to modeling the effect of the supplementary variable on the distribution function $F(t)$ is through its hazard function $h(t)$. As a result, we have

$$h(t) = \psi(S)h_0(t) \tag{2.48}$$

where $h_0(t)$ is called the baseline hazard. The only restriction on the scalar function $\psi\,(\,)$ is that it be a positive. Many different forms for $\psi\,(\,)$ have been proposed. One such is the following:

$$\psi(S) = \exp\left(b_0 + \sum_{i=1}^{k} b_i s_i\right) \tag{2.49}$$

These models are also called the *proportional hazard models* [see Nelson (1990)].

2.6.4 Model with Time and Parameter Change

This model is a combination of IV(a) and IV(b) type models.

2.6.5 Random Parameters

Here the scale parameter (α) in the standard Weibull model is a random variable with distribution function $F_\alpha(\cdot)$. Let Y be a random variable from a distribution given by (1.3) conditional on $\alpha = u$. Then, the unconditional distribution of Y is given by

$$G(t) = \int F(t|\alpha = u)\, dF_\alpha(u) \tag{2.50}$$

Note that this is a *continuous mixture model* and is called as the *compound Weibull model* by Dubey (1968). Different forms for $F_\alpha(\cdot)$ lead to different forms for $G(t)$.

2.7 TYPE V MODELS: DISCRETE WEIBULL MODELS

These are discrete versions of the standard Weibull model. As a result, the model involves a discrete distribution. Three models that have appeared in the literature are as follows:

Model V-1

The distribution function is given by

$$F(t) = \begin{cases} 1 - q^{t^\beta} & t = 0, 1, 2, 3, \ldots \\ 0 & t < 0 \end{cases} \tag{2.51}$$

This model is due to Nakagawa and Osaki (1975).

Model V-2

The cumulative hazard function is given by

$$H(t) = \begin{cases} ct^{\beta-1} & t = 1, 2, \ldots, m \\ 0 & t < 0 \end{cases} \tag{2.52}$$

where m is given by

$$m = \begin{cases} \text{int}\{c^{-[1/(\beta-1)]}\} & \text{if } \beta > 1 \\ \infty & \text{if } \beta \le 1 \end{cases} \tag{2.53}$$

and int $\{\cdot\}$ represents the integer part of the quantity inside the brackets. This model is due to Stein and Dattero (1984).

Model V-3

The distribution function is given by

$$F(t) = 1 - \exp\left[-\sum_{i=1}^{t} r(i)\right] = 1 - \exp\left[-\sum_{i=1}^{t} c i^{\beta-1}\right] \qquad t = 0, 1, 2, \ldots \quad (2.54)$$

This model is due to Padgett and Spurrier (1985).

2.8 TYPE VI MODELS: MULTIVARIATE WEIBULL MODELS

This group of models is multivariate extensions of the univariate case so that the distribution function is given by an n-dimensional distribution function $F(t_1, t_2, \ldots, t_n)$. The models can be grouped into two groups based on basis of the extension. These are as follows:

Type VI(a) These are transforms of the multivariate exponential distributions using the power law relationship.

Type VI(b) These are multivariate distributions that have the marginal distributions as univariate Weibull distributions.

We first consider the bivariate case ($n = 2$) and later on we look at the multivariate case ($n > 2$). For the models in the remainder of the section, the ranges of the independent variables are $[0, \infty)$.

2.8.1 Bivariate Models

Model VI(a)-1

This is obtained from the power law transformation of the bivariate exponential distribution studied in Marshall and Olkin (1967) with the survivor function

$$\bar{F}(t_1, t_2) = \exp\left\{-\left[\lambda_1 t_1^{\beta_1} + \lambda_2 t_2^{\beta_2} + \lambda_{12} \max\left(t_1^{\beta_1}, t_2^{\beta_2}\right)\right]\right\} \quad t_1 \geq 0, t_2 \geq 0 \quad (2.55)$$

This reduces to the bivariate exponential distribution when $\beta_1 = \beta_2 = 1$.

Model VI(a)-2

A related model due to Lee (1979) involves the transformation $Z_i = T_i / c_i$ and $\beta_1 = \beta_2 = \beta$. The model is given by

$$\bar{F}(t_1, t_2) = \exp\left\{-\left[\lambda_1 c_1^{\beta} t_1^{\beta} + \lambda_2 c_2^{\beta} t_2^{\beta} + \lambda_{12} \max\left(c_1^{\beta} t_1^{\beta}, c_2^{\beta} t_2^{\beta}\right)\right]\right\} \quad (2.56)$$

Model VI(a)-3

Yet another related model due to Lu (1989) has the following survival function:

$$\bar{F}(t_1, t_2) = \exp\left[-\lambda_1 t_1^{\beta_1} - \lambda_2 t_2^{\beta_2} - \lambda_0 \max(t_1, t_2)^{\beta_0}\right] \tag{2.57}$$

This can be seen as a slight modification (or generalisation) of the Marshall and Olkin's bivariate exponential distribution due to the exponent in the third term having a new parameter.

Model VI(a)-4

A general model proposed by Lu and Bhattacharyya (1990) has the following form:

$$\bar{F}(t_1, t_2) = \exp\left[-\left(\frac{t_1}{\alpha_1}\right)^{\beta_1} - \left(\frac{t_2}{\alpha_2}\right)^{\beta_2} - \delta\psi(t_1, t_2)\right] \tag{2.58}$$

Different forms for the function of $\psi(t_1, t_2)$ yield a family of models. One form for $\psi(t_1, t_2)$ is the following:

$$\psi(t_1, t_2) = \left[\left(\frac{t_1}{\alpha_1}\right)^{\beta_1/m} + \left(\frac{t_2}{\alpha_2}\right)^{\beta_2/m}\right]^m \tag{2.59}$$

This yields the following survival function for the model:

$$\bar{F}(t_1, t_2) = \exp\left\{-\left(\frac{t_1}{\alpha_1}\right)^{\beta_1} - \left(\frac{t_2}{\alpha_2}\right)^{\beta_2} - \delta\left[\left(\frac{t_1}{\alpha_1}\right)^{\beta_1/m} + \left(\frac{t_2}{\alpha_2}\right)^{\beta_2/m}\right]^m\right\} \tag{2.60}$$

Model VI(a)-5

This is another case of (2.58) with

$$\begin{aligned}\bar{F}(t_1, t_2) = \exp\Bigg(&-\left(\frac{t_1}{\alpha_1}\right)^{\beta_1} - \left(\frac{t_2}{\alpha_2}\right)^{\beta_2} - \delta\left\{1 - \exp\left[-\left(\frac{t_1}{\alpha_1}\right)^{\beta_1}\right]\right\} \\ &\times \left\{1 - \exp\left[-\left(\frac{t_2}{\alpha_2}\right)^{\beta_2}\right]\right\}\Bigg)\end{aligned} \tag{2.61}$$

Model VI(a)-6

Finally, the Morgenstern–Gumbel–Farlie system of distributions (Johnson and Kotz, 1970b) is given by

$$\bar{F}(t_1, t_2) = \bar{F}_1(t_1)\bar{F}_2(t_2)\{1 + \gamma[1 - \bar{F}_1(t_1)][1 - \bar{F}_2(t_2)]\} \tag{2.62}$$

With $\bar{F}_i(t_i) = \exp\{-t_i^{c_i}\}$, this yields a special case of the model given by (2.58).

Model VI(a)-7

Sarkar (1987) proposed a new bivariate exponential distribution and mentions that with simple power transformation, a bivariate Weibull distribution can be obtained. The survival function is given by

$$\bar{F}(t_1, t_2) = \begin{cases} \exp\left[-(\lambda_1 + \lambda_{12})t_1^{\beta_1}\right]\left\{1 - \left[A\left(\lambda_2 t_1^{\beta_1}\right)\right]^{-\gamma}\left[A\left(\lambda_2 t_2^{\beta_2}\right)\right]^{1+\gamma}\right\} & t_1 \geq t_2 > 0 \\ \exp\left\{-(\lambda_2 + \lambda_{12})t_2^{\beta_2}\right\}\left\{1 - \left[A\left(\lambda_1 t_2^{\beta_2}\right)\right]^{-\gamma}\left[A\left(\lambda_1 t_1^{\beta_1}\right)\right]^{1+\gamma}\right\} & t_2 \geq t_1 > 0 \end{cases} \tag{2.63}$$

where $\gamma = \lambda_{12}/(\lambda_1 + \lambda_2)$ and $A(z) = 1 - e^{-z}, z > 0$.

Model VI(b)-1

A different type of bivariate Weibull distributions due to Lu and Bhattacharyya (1990) is given by

$$\bar{F}(t_1, t_2) = \left[1 + \left(\left\{\exp\left[\left(\frac{t_1}{\alpha_1}\right)^{\beta_1}\right] - 1\right\}^{1/\gamma} + \left\{\exp\left[\left(\frac{t_2}{\alpha_2}\right)^{\beta_2}\right] - 1\right\}^{1/\gamma}\right)^{\gamma}\right]^{-1} \tag{2.64}$$

Model VI(b)-2

Lee (1979) proposed the following bivariate Weibull distribution:

$$\bar{F}(t_1, t_2) = \exp\left[-\left(\lambda_1 t_1^{\beta_1} + \lambda_2 t_2^{\beta_2}\right)^{\gamma}\right] \tag{2.65}$$

where $\beta_i > 0, 0 < \gamma \leq 1, \lambda_i > 0, t_i \geq 0, i = 1, 2$.

A similar model has been proposed by Hougaard (1986) and studied in Lu and Bhattacharyya (1990).

2.8.2 Multivariate Models

We use the following notation: **t** is a n-dimensional vector with components t_i, $1 \leq i \leq n$.

Model VI(a)-8

This is a multivariate version of Model VI(a)-1 and the survivor function is given by

$$\bar{G}(\mathbf{t}) = \exp\left[-\sum_{S} \lambda_S \max_{i \in S} (t_i^{\alpha})\right] \qquad \mathbf{t} > 0 \tag{2.66}$$

where S denote the set of vectors $(s_1, s_2, \ldots s_n)$ where each element takes value of 0 or 1, but not all equal to 0.

Model VI(b)-3

Roy and Mukherjee (1998) proposed a general class of multivariate distribution with Weibull marginals and the survivor function is given by

$$\bar{G}(\mathbf{t}) = \exp\left\{-\left[\sum_{i=1}^{p}(\lambda_i^{\nu} t_i^{\alpha\nu})\right]^{1/\nu}\right\} \qquad \mathbf{t} > 0 \tag{2.67}$$

A similar model can also be found in Hougaard (1986) and Crowder (1989).

Model VI(b)-4

The following model is from Patra and Dey (1999). It is based on the study of r-component system for which the lifetime of each component is assumed to be a mixture of Weibull distribution. The survivor function is given by

$$\bar{G}(\mathbf{t}) = \prod_{i=1}^{r}\sum_{j=1}^{k} a_{ij} \exp\left[-\left(\theta_{ij} t_i^{\alpha_{ij}} + \frac{\theta_0 t_0}{r}\right)\right] \qquad \mathbf{t} > 0 \tag{2.68}$$

where $t_0 = \max(t_1, \ldots, t_r)$.

2.9 TYPE VII MODELS: STOCHASTIC POINT PROCESS MODELS

These are stochastic point process models with some link to the standard Weibull model.

2.9.1 Weibull Intensity Function Models

Model VII(a)-1: Power Law Process

This is a point process model with the intensity function given by

$$\lambda(t) = \frac{\beta t^{\beta-1}}{\alpha^{\beta}} \tag{2.69}$$

This is identical to the hazard function for the standard Weibull model. This model has been called by many different names: *power law process* (Bassin, 1973; Rigdon and Basu, 1989; Klefsjo and Kumar, 1992); *Rasch–Weibull process* (Moller, 1976); *Weibull intensity function* (Crow, 1974); *Weibull–Poisson process* (Bain and Englehardt, 1980); and *Weibull process* (Bain, 1978). Ascher and Feingold (1984) address the confusion and misconception that has resulted from this plethora of terminology.

Model VII(a)-2: Modulated Power Law Process

The joint density function for the time instants ($t_1 < t_2 < \cdots < t_n$) of the first n events for the modulated power law process is given by

$$f(t_1, t_2, \ldots, y_n) = \left\{ \prod_{i=1}^{n} \lambda(t_i)[\Lambda(t_i) - \Lambda(t_{i-1})]^{k-1} \right\} \frac{\exp[-\Lambda(t_n)]}{[\Gamma(k)]^n} \qquad (2.70)$$
$$0 < t_1 < t_2 < \cdots < t_n < \infty$$

where $\lambda(t)$ is given by (2.69) and

$$\Lambda(t) = \int_0^t \lambda(x)\, dx \qquad (2.71)$$

This model is due to Lakey and Rigdon (1993) and is a special case of the modulated gamma process proposed by Berman (1981).

Model VII(a)-3: Proportional Intensity Model

The intensity function is given by

$$\lambda(t; S) = \lambda_0(t)\psi(S) \qquad t \geq 0 \qquad (2.72)$$

where $\lambda_0(t)$ is of the form given by (2.69) and $\psi(S)$ is a function of explanatory variables S. The only restriction on $\psi(S)$ is that $\psi(S) > 0$. Various forms of $\psi(S)$ have been studied; see Cox and Oaks (1984) and Prentice et al. (1981).

2.9.2 Weibull Renewal Process Models

Here we have a variety of models.

Model VII(b)-1: Ordinary Renewal Process

Here the point process is a renewal process with the time between events being independent and identically distributed with the distribution function given by (2.1).

Model VII(b)-2: Modified Renewal Process

Here the distribution function for the time to first event, $F_0(t)$, is different from that for the subsequent interevent times, which are identical and independent random variables with distribution function $F(\cdot)$. Note that $F_0(t)$ and/or $F(t)$ are standard Weibull distributions. When $F_0(t) = F(t)$, this reduces to the ordinary renewal process.

Model VII(b)-3: Alternating Renewal Process

Here the odd-numbered interevent times are a sequence of independent and identically distributed random variables with a distribution function $F_1(t)$. The even-numbered interevent times are another sequence of independent and identically distributed random variables with a distribution function $F_2(t)$. As before, either $F_1(t)$ and/or $F_2(t)$ are standard Weibull distributions. For one such example, see, for example, Dickey (1991).

2.9.3 Type VII(c) Model: A Generalized Model

This model combines the power law intensity model and the Weibull renewal model. Let $(t_1 < t_2 < \cdots < t_n)$ be the time instants for the first n events. For $t_i \le t < t_{i+1}$, define $u(t) = t - t_i$. In this model, the occurrence of events is given by intensity function

$$\lambda(t) = \frac{\beta + \delta - 1}{\alpha^{\beta+\delta-1}} t^{\beta-1} [u(t)]^{\delta-1} \qquad t \ge 0 \tag{2.73}$$

with $\alpha > 0$ and $\beta + \delta > 1$. Note that when $\delta = 1$, the model reduces to the simple Weibull intensity model [Model VII(a)-1] and when $\beta = 1$, it reduces to an ordinary Weibull renewal process model [Model VII(b)-1]. This model is a special case of a more general class of processes introduced by Lawless and Thiagarajah (1996).

EXERCISES

2.1. Is the taxonomy proposed for the Weibull models in Section 2.2 valid for models based on some distribution other than the Weibull distribution?

2.2. Propose some new nonlinear transformations that will yield new Type I(b) Weibull models.

2.3. Propose some new Type II(a) and Type II(b) Weibull models. (Hint: Combine the features of one or more models discussed in Section 2.4 to derive new models.)

2.4. Propose four different models (involving a Weibull and a non-Weibull distribution) belonging to Type III(a)-3. Give the details of the model structure.

2.5. Repeat Exercise 2.4 with Type III(a)-3 replaced by Type III(b)-3.

2.6. Repeat Exercise 2.4 with Type III(a)-3 replaced by Type III(c)-3.

2.7. Can one use a Type IV(b) model to model the lifetime of Exercise 1.5? If so, indicate the possible relationship between parameters of the model and the relevant factors.

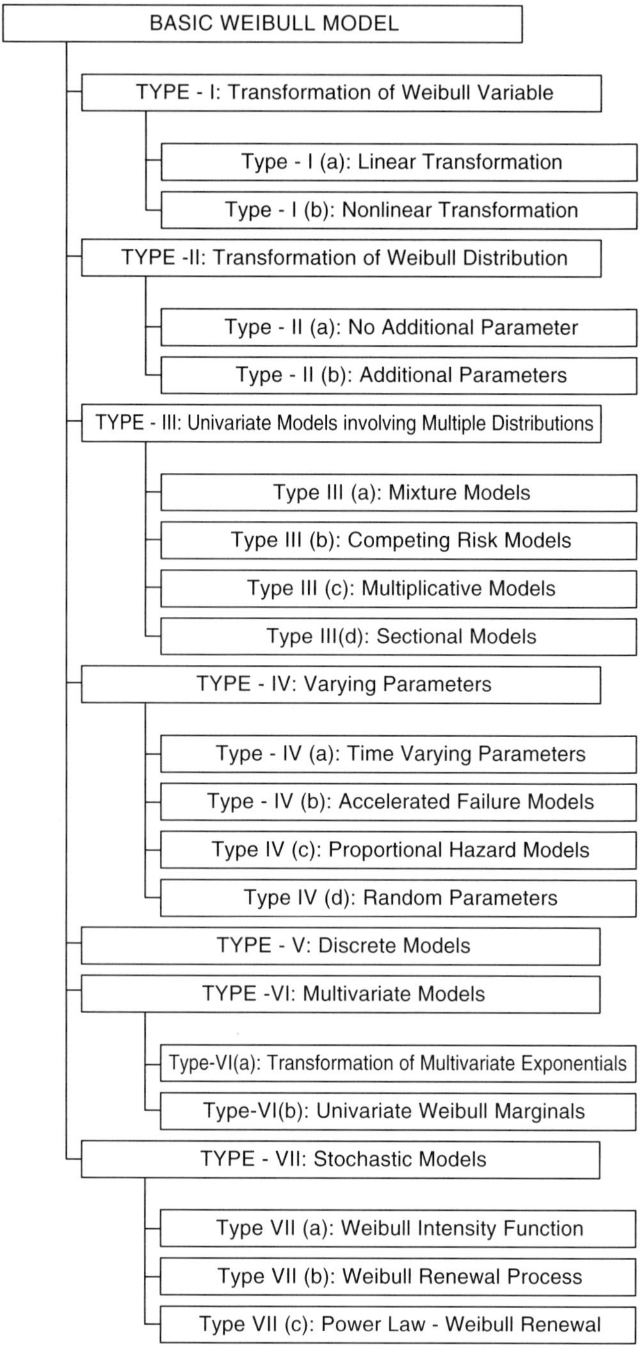

Figure 2.1 Taxonomy for Weibull Models.

2.8. Describe three real-world problems where one or more variables can be modeled by a Type V model.

2.9. For many components the degradation and failure is a function of age and usage. Describe some real-world examples where this is true and discuss how one can use Type VI bivariate models to model the failures.

2.10. The variables $(t_1, t_2, \ldots, t_n)$ for Type VI models in Section 2.8 are continuous valued. How would the formulation change if one or more of them are discrete valued?

2.11. The cracks in a long pipeline can be viewed as random points along the length. One can use a Type VII(b) model to model this. Give the details of your model formulation.

PART B

Basic Weibull Model

CHAPTER 3

Model Analysis

3.1 INTRODUCTION

A thorough understanding of model properties is important in deciding on the appropriateness of a model to model a given data set or for its application in a particular context. Model analysis deals with this topic. In this chapter we introduce various concepts and techniques needed for model analysis. Following this, we carry out an analysis of the standard Weibull model [given by (1.3)] and the three-parameter Weibull model [given by (1.2)]. The techniques and concepts discussed will be used in later chapters for the analysis of other models derived from the standard Weibull model.

The outline of the chapter is as follows. In Section 3.2 we discuss some basic concepts needed for the analysis of models in general. Section 3.3 deals with the analysis of the standard Weibull model, and Section 3.4 deals with the analysis of the three-parameter Weibull model. The main focus in these sections is a study of the density function, hazard function, and Weibull transformation.

3.2 BASIC CONCEPTS

3.2.1 Density, Hazard, and Cumulative Hazard Functions

In Section 2.2.8, we defined the density function $f(t)$, the survivor function $\bar{F}(t)$, the hazard function $h(t)$, and the cumulative hazard function $H(t)$ associated with a distribution function $F(t)$. These are related to each other as indicated below:

$$F(t) = 1 - e^{-H(t)} \qquad f(t) = h(t)e^{-H(t)} \quad \text{and} \quad H(t) = -\ln[1 - F(t)] \tag{3.1}$$

The survivor function $\bar{F}(t)$ is of special importance in reliability theory as it represents the reliability of a component.

Weibull Models, by D.N.P. Murthy, Min Xie, and Renyan Jiang.
ISBN 0-471-36092-9

The shapes of the density and hazard functions depend on the model parameters. Weibull models exhibit a wide variety of shapes for the density and hazard functions. An understanding of these different shapes is part of model analysis and of critical importance in the context of model selection, and this is discussed further in Chapter 16.

The different possible shapes for the density function are as follows:

- Type 1: Decreasing
- Type 2: Unimodal
- Type 3: Decreasing followed by unimodal
- Type 4: Bimodal
- Type $(2k - 1)$: Decreasing followed by $k - 1$ modal $(k > 2)$
- Type $(2k)$: k modal $(k > 2)$

Figure 3.1 shows types 1 to 4. A further subdivision is possible based on whether $f(0)$ is zero, finite, or infinite.

The different possible shapes for the hazard function are as follows:

- Type 1: Decreasing
- Type 2: Constant
- Type 3: Increasing
- Type $4k$: k reverse modal $(k \geq 1)$

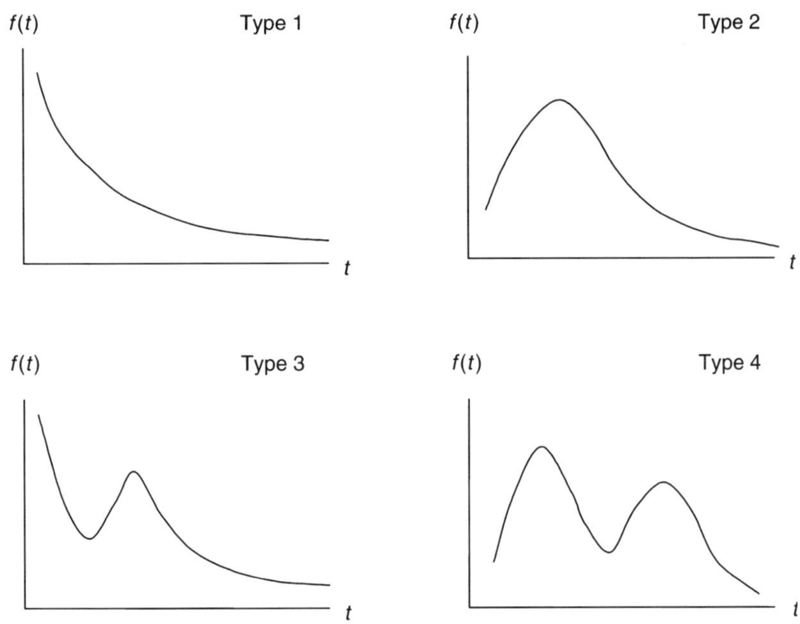

Figure 3.1 Different shapes for the density function.

[Note: Type 4 ($k = 1$) is Bathtub shaped and type 8 ($k = 2$) is W shape]

- Type $4k + 1$: k modal ($k \geq 1$)

[Note: Type 5 ($k = 1$) is unimodal and type 9 ($k = 2$) is bimodal]

- Type $4k + 2$: k modal followed by increasing ($k \geq 1$)
- Type $4k + 3$: Decreasing followed by k modal ($k \geq 1$)

Figure 3.2 shows types 1 to 8. A further subdivision is based on whether $h(\infty)$ is finite or infinite.

The possible shapes for the density and hazard functions will be discussed in detail for the different Weibull models in the remainder of the book.

A *mode* of a density or hazard function is the value of t which corresponds to a local maximum of the function. There can be none, one, or more modes depending on the model and its parameters.

3.2.2 Moments

The term $M_j(\theta)$, the jth moment about the origin, of a continuous random variable T from a distribution $F(t;\theta)$ (also called the jth moment of the probability distribution), $j \geq 1$, is given by $M_j(\theta) = E[T^j]$ (where $E[\cdot]$ is the *expectation* of the random variable). It is obtained as follows:

$$M_j(\theta) = \int_{-\infty}^{\infty} t^j f(t;\theta)\, dt. \tag{3.2}$$

The first moment $M_1(\theta)$ is also denoted μ.

The term $\mu_j(\theta)$, the jth central moment of a continuous random variable T, $j \geq 1$, is given by $\mu_j(\theta) = E[(T - \mu)^j]$ and is calculated as

$$\mu_j(\theta) = \int_{-\infty}^{\infty} (t - \mu)^j f(t;\theta)\, dt \tag{3.3}$$

Note that $M_j(\theta)$ and $\mu_j(\theta)$ ($j \geq 1$) are functions of the parameters of the distribution function $F(t;\theta)$. This will be used in Chapter 4 for estimation of model parameters:

1. The first moment μ is called the *mean*. It is a measure of central tendency.
2. The second central moment, $\mu_2(\theta)$, is called the *variance* and is also denoted σ^2. The square root of the variance (σ) is called the *standard deviation*. Both are measures of dispersion or spread of a probability distribution.
3. The expression $\gamma_1 = \mu_3(\theta)/[\mu_2(\theta)]^{3/2}$ is called the *coefficient of skewness*. It is measure of departure from symmetry of the density function $f(x;\theta)$. If

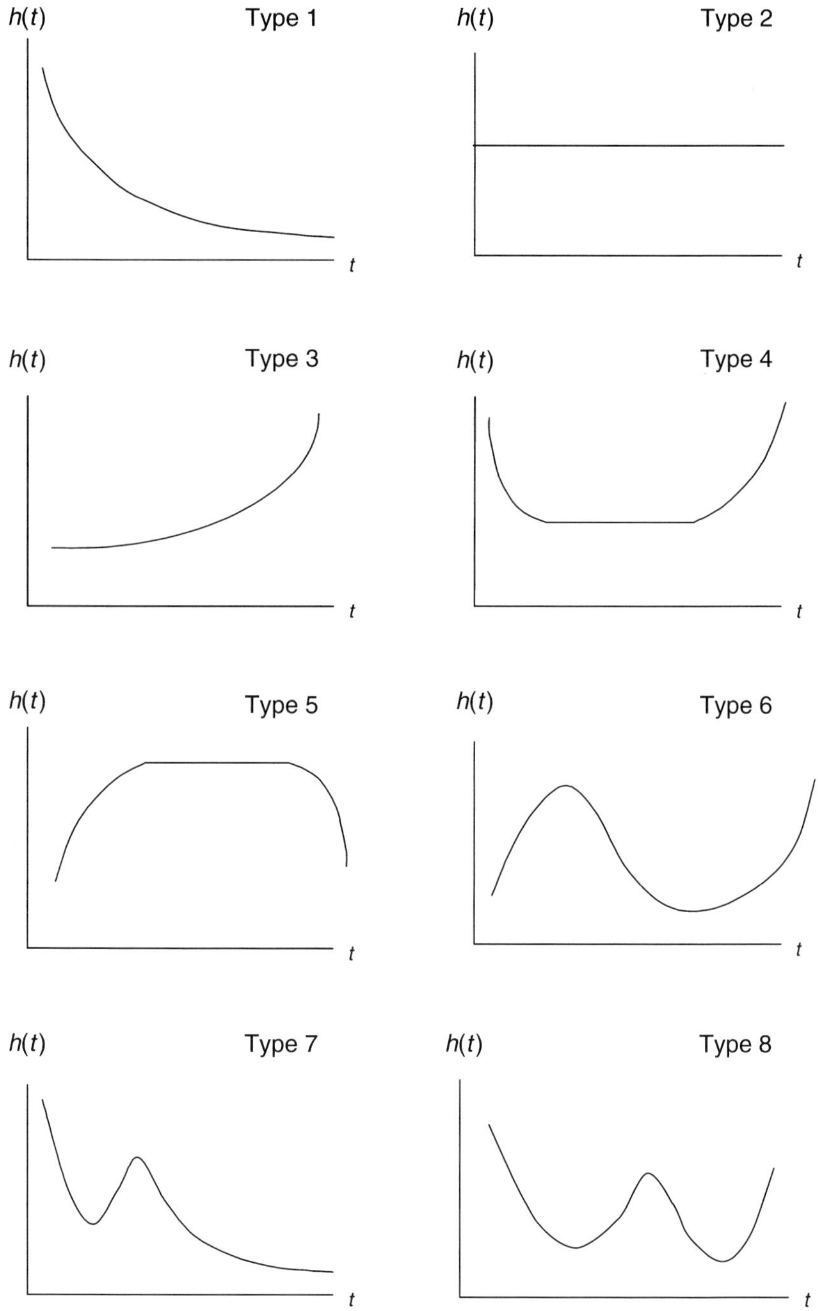

Figure 3.2 Different shapes for the hazard function.

$\gamma_1 > 0$, it is positively skewed (or skewed right), and if $\gamma_1 < 0$, it is negatively skewed (or skewed left).

4. The expression $\gamma_2 = \mu_4(\theta)/[\mu_2(\theta)]^2$ is called the *coefficient of excess or kurtosis*. It is a measure of the degree of the flatness (or peakedness) of the density function. If $\gamma_2 = 3$, the density function is called mesokurtic. If $\gamma_2 > (<)3$, it is called leptokurtic (platykurtic).
5. The expression $\rho = \sigma/\mu$ is called the *coefficient of variation*. It is a standardized (unit free) measure of dispersion.

The moments of a distribution are easily derived from the *moment generating function*, which is given by

$$\psi(s) = E(e^{sT}) = \int_{-\infty}^{\infty} e^{st} f(t)\, dt \tag{3.4}$$

with

$$M_j = E(T^j) = \left.\frac{d^j \psi(s)}{ds^j}\right|_{s=0} \tag{3.5}$$

3.2.3 Percentile and Median

For a continuous distribution, the p percentile (also referred to as fractile or quantile), t_p, for a given $p, 0 < p < 1$, is a number such that

$$P\{T \le t_p\} = F(t_p) = p \tag{3.6}$$

The percentile for $p = 0.25$ and 0.75 are called first and third *quartiles* and the 0.50 percentile is called the *median*.

3.2.4 Order Statistics

Let $T_1, T_2, \ldots, T_n$ denote n-independent random variables from a distribution function $F(t)$. The ordered sample is the arrangement of the sample values from the smallest to the largest and denoted by $T_{(1)}, T_{(2)}, \ldots, T_{(n)}$. The kth value of this ordered sample is then called the kth order statistic of the sample. The joint probability density function of $T_{(1)}, T_{(2)}, \ldots, T_{(r)}; r \le n$, is given by

$$f_r(t_{(1)}, t_{(2)}, \ldots, t_{(r)}) = \frac{n!}{(n-r)!}\left[\prod_{i=1}^{r} f(t_{(i)})\right] \cdot [1 - F(t_{(r)})]^{n-r} \tag{3.7}$$

Order statistics play an important role in data analysis and are useful in statistical inference. For further details about order statistics, see Balakrishnan and Rao (1998) and Harter (1978).

3.2.5 WPP Plot

The Weibull transformation involves the transformations given in (1.7). A plot of y versus x is called the WPP plot and, as mentioned in Section 1.5, plays an important role in model selection. This will be discussed further in the remainder of the book.

3.2.6 TTT Plot

For a distribution $F(t)$, define $F^{-1}(t) = \inf\{x : F(x) = t\}$. The total time on test (TTT) transform of $F(t)$, denoted by $\Psi_F^{-1}(t)$, is defined (Barlow and Campo, 1975; Bergman and Klefsjo, 1984) as

$$\Psi_F^{-1}(t) = \int_0^{F^{-1}(t)} \bar{F}(x)\,dx \qquad 0 \le t \le 1 \tag{3.8}$$

The TTT transform and the hazard function are related as indicated below:

$$\left.\frac{d}{dx}\Psi_F^{-1}(x)\right|_{x=F(t)} = \frac{1}{h(t)} \tag{3.9}$$

An important use of the TTT transform is the development of the TTT plot. The TTT plot provides a useful graphical technique for model identification.

3.3 STANDARD WEIBULL MODEL

The standard Weibull model is given by (1.3). The survivor function is given by

$$\bar{F}(t) = 1 - F(t) = \exp\left[-\left(\frac{t}{\alpha}\right)^{\beta}\right] \qquad t \ge 0 \tag{3.10}$$

3.3.1 Parametric Study of Density Function

The density function is given by

$$f(t) = \frac{dF(t)}{dt} = \frac{\beta t^{\beta-1}}{\alpha^{\beta}} \exp\left[-\left(\frac{t}{\alpha}\right)^{\beta}\right] \tag{3.11}$$

The shape of the density function depends on the model parameters. The two possible shapes for the density function are as follows:

- Monotonically decreasing
- Unimodal

A parametric study is a characterization of the shapes in the two-dimensional parameter plane. It is easily seen that the shape of the density function depends only on the shape parameter β and the scale parameter has no effect. The results are as follows:

- For $\beta \leq 1$ the density function is monotonically decreasing (type 1 in Fig. 3.1).
- For $\beta > 1$ the density function is unimodal with the mode at $t_m = \alpha[(\beta - 1)/\beta]^{1/\beta}$ (type 2 in Fig. 3.1).

3.3.2 Parametric Study of Hazard Function

The hazard and cumulative hazard functions are given by

$$h(t) = \frac{f(t)}{1 - F(t)} = \frac{\beta t^{\beta-1}}{\alpha^{\beta}} \tag{3.12}$$

and

$$H(t) = \int_0^t h(x)\,dx = \left(\frac{t}{\alpha}\right)^{\beta} \tag{3.13}$$

respectively.

As with the density function, the shape depends only on the shape parameter and the scale parameter has no effect. The three possible shapes are as follows:

- For $\beta < 1$ the hazard function is decreasing (type 1 in Fig. 3.2).
- For $\beta = 1$ the hazard function is a constant (type 2 in Fig. 3.2).
- For $\beta > 1$ the hazard function is increasing (type 3 in Fig. 3.2).

3.3.3 Moments

The moment generating function is given by

$$\psi(s) = \alpha^s \Gamma(1 + s/\beta) \tag{3.14}$$

As a result, the kth moment about the origin is given by

$$M_k = \alpha^k \Gamma(1 + k/\beta) \tag{3.15}$$

where $\Gamma(\cdot)$ is the gamma function [see, Abramowitz and Stegun (1964) for more details]. The mean, μ, is given by

$$\mu = E(T) = \alpha \Gamma(1 + 1/\beta) \tag{3.16}$$

The kth moment of the Weibull distribution about the mean is given by

$$\mu_k = \alpha^k \sum_{j=0}^{k} (-1)^j \binom{k}{j} \Gamma\left(\frac{k-j}{\beta} + 1\right) \left[\Gamma\left(\frac{1}{\beta} + 1\right)\right]^j \tag{3.17}$$

The variance, σ^2, is given by

$$\sigma^2 = \alpha^2 \left[\Gamma\left(1 + \frac{2}{\beta}\right) - \left[\Gamma\left(1 + \frac{1}{\beta}\right)\right]^2\right] \tag{3.18}$$

The third moment, μ_3, is given by

$$\mu_3 = \alpha^3 \left\{\Gamma\left(1 + \frac{3}{\beta}\right) - 3\Gamma\left(1 + \frac{2}{\beta}\right)\Gamma\left(1 + \frac{1}{\beta}\right) + 2\left[\Gamma\left(1 + \frac{1}{\beta}\right)\right]^3\right\} \tag{3.19}$$

From this, one can derive the skewness. The fourth moment, μ_4, is given by

$$\begin{aligned} \mu_4 = \alpha^4 \Bigg\{ & \Gamma\left(1 + \frac{4}{\beta}\right) - 4\Gamma\left(1 + \frac{3}{\beta}\right)\left[\Gamma\left(1 + \frac{1}{\beta}\right)\right] \\ & + 6\Gamma\left(1 + \frac{2}{\beta}\right)\left[\Gamma\left(1 + \frac{1}{\beta}\right)\right]^2 - 3\left[\Gamma\left(1 + \frac{1}{\beta}\right)\right]^4\Bigg\} \end{aligned} \tag{3.20}$$

From this, one can derive the kurtosis. Finally, the coefficient of variation is given by

$$\rho = \left[\frac{\Gamma(1 + 2/\beta)}{[\Gamma(1 + 1/\beta)]^2} - 1\right]^{1/2} \tag{3.21}$$

When the shape parameter is large, the mean and variance can be approximated by the following expressions (McEwen and Parresol, 1991):

$$\mu \approx 1 - \frac{\gamma}{\beta} + \frac{\pi^2/6 + \gamma^2}{2\beta^2} \quad \text{and} \quad \sigma^2 \approx \frac{\pi^2}{6\beta^2} \tag{3.22}$$

where $\gamma \approx 0.57721566$ is the Euler constant.

For $\beta > 3.6$, the Weibull distribution has a form very similar to a normal distribution (Dubey, 1967a). The coefficient of skewness is zero for $\beta = 3.6$, and the coefficient of kurtosis has a minimum value of 2.71 for $\beta = 3.35$ [see, Mudholkar and Kollia (1994)].

3.3.4 Median, Mode, and Percentile

The median, t_{median}, is given by

$$t_{\text{median}} = \alpha(\ln 2)^{1/\beta} \tag{3.23}$$

When $\beta > 1$, there is a single mode, t_{mode}, and it is given by

$$t_{\text{mode}} = \alpha(1 - 1/\beta)^{1/\beta} \tag{3.24}$$

When $\beta \leq 1$ the mode is at $t = 0$.

The p percentile (or p quantile) is given by

$$t_p = \alpha[-\ln(1 - p)]^{1/\beta} \tag{3.25}$$

3.3.5 Order Statistics

The distribution function for the kth order statistic is given by (3.7) with $F(t)$ given by (1.3). As indicated in Chapter 2, the standard Weibull model is related to the extreme-value distribution. Lieblein (1955) gives the expression for the noncentral moments in terms of gamma functions and for the covariances in terms of gamma and incomplete beta functions. Tables for these values have been tabulated. For a related discussion, see Harter (1978).

3.3.6 Weibull Random Variable

Weibull random variables can be generated from the uniform distribution. Let U be uniform $[0, 1]$ random variable, then $\alpha(-\ln U)^{1/\beta}$ is Weibull distributed with parameter (α, β).

3.3.7 WPP Plot

Under the Weibull transformation (1.7), (1.3) gets transformed into

$$y = \beta x - \beta \ln(\alpha) \tag{3.26}$$

This is an equation to a straight line with the slope β (the shape parameter). The intercept with the y axis is $-\beta \ln(\alpha)$ and with the x axis is $\ln(\alpha)$.

3.3.8 TTT Plot

The TTT plot is a straight line for $\beta = 1$. It is convex for $\beta < 1$ and concave for $\beta > 1$. For further discussion, see Bergman and Klefsjo (1984).

3.3.9 Characterization Results

There are several characterization results for the Weibull distribution in the literature. The Notes at the end of the chapter give additional references dealing with this topic.

3.4 THREE-PARAMETER WEIBULL MODEL

The three-parameter Weibull model is given by (1.2).

3.4.1 Parametric Study of Density Function

The density function for the three-parameter Weibull model is given by

$$f(t) = \frac{\beta}{\alpha}\left(\frac{t-\tau}{\alpha}\right)^{\beta-1} \exp\left[-\left(\frac{t-\tau}{\alpha}\right)^{\beta}\right] \qquad t \geq \tau \tag{3.27}$$

The shapes of the density function depend only on the shape parameter β and the parametric characterization is similar to that for the two-parameter case.

3.4.2 Parametric Study of Hazard Function

The hazard function is given by

$$h(t) = \frac{\beta}{\alpha}\left(\frac{t-\tau}{\alpha}\right)^{\beta-1} \qquad t \geq \tau \tag{3.28}$$

The shapes of the hazard function are dependent only on the shape parameter β, and the parametric characterization is similar to that for the two-parameter case.

3.4.3 Moments, Median, Mode, and Percentiles

The mean is given by

$$\mu = E(T) = \tau + \alpha\Gamma(1 + 1/\beta) \tag{3.29}$$

and the variance is given by (3.18).

The median is given by

$$t_{\text{median}} = \tau + \alpha(\ln 2)^{1/\beta} \tag{3.30}$$

When $\beta > 1$, there is a single mode and it is given by

$$t_{\text{mode}} = \tau + \alpha(1 - 1/\beta)^{1/\beta} \tag{3.31}$$

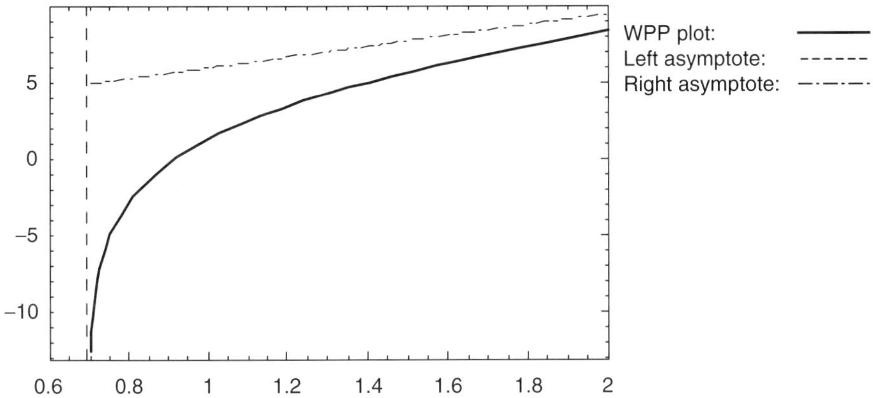

Figure 3.3 WPP plot for three-parameter Weibull model ($\alpha = 0.5, \beta = 3.5, \tau = 2.0$).

The p percentile is given by

$$t_p = \tau + \alpha[-\ln(1-p)]^{1/\beta} \tag{3.32}$$

Note that the expressions for the mean, median, mode, and p percentile for the three-parameter Weibull model depend on the location parameter τ. In contrast, the variance does not depend on location parameter.

3.4.4 WPP Plot

Under the Weibull transformation (1.7), (1.2) gets transformed into

$$y = \beta \ln(e^x - \tau) - \beta \ln(\alpha) \tag{3.33}$$

Note that y is a nonlinear function of x, and as a result the WPP plot is a smooth curve shown in bold in Figure 3.3. This is in contrast to the standard Weibull model where the relationship is linear and the WPP plot is a straight line.

The asymptotes of the WPP plot are as follows:

- As $x \to \infty$, the asymptote is a straight line given by (3.26), the WPP plot for the standard Weibull model.
- As $x \to \ln(\tau)$, the asymptote is a vertical line that intersects the x axis at $\ln(\tau)$.

The two asymptotes are shown (as dotted lines) in Figure 3.3.

3.5 NOTES

Many books on statistical distributions and on reliability (e.g., Bain and Englehardt, 1991; Zacks, 1992; Leemis, 1995) deal with the analysis of the two- and three-parameter Weibull models.

As mentioned in the text, there are many studies dealing with the characterization of the Weibull distribution.

Scholz (1990) gives the characterization in terms of the quantiles. Gertsbakh and Kagan (1999) proposed a characterization based on the Fisher information and censored experiments. For additional results on characterization, see Nigm and El-Hawary (1996), Mohie El-Din et al. (1991), and Chaudhuri and Chandra (1990).

Roy and Mukherjee (1986) deal with four miscellaneous characterizations of the two-parameter Weibull distribution. Khan and Beg (1987) and Khan and Abu-Salih (1988) study the characterization of Weibull through some conditional moments. El-Arishy (1993) uses truncated moments for the characterization of Weibull distribution. Ali (1997) showed that Weibull distributions can be characterized by the product moments of any two order statistics of their respective distributions.

Other references dealing with the characterization results are Seshadri (1968), Wang (1976), Shimizu and Davies (1981), Eldin et al. (1991), Yehia (1993), Janardan and Schaffer (1978), Janardan and Taneja (1979a, 1979b), and Zheng (2001).

Groeneveld (1986) discusses the skewness of Weibull distribution. Lochner et al. (1974) study the covariance of order statistics. Makino (1984) studies the mean hazard rate for the Weibull distribution. Patel (1975) deals with bounds on moments of order statistics. Patel and Read (1975) present some additional bounds based on conditional moments, while Rao and Talwalker (1989) derive some bounds on the life expectancy.

Lieblein (1955) expresses the covariance of order statistics drawn from Weibull distribution in terms of incomplete beta and gamma distributions. For other references dealing with this topic, see White (1969), Balakrishnan and Joshi (1981), Malik and Trudel (1982), Menon and Daghel (1987), Mohie El-Din et al. (1991), Marohn (1994), Nigm and El-Hawary (1996), and Barakat and Abdelkader (2000).

EXERCISES

3.1. Derive (3.1).

3.2. Derive (3.5) from (3.4).

3.3. Derive (3.9).

3.4. Derive (3.14) and (3.15).

3.5. Derive expressions for the covariance of order statistics drawn from Weibull distribution. [See, Lieblein (1955).]

3.6. Consider a two-parameter Weibull distribution with $\alpha = 10$ and $\beta = 1.5$. Compute the percentile t_p for p varying from 0.1 to 0.9 in steps of 0.1. How does the difference between these computed values change with p?

3.7. For a two-parameter Weibull distribution prove that the TTT plot is convex for $\beta > 1$ and concave for $\beta < 1$.

3.8. Suppose that the lifetime, T, of a component can be adequately modeled by a two-parameter Weibull distribution with shape parameter $\beta > 1$. The component has been operating for a period t_1. Obtain the distribution function for the residual life $(T - t_1)$ and derive an expression for the mean residual life. How does this change with t_1 and β?

3.9. The lifetime distribution of a device is known to be Weibull with parameters $(\alpha, \beta) = (10\,\text{years}, 1.5)$. Compute the mean, median, and standard deviation of the lifetime. What is the probability that the device will survive for 10 years?

3.10. The life of a bearing can be modeled by a three-parameter Weibull distribution with $\tau > 10$ (months). Describe the physical basis why this should be so?

3.11. Show that the asymptote for the WPP plot given by (3.33) is given by (3.26) as $x \to \infty$.

3.12. Consider a Weibull distribution with shape parameter β and scale parameter 1 and a normal distribution with the same mean and variance. Study the difference between the two density functions as function of the independent variable. For what value of the independent variable is the difference a maximum? How does β affect this value?

3.13. Let $F_1(t)$ and $F_2(t)$ be two Weibull distributions with common shape parameter β and different scale parameters with $\alpha_1 < \alpha_2$. Show that the sign of $\{F_1(t) - F_2(t)\}$ does not change with t? What is the implication of this in the reliability context?

3.14. Suppose that in Exercise 3.13 the scale parameters are the same but the shape parameters are different. Show that $F_1(t)$ and $F_2(t)$ cross each other at least once.

3.15. Section 3.5 lists studies dealing with alternate characterization of the Weibull distribution. Review the works and write a report that summarizes the results.

CHAPTER 4

Parameter Estimation

4.1 INTRODUCTION

As indicated in Section 1.3, model building involves five steps with the last three being the following:

3. Model selection
4. Estimating model parameters
5. Model validation

In this chapter we focus on the estimation of model parameters (step 4) assuming that the model selected is appropriate. The remaining two steps (steps 3 and 5), which are related to each other, are discussed in the next chapter.

The estimate obtained depends on the type of data available and the method used. We first discuss these two topics in a general setting, following which we look at parameter estimation for the standard Weibull and the three-parameter Weibull models.

The outline of the chapter is as follows. In Section 4.2 we look at different data structures. In Section 4.3 we discuss parameter estimation and some related issues. The aim is to highlight the underlying concepts. Section 4.4 deals with different estimation methods and some properties of the estimators. Section 4.5 deals with graphical methods, and Section 4.6 with statistical methods, for estimating the parameters of the standard Weibull model. The estimation of the parameters for the three-parameter model is discussed in Section 4.7.

4.2 DATA TYPES

Data may be classified as (i) complete and (ii) censored (or incomplete). The latter can be further subdivided into different categories. In this section we discuss briefly both these types of data.

Weibull Models, by D.N.P. Murthy, Min Xie, and Renyan Jiang.
ISBN 0-471-36092-9

Let $T_1, T_2, \ldots, T_n$ denote a sample of n-independent random variables from a distribution function $F(t, \theta)$, where $\theta = (\theta_1, \ldots, \theta_k)$ is a k-dimensional parameter. (These correspond to the life times for n different items in the reliability context.) Let $t_1, t_2, \ldots, t_n$ denote the actual realized values of the T_i's.

4.2.1 Complete Data

The data set available for estimation is the set $\{t_1, t_2, \ldots, t_n\}$. In other words, the actual realized values are known for each observation in the data set.

4.2.2 Censored Data

In this case the actual realized values for some or all of the variables are not known, and this depends on the kind of censoring. There are many different kinds of censoring. These include

- Right, left, or interval censoring
- Type I or type II censoring
- Single or multiply censoring

In reliability context, the following types of censored data are of particular interest.

Right Type I Censoring
Let v denote a variable (deterministic or random). Under type I right censoring the data available for estimation is as follows: For item i, the actual realized value of T_i is known only if $t_i \leq v$. When $t_i > v$ the only information available is that $T_i > v$.

Right Type II Censoring
Let r denote a predetermined number such that $r < n$. Under type II right censoring the data available for estimation is given by t_i (the actual realized value of T_i) for r data and that $T_i > t_{\max}$ for the remaining $n - r$ data where $t_{\max}$ is the maximum of the r observed t_i's.

Random Censoring
The general formulation of random censoring is as follows. Let s_i be a random variable independent of T_i. The observed value is given by $\min\{t_i, s_i\}$ so that it is censored if $t_i > s_i$.

For further details of the different types of censoring, see, for example, Lawless (1982), Nelson (1982), and Leemis (1995).

Grouped Data: Interval Censoring
Grouped data are data that have been categorized into classes (usually nonoverlapping intervals), with only class frequencies known. As a result, in this case one only knows that t_i is in some interval but its actual value is unknown.

4.3 ESTIMATION: AN OVERVIEW

4.3.1 Basic Concepts

The parameter to be estimated is θ, and we assume that it is a k-dimensional vector with components $\theta_1, \theta_2, \cdots, \theta_k$. There are two approaches to estimation and several methods (discussed in the next section) for each of them. The two approaches are *point estimation* and *interval* (or *confidence interval*) *estimation.*

In point estimation, a numerical value for θ is calculated. In interval estimation, a k-dimensional region is determined in such a way that the probability that the region contains the true parameter θ is a specified, predetermined value. (If $k = 1$, this region is an interval, hence the name.)

In point estimation, $\hat{\theta}$ is a function of $T_1, T_2, \ldots, T_n$ or $t_1, t_2, \ldots, t_n$ in the case of complete data. The expression $\hat{\theta} = \hat{\theta}(T_1, T_2, \ldots, T_n)$ is called an *estimator* and is a random variable; $\hat{\theta} = \hat{\theta}(t_1, t_2, \ldots, t_n)$ is called an *estimate* and is the numerical value obtained using the data for estimation. In the case of censored or grouped data, the estimate is a function of the observed data and the censoring or grouping values.

4.3.2 Properties of Estimators

Unbiasedness An estimator $\hat{\theta}_i$ of θ_i is said to be *unbiased* if $E(\hat{\theta}_i) = \theta_i$ for all possible values of θ_i. The term $\hat{\theta}$ is an unbiased estimator of θ if $\hat{\theta}_i$ is unbiased for $i = 1, \ldots, k$. An estimator for which $E(\hat{\theta}_i) \neq \theta_i$ is said to be *biased*. The *bias* $b(\hat{\theta}_i)$ of an estimator $\hat{\theta}_i$ is given by $b(\hat{\theta}_i) = E(\hat{\theta}_i) - \theta_i$.

Asymptotic Unbiasedness An estimator $\hat{\theta}_i$ of θ_i is said to be *asymptotically unbiased* if $E(\hat{\theta}_i) \to \theta_i$ as $n \to \infty$ for all possible values of θ_i. The term $\hat{\theta}$ is an asymptotically unbiased estimator of θ if $\hat{\theta}_i$ is asymptotically unbiased for $i = 1, \ldots, k$.

Consistency An estimator $\hat{\theta}_i$ of θ_i is said to be *consistent* if for any $\varepsilon > 0$ and all possible values of θ_i, $P(|\hat{\theta}_i - \theta_i| > \varepsilon) \to 0$ as $n \to \infty$. Estimator $\hat{\theta}$ is a consistent estimator of θ if $\hat{\theta}_i$ is consistent for $i = 1, \ldots, k$.

Efficiency Efficiency of an estimator may be assessed relative to another estimator or estimators (*relative efficiency*) or relative to an absolute standard. The results given below involve the variance of the estimator, denoted $V(\cdot)$.

Minimum Variance Unbiased Estimators An unbiased estimator $\hat{\theta}$ of a parameter θ is *minimum variance unbiased* if $V(\hat{\theta}) \leq V(\theta^*)$ for any other unbiased estimator θ^* and for all possible values of θ.

Estimators that are efficient in this sense are also called UMVUE (for uniformly minimum variance unbiased estimator), and, if $\hat{\theta}$ is a linear function of the T_i's, BLUE (for best linear unbiased estimator).

Cramér-Rao Inequality Suppose that $\hat{\psi} = g(T_1, T_2, \ldots, T_n)$ is an unbiased estimator of $\psi(\theta)$ (a scalar function of the parameter). Then under certain regularity conditions [see Stuart and Ord (1991)]

$$V(\hat{\psi}) \geq \frac{[\psi'(\theta)]^2}{E\left(\frac{\partial \ln[L(T_1,\ldots,T_n;\theta)]}{\partial\theta}\right)^2} \tag{4.1}$$

where $L(\cdot)$ is the likelihood function. The vector case is straightforward and involves the inverse of the covariance matrix.

The likelihood function $L(\cdot)$ is given by

$$L(\theta) = \prod_{i=1}^{n} f(t_i; \theta) \tag{4.2}$$

for the case of complete data. For other kinds of data, the expression for $L(\cdot)$ would be different and this will be discussed later in this chapter.

4.4 ESTIMATION METHODS AND ESTIMATORS

Many different estimation methods have been developed for estimating model parameters. They can be broadly grouped into three categories:

- Graphical methods
- Statistical methods
- Combination of the above two methods

In graphical methods, the estimates are obtained from plotting the data. The plot depends on the model selected, and hence each needs to be treated separately. The main drawback of these methods is that there is no well-developed statistical theory for determining the small sample or asymptotic properties. However, they are useful in providing an initial estimate for statistical methods of estimation. We will discuss the use of WPP plots to estimate the parameters of the standard Weibull and three-parameter Weibull models later in the chapter.

The statistical methods, in contrast, are more general and applicable to all kinds of models and data types. The asymptotic properties of the estimators are well understood. We first discuss some of the common statistical methods for point estimation and then briefly discuss the interval estimation.

4.4.1 Moment Estimator

In Section 3.2.2, we discussed the relationship between the moments of a random variable T from a distribution function $F(t; \theta)$ as a function of the parameters θ (of dimension k) of the distribution function. The *method of moments* is based on expressing k moments in terms of the parameters of the distribution. Using sample moments (obtained from the data available for estimation) in place of the moments in these relationships yields k equations containing the k unknown parameters. Solving these yields the estimates. Note that in most cases, the estimates need to be obtained using numerical techniques.

One can use the moments $M_r = E[T^r]$ [given by (3.2)], the central moments $\mu_r = E[(T - \mu)^r]$ [given by (3.3)] or a combination of the two. In the case of complete data $\{t_1, t_2, \ldots, t_n\}$, the sample moments $\hat{M}_r$ $(r = 1, 2, 3)$ and sample central moments are obtained as follows:

$$\hat{M}_r = \frac{1}{n}\sum_{i=1}^{n} t_i^r \tag{4.3}$$

and

$$\hat{\mu}_r = \frac{1}{n}\sum_{i=1}^{n} (t_i - \bar{t})^r \tag{4.4}$$

where $\bar{t} = \hat{M}_1$.

For most models, moment estimators are asymptotically consistent and are normally distributed. In general, moment estimators are usually not efficient.

4.4.2 Percentile Estimator

In Section 3.2.3 we discussed the p percentile. The percentile estimator is obtained by first obtaining the expression for the percentile in terms of the model parameters.

From (3.6) we have

$$p = F(t_p, \theta) \tag{4.5}$$

Let $0 < p_1 < p_2 < \cdots p_k < 1$. Estimates of the corresponding k p-percentiles are obtained from the empirical distribution function (EDF) using the data. Using these in (4.5) yields k equations containing the k unknown parameters. Solving these yields the parameter estimates.

The EDF plot is a plot of the cumulative proportion of the data that lie below t, against $t(-\infty < t < \infty)$ and denoted by $\hat{F}(\cdot)$. It is also called the sample cumulative distribution function. There are many different ways of computing this. One such, for the case of complete data comprising of n data points, is a *step function* having steps of height $1/n$ at each data point x_i. This is most easily done by use of the order statistics $t_{(1)}, t_{(2)}, \ldots, t_{(n)}$ with $t_{(0)} = 0$ and $t_{(n+1)} = \infty$ so that $\hat{F}(t_{(i)}) = i/n$.

There are other approaches to computing the EDF for the case of complete data. In the *mean rank* approach, $\hat{F}(t_{(i)}) = i/(n+1)$. Three other alternatives are as follows:

- $\hat{F}(t_{(i)}) = (i - 0.5)/n$ (called the *median rank* estimator).
- $\hat{F}(t_{(i)}) = (i - 0.3)/(n + 0.4)$.
- $\hat{F}(t_{(i)}) = (i - \frac{3}{8})/(n + \frac{1}{4})$.

For further discussion on computing the EDF with censored data, see King (1971), Nelson (1982), and Lawless (1982).

For most models, the percentile estimators are asymptotically consistent and are normally distributed. However, they are usually not efficient.

4.4.3 Maximum-Likelihood Estimator

For the case of complete data, the *likelihood function* L is given by (4.2). The *maximum-likelihood estimate* (MLE) of θ is the value $\hat{\theta}$ that maximizes the likelihood function given by (4.2). As a result, the estimate is a function [say, $\psi(t_1, t_2, \ldots, t_n)$] of the data. The expression $\psi(T_1, T_2, \ldots, T_n)$ is called the *maximum-likelihood estimator* and plays an important role in study of the estimate.

Under certain regularity conditions, maximum-likelihood estimators are consistent, asymptotically unbiased, efficient, and normally distributed. Asymptotic efficiency here means that as $n \to \infty$, the covariance matrix of the estimator achieves the lower bound of the Cramér–Rao inequality discussed earlier. For a rigorous treatment of the theory of maximum likelihood, asymptotic results, and related topics, see Stuart and Ord (1991).

Censored Data

Right Type I Censoring

For right Type I censored data the likelihood function is given by

$$L(\theta) = \prod_{i=1}^{n} [f(t_i)]^{\delta_i} [1 - F(v)]^{1-\delta_i} \tag{4.6}$$

where $\delta_i = 1$ if the ith obvervation is uncensored $(T_i = t_i < v)$ and $= 0$ if censored $(T_i > v)$.

Right Type II Censoring

For right Type II censored data (with r failures) likelihood function is given by

$$L(\theta) = \left[\prod_{i=1}^{r} f(t_i)\right] [1 - F(t_r)]^{n-r} \tag{4.7}$$

where the data has been reordered so that the first r corresponds to uncensored and the remaining to censored.

Random Censoring
For random censoring let D denote the set of uncensored observations ($t_i \leq v_i$) and C denote the set of censored observations ($t_i > v_i$). The likelihood function is given by

$$L(\theta) = \left[\prod_{i \in D} f(t_i)\right]\left[\prod_{i \in C}[1 - F(v_i)]\right] \tag{4.8}$$

Grouped Data
Here the time axis is divided into $(k + 1)$ intervals. The jth interval is given by $[a_j, a_{j+1})$ with $a_0 = 0$ and $a_{k+1} = \infty$. Let d_j denote the number of uncensored observations falling in $[a_{j-1}, a_j)$ and w_j denote the number of observations censored at a_j. In this case the likelihood function is given by

$$L(\theta) = \prod_{j=1}^{k+1} [F(a_j) - F(a_{j-1})]^{d_j}[1 - F(a_j)]^{w_j} \tag{4.9}$$

and the maximum-likelihood estimates are obtained by maximizing this function.

For further discussion on the method of maximum likelihood, see Lawless (1982) or Leemis (1995).

4.4.4 Bayesian Estimator

The Bayesian approach starts with a prior information regarding the parameters, and estimates of the parameters are updated as new data is obtained. The parameters are assumed to be random variables with a *prior* density function, which we denote by $g(\theta)$. We assume that θ is one-dimensional (one can easily extend the results to the multidimensional case) and $g(\theta)$ a continuous function. Let $f(t|\theta)$ be the conditional density function of T given θ. The joint density function for θ and $\{T_1, T_2, \ldots, T_n\}$ is given by

$$f_{t\theta}(t_1, \ldots, t_n, \theta) = \left[\prod_{i=1}^{n} f(t_i|\theta)\right] g(\theta) \tag{4.10}$$

Note that $f_{t\theta}$ is the product of the likelihood function (given the parameter) and the prior density function. The marginal density function of $\{T_1, T_2, \ldots, T_n\}$ is given by

$$f(t_1, \ldots, t_n) = \int_{-\infty}^{\infty} f_{t\theta}(t_1, \ldots, t_n, \theta)\, d\theta = \int_{-\infty}^{\infty} f(t_1, \ldots, t_n|\theta) g(\theta)\, d\theta \tag{4.11}$$

Using Bayes' theorem (Martz and Waller, 1982), the *posterior* density function of θ, given the data $t_1, t_2, \ldots, t_n$, is given by

$$g(\theta|t_1, \ldots, t_n) = \frac{f_{t\theta}(t_1, \ldots, t_n, \theta)}{f(t_1, \ldots, t_n)} \tag{4.12}$$

The Bayesian point estimate $\hat{\theta}_b$ of a parameter θ is the conditional mean and is given by

$$\hat{\theta}_b = E(\theta|t_1, \ldots, t_n) = \int \theta g(\theta|t_1, \ldots, t_n)\, d\theta \tag{4.13}$$

If the prior and posterior distributions belong to the same family, then we have a conjugate family of distributions. In general, it is difficult to achieve this.

4.4.5 Interval Estimation

In the case where θ is scalar, a confidence interval based on a sample of size n, $T_1, T_2, \ldots, T_n$, is an interval defined by two limits, the lower limit $L_1(T_1, T_2, \ldots, T_n)$ and the upper limit $L_2(T_1, T_2, \ldots, T_n)$ having the property that

$$P[L_1(T_1, \ldots, T_n) < \theta < L_2(T_1, \ldots, T_n)] = \gamma \tag{4.14}$$

where $\gamma (0 < \gamma < 1)$ is called the *confidence coefficient*.

Confidence is usually expressed in percent, for example, if $\gamma = .95$, the result is a "95% confidence interval" for θ. Note that the random variables in expression (4.14) are L_1 and L_2, not θ, that is, this is not a probability statement about θ, but about L_1 and L_2. Hence we use the term *confidence* rather than *probability* when discussing this as a statement about θ. The proper interpretation is that the procedure gives a correct result $100\gamma\%$ of the time.

The confidence interval defined above is a *two-sided interval*; if a fraction of the remaining probability, $(1 - \gamma)$, is below L_1 and the rest above L_2 [usually $(1 - \gamma)/2$ on each side]. A *lower* one-sided confidence interval is obtained by omitting L_2 in the above equation and modifying L_1 accordingly; the interpretation is that we are $100\gamma\%$ confident that the true value is at least L_1. Similarly, one can define an *upper* one-side confidence interval.

Confidence intervals for functions $\psi(\theta)$ are defined by replacing θ by $\psi(\theta)$ in (4.14). If $\psi(\theta)$ is a monotonic function, the interval is simply $(\psi(L_1), \psi(L_2))$. Construction of confidence intervals, that is, derivation of the limits L_1 and L_2, may be done in several ways [see, e.g., Mood et al. (1974)]. Usually the interval is based on the distribution of the best point estimator, if such exists. If not, it is based on the distribution of a "good" point estimator. If the distribution is unknown or is mathematically intractable, approximations may be used. Also, asymptotic results are often used to obtain approximate confidence intervals.

4.5 TWO-PARAMETER WEIBULL MODEL: GRAPHICAL METHODS

In this section we discuss various graphical methods for estimation of the two parameters of the standard Weibull model given by (1.3).

4.5.1 WPP Plot

As indicated in Section 3.3.1, under the Weibull transformation, the WPP plot for the model is given by (3.26). This is a linear relationship between y and x.

The approach to estimation involves two parts (denoted as parts A and B, respectively). Part A involves plotting the data to obtain the WPP plot. Part B involves fitting the straight line to the plot and estimating the parameters from this fit. Part A depends on the type of data available. Once part A is executed, part B is the same for all types of data.

Complete Data
In this case the data is given by $t_1, t_2, \ldots, t_n$. The estimation method is as follows:

Part A: Plotting

1. Reorder the data from the smallest to the largest so that $t_{(1)} \le t_{(2)} \le \cdots \le t_{(n)}$.
2. Compute $\hat{F}(t_{(i)})$ for $1 \le i \le n$. (As indicated in Section 4.4.2, one can use many different approaches to compute the EDF.)
3. Compute $y_i = \ln\{-\ln[1 - \hat{F}(t_{(i)})]\}$ for $1 \le i \le n$.
4. Compute $x_i = \ln(t_{(i)})$ for $1 \le i \le n$.
5. Plot y_i versus x_i for $1 \le i \le n$.

Part B: Estimation

6. Determine the best straight-line fit using regression or least-squares method.
7. The slope of this line yields $\hat{\beta}$, the estimate of β.
8. Compute y_0, the y intercept of the fitted line; $\hat{\alpha}$, the estimate of α, is given by $\hat{\alpha} = \exp(-y_0/\hat{\beta})$.

Censored Data
The procedure is essentially the same as that for the complete data case with a modification to step 2 in part A. For further details, see Nelson (1982), Lawless (1982), and Kececioglu (1991).

Grouped Data
For grouped data, the procedure is similar to the case of complete data with a modification to step 2. See Dodson (1994) for the details of the procedure.

4.5.2 Hazard Plot

The cumulative hazard function, $H(t)$, for the standard Weibull model is given by (3.13) and is a nonlinear function of t. Under the transformation

$$x = \ln(t) \quad \text{and} \quad y = \ln[H(t)] \tag{4.15}$$

the cumulative hazard function gets transformed into a linear relationship

$$x = \ln(\alpha) + (1/\beta)y \tag{4.16}$$

The approach to estimation involves two parts (A and B) similar to the WPP plot. The execution of part A depends on the type of data available for estimation and of part B is as before. As such, we only discuss part A.

Complete Data

Part A: Plotting

1. Reorder the data from the smallest to the largest so that $t_{(1)} \leq t_{(2)} \leq \cdots \leq t_{(n)}$.
2. Compute a hazard value for each failure as 100/k, where k is its reverse rank of each failure time.
3. Compute the cumulative hazard by summing up the previous hazard values.
4. Compute $x_i = \ln(t_{(i)})$ for $1 \leq i \leq n$.
5. Plot y_i, the logarithm of the cumulative hazard, versus $t_{(i)}$ for $1 \leq i \leq n$.

Censored Data and Grouped Data

The procedure is essentially the same as that for the complete data case with a modification to step 2 in part A. For further details, see Nelson (1982).

4.6 STANDARD WEIBULL MODEL: STATISTICAL METHODS

In this section we discuss a number of common and some specialized statistical methods for estimating the parameters of the standard Weibull model. We focus our attention on point estimates and briefly discuss interval estimates toward the end of the section.

4.6.1 Method of Moments

Since the basic Weibull model has two parameters, estimates of the model parameters can be obtained using the sample mean $\bar{t}$ and sample variance s^2. In the case of complete data $\{t_1, t_2, \ldots, t_n\}$ these are given by

$$\bar{t} = \sum_{i=1}^{n} \frac{t_i}{n} \tag{4.17}$$

and

$$s^2 = \sum_{i=1}^{n} \frac{(t_i - \bar{t})^2}{n-1} \tag{4.18}$$

($n - 1$ is used as it yields an unbiased estimate of the sample variance).

Using the expressions for the mean and variance [given by (3.16) and (3.18), respectively], $\hat{\beta}$ is obtained as the solution of

$$\frac{s^2}{\bar{t}^2} = \frac{\Gamma(1+2/\hat{\beta})}{\Gamma^2(1+1/\hat{\beta})} - 1 \tag{4.19}$$

and $\hat{\alpha}$ is given by

$$\hat{\alpha} = \frac{\bar{t}}{\Gamma(1+1/\hat{\beta})} \tag{4.20}$$

Note that solving (4.19) requires a numerical procedure and involves computing the gamma function.

4.6.2 Method of Percentiles

From (3.25) the 63.2th percentile is equal to α, and this can be used as a percentile estimator of α so that

$$\hat{\alpha} = t_{(1-e^{-1})} \tag{4.21}$$

where the percentile is calculated from the empirical plot of the distribution function.

A percentile estimate of β is given by

$$\hat{\beta} = \frac{\ln[-\ln(1-p)]}{\ln(t_p/t_{0.632})} \tag{4.22}$$

for $0 < t_p < t_{0.632}$ and it can be obtained from the EDF.

Seki and Yokoyama (1993) suggest $p = 0.31$ to obtain an estimate of β. Wang and Keats (1995) suggest (based on simulation studies) that the optimal p for minimum bias in the estimate is $p = 0.15$. The corresponding t_p value is obtained using the following linear interpolation:

$$t_p = \frac{p - F(t_{(r)})}{F(t_{(r+1)}) - F(t_{(r)})}\left(t_{(r+1)} - t_{(r)}\right) + t_{(r)} \tag{4.23}$$

where $t_{(r)}$ is the time of the rth ordered failure time and $F(t_{(r)})$ is the proportion of the population that failed by $t_{(r)}$. Different estimators of $F(t_{(r)})$ can also be used and Wang and Keats (1995) compare the results obtained by their method with other methods such as the method of maximum likelihood.

4.6.3 Method of Maximum Likelihood

The maximum-likelihood estimates (MLEs) depend on the type of data available and we consider several different cases.

Complete Data

The likelihood function is given by

$$L(\alpha, \beta) = \prod_{i=1}^{n} \left(\frac{\beta t_i^{(\beta-1)}}{\alpha^{\beta}} \right) \exp\left[-\left(\frac{t_i}{\alpha} \right)^{\beta} \right] \tag{4.24}$$

The maximum-likelihood estimates are obtained by solving the equations resulting from setting the two partial derivatives of $L(\alpha, \beta)$ to zero. As a result, $\hat{\beta}$ is obtained as the solution of

$$\frac{\sum_{i=1}^{n} (t_i^{\hat{\beta}} \ln t_i)}{\sum_{i=1}^{n} t_i^{\hat{\beta}}} - \frac{1}{\hat{\beta}} - \frac{1}{n} \sum_{i=1}^{n} \ln t_i = 0 \tag{4.25}$$

Although analytical solution is not available, $\hat{\beta}$ can be easily solved using a computational approach. Once the shape parameter is estimated, the scale parameter can then be estimated as follows:

$$\hat{\alpha} = \left(\frac{1}{n} \sum_{i=1}^{n} t_i^{\hat{\beta}} \right)^{1/\hat{\beta}} \tag{4.26}$$

Right Type I Censoring

The likelihood function is given by

$$L(\alpha, \beta) = \frac{\beta^k}{\alpha^{\beta k}} \left[\prod_{i=1}^{k} (t_i)^{\beta-1} \right] \exp\left\{ -\frac{1}{\alpha^{\beta}} \left[\sum_{i=1}^{k} t_i^{\beta} + (n-k) v^{\beta} \right] \right\} \tag{4.27}$$

where the data has been reordered so that the first k are uncensored data and the remaining are censored data. The maximum-likelihood estimate $\hat{\beta}$ is obtained as the solution of

$$\frac{\sum_{i=1}^{n} t_i^{\hat{\beta}} \ln t_i}{\sum_{i=1}^{n} t_i^{\hat{\beta}}} - \frac{1}{\hat{\beta}} - \frac{1}{k} \sum_{i=1}^{k} \ln t_i = 0 \tag{4.28}$$

and the estimate $\hat{\alpha}$ is given by

$$\hat{\alpha} = \left\{ \frac{1}{n} \left[\sum_{i=1}^{k} t_i^{\hat{\beta}} + (n-k) v^{\hat{\beta}} \right] \right\}^{1/\hat{\beta}} \tag{4.29}$$

Sirvanci and Yang (1984) discuss the MLEs based on an alternate characterization of the two-parameter Weibull distribution.

Right Type II Censoring

The likelihood function is given by

$$L(\alpha, \beta) = \frac{\beta^r}{\alpha^{\beta r}} \left(\prod_{i=1}^{r} t_i^{\beta-1} \right) \cdot \exp\left\{ -\frac{1}{\alpha^\beta} \left[\sum_{i=1}^{r} t_i^\beta + (n-r) t_r^\beta \right] \right\} \tag{4.30}$$

where the data has been reordered so that the first r in the set are the uncensored observations. The maximum-likelihood estimate $\hat{\beta}$ is obtained by solving

$$\frac{\sum_{i=1}^{r} t_i^{\hat{\beta}} \ln t_i + (n-r) t_r^{\hat{\beta}} \ln t_r}{\sum_{i=1}^{r} t_i^{\hat{\beta}} + (n-r) t_r^{\hat{\beta}}} - \frac{1}{\hat{\beta}} - \frac{1}{r} \sum_{i=1}^{r} \ln t_i = 0 \tag{4.31}$$

and using this, the estimate $\hat{\alpha}$ is obtained from

$$\hat{\alpha}^{\hat{\beta}} = \frac{1}{r} \left[\sum_{i=1}^{r} t_i^{\hat{\beta}} + (n-r) t_r^{\hat{\beta}} \right] \tag{4.32}$$

Iterative methods, such as the Newton–Raphson method, need to be used to obtain the estimates. Qiao and Tsokos (1994) propose an iterative procedure that is shown to be more effective than the Newton–Raphson method. A FORTRAN program for point and interval estimates is given in Keats et al. (1997).

Random Censoring

For randomly censored data, the likelihood function is a bit more complicated. Let $u_i = \min\{t_i, v_i\}$ where t_i is actual realized value and v_i is the censoring time for T_i. In this case, the MLEs are given by

$$\hat{\alpha}^{\hat{\beta}} = \frac{1}{n} \left(\sum_{i=1}^{n} u_i^{\hat{\beta}} \right) \tag{4.33}$$

and

$$\frac{\sum_{i=1}^{n} u_i^{\hat{\beta}} \ln u_i}{\sum_{i=1}^{n} u_i^{\hat{\beta}}} - \frac{1}{\hat{\beta}} - \frac{1}{k} \sum_{i \in D} \ln u_i = 0 \tag{4.34}$$

For further details, see Leemis (1995).

Grouped Data

Let d_j denote the number of uncensored observations falling in the interval $I_j = (a_{j-1}, a_j]$ with $a_0 = 0$ and $a_{k+1} = \infty$ and let w_j denote the number of censored observations with censoring at a_j. The likelihood function is

$$L(\alpha, \beta) = \prod_{j=1}^{k+1} \left\{ \exp\left[-\left(\frac{a_{j-1}}{\alpha}\right)^{\beta} \right] - \exp\left[-\left(\frac{a_j}{\alpha}\right)^{\beta} \right] \right\}^{d_j} \left\{ \exp\left[-\left(\frac{a_j}{\alpha}\right)^{\beta} \right] \right\}^{w_j} \tag{4.35}$$

The maximum-likelihood estimates are obtained by maximizing the likelihood function. For further details, see Nelson, (1982), Cheng and Chen (1988), and Rao et al. (1994).

Comments

The method of maximum likelihood for the estimation of the Weibull parameters has received a lot of attention in the literature. Although MLEs have nice asymptotic properties, they are biased for small and censored samples. Corrections to compensate for the bias have received some attention in the literature. Watkins (1996), Ross (1996), and Montanari et al. (1997) investigate the bias of the MLE for the two-parameter Weibull distribution and Hirose (1999) presentes a simple method to correct the bias.

Fei and Kong (1995) consider the estimation with type II censored data and compare the MLE with some approximate MLE and best linear estimators (discussed later in the section). See also Seki and Yokoyama (1996) for a comparison between the MLE and a bootstrap robust estimator.

4.6.4 Bayesian Method

In this case both the scale and shape parameters are assumed to be random variables and characterized through a prior density function. In this section we review the relevant literature where interested readers can find the details.

Soland (1968,1969) uses a discrete prior distribution for the shape parameter, and conditional on this, a gamma distribution is used as the prior distribution for the scale parameter. This results in the joint posterior distribution to be in the same family as the joint prior distribution. Papadopoulos and Tsokos (1975a) derive some Bayesian confidence bounds for the standard Weibull model.

Canavos and Tsokos (1973) and Erto and Rapone (1984) assume an inverse Gaussian distribution for the scale parameter and a uniform distribution for the shape parameter. Lamberson and De Souza (1987) use inverse Weibull and beta distributions as the priors for the scale and shape parameters, respectively. De Souza and Lamberson (1995) use the inverse Weibull as the prior distribution for the scale parameter and the negative log-Gamma distribution as the prior distribution for the shape parameter.

Often approximations and/or numerical methods have to be used for obtaining the posterior distribution. Dellaportas and Wright (1991) discuss this problem in detail and propose a numerical approach for the two-parameter Weibull model. Nigm (1990) deals with Bayesian approach to obtain prediction bounds under type 1 censoring. Canavos and Tsokos (1973) and Dellaportas and Wright (1991) discuss some numerical problems associated with the Bayesian estimation of Weibull parameters.

When the shape parameter is known, Bayesian estimation of the scale parameter can be reduced to the Bayesian estimation of the parameters of an exponential distribution by a simple power transformation discussed in Section 2.3.2. This approach has been studied extensively by Martz and Waller (1982). An early and interesting reference is Harris and Singpurwalla (1969), which looks at the Bayesian method with the shape parameter treated as a random variable. For exponential distribution (or the Weibull distribution with shape parameter one) the simplest case is a gamma prior distribution. Papadopoulos and Tsokos (1975b) consider the case where the scale parameter has either a uniform distribution or an inverse Gaussian distribution as the prior distribution.

4.6.5 Linear Estimators

Linear estimators, because of their simplicity, have been studied for both the two- and three-parameter Weibull distributions. Here the estimators are linear combinations of data (for the case of complete data), and the coefficients used need to be obtained from special tables.

The problem of estimating the parameters for the standard Weibull distribution is converted into a problem of estimating the parameters of an extreme value distribution [see Model I(b)-2 in Section 2.3.2] with parameters $\lambda[\equiv \ln(\alpha)]$ and $\delta(\equiv 1/\beta)$. The best linear unbiased estimators for the parameters of the extreme value distribution are given by

$$\lambda^* = \sum_{i=1}^{r} a(i;n,r)X_i \tag{4.36}$$

and

$$\delta^* = \sum_{i=1}^{r} b(i;n,r)X_i \tag{4.37}$$

where $X_i = \ln(T_i)$ and the coefficients in (4.36) and (4.37) are selected so as to yield unbiased estimators. The variances of the estimators are given by $\mathrm{Var}(\lambda^*) \approx A_{n,r}/\beta^2$ and $\mathrm{Var}(\delta^*) \approx B_{n,r}/\beta^2$, respectively, where the coefficients are again available in tabulated form. The covariance of (λ^*, δ^*) is given by $\mathrm{Cov}(\lambda^*, \delta^*) = C_{n,r}/\beta^2$.

Tables for $a(i;n,r)$, $b(i;n,r)$, $A_{n,r}$, $B_{n,r}$, and $C_{n,\gamma}$ can be found in Nelson (1982), White (1964b), Lieblein and Zelen (1956), and Mann et al. (1974), where references to earlier literature can also be found. See also Johnson et al. (1995).

Note that $\alpha^* = \exp(\lambda^*)$ is only asymptotically unbiased. When r becomes large, the distribution of α^* is approximately normal with variance given by

$$\mathrm{Var}(\alpha^*) \approx A_{n,r}(\alpha/\beta)^2 \tag{4.38}$$

Similarly, the estimator $\beta^* = 1/\delta^*$ is only asymptotically unbiased. When r becomes large, the distribution of β^* is approximately normal with variance given by

$$\mathrm{Var}(\beta^*) \approx B_{n,r}\beta^2 \tag{4.39}$$

4.6.6 Interval Estimation

The asymptotic results from the theory of maximum-likelihood estimation are used in the interval estimation of parameters. The common procedure is to use the Fisher information matrix given by

$$I = \begin{bmatrix} -\dfrac{\partial^2 L}{\partial \alpha^2} & -\dfrac{\partial^2 L}{\partial \beta \partial \alpha} \\ -\dfrac{\partial^2 L}{\partial \beta \partial \alpha} & -\dfrac{\partial^2 L}{\partial \beta^2} \end{bmatrix} \tag{4.40}$$

For the case of complete data the various second derivatives of the log-likelihood function are given by

$$\frac{\partial^2 \ln L}{\partial \alpha^2} = \frac{\beta}{\alpha^2} - \sum_{j=1}^{n} \left(\frac{t_j}{\alpha}\right)^{\beta} \left(\frac{\beta}{\alpha^2}\right)(\beta + 1) \tag{4.41}$$

$$\frac{\partial^2 \ln L}{\partial \beta^2} = -\frac{1}{\beta^2} - \sum_{j=1}^{n} \left(\frac{t_j}{\alpha}\right)^{\beta} \left[\ln\left(\frac{t_j}{\alpha}\right)\right]^2 \tag{4.42}$$

and

$$\frac{\partial^2 \ln L}{\partial \alpha\, \partial \beta} = -\frac{1}{\alpha} + \sum_{j=1}^{n} \left(\frac{t_j}{\alpha}\right)^{\beta} \left(\frac{1}{\alpha}\right) \left[\beta \ln\left(\frac{t_j}{\alpha}\right) + 1\right] \tag{4.43}$$

The above is the sample estimate of the information matrix I. The actual information matrix is the expected values of the estimator.

The upper confidence limit for the scale parameter α is given by

$$\alpha_{[\mathrm{UL},1-(\delta/2)]} = \hat{\alpha} \exp\left(\frac{k\sqrt{I^{-1}(1,1)}}{\hat{\alpha}}\right) \tag{4.44}$$

The lower and upper confidence limits for the shape parameter β are given by

$$\beta_{[LL,1-(\delta/2)]} = \frac{\hat{\beta}}{\exp\left(k\sqrt{I^{-1}(2,2)}/\hat{\beta}\right)} \tag{4.45}$$

and

$$\beta_{[UL,1-(\delta/2)]} = \hat{\beta}\exp\left(\frac{k\sqrt{I^{-1}(2,2)}}{\hat{\beta}}\right) \tag{4.46}$$

respectively, where k is the $[100(1-\delta/2)]$ percentile of a standard normal distribution. For further discussion, see Lawless (1982).

Lawless (1978) presents a review of the interval estimation for Weibull distribution. Some other relevant references are Mann (1968), Mann and Fertig (1975), Fertig et al. (1980), Schneider and Weissfield (1989), and Kotani et al. (1997).

Stone and Rosen (1984) discuss the use of graphical techniques for the estimation of statistical confidence interval. O'Connor (1997) discusses the confidence limits in the context of the WPP plot.

Although we have confined our discussion to the case of complete data, the approach can be extended, with some modification of the likelihood function, to the censored data case.

4.7 THREE-PARAMETER WEIBULL MODEL

In this section we discuss the graphical and statistical methods for estimating the parameters of the three-parameter Weibull model given by (1.2).

4.7.1 Graphical Methods

The WPP plot for this model is discussed in Section 3.4.4 and the relationship between y and x [given by (3.33)] is nonlinear. Five different methods have been proposed to estimate the parameters from the plot.

The first and simplest method is as follows. As estimate of τ is given by $\hat{\tau} = t_{(1)}$ (a better estimate is $t_{(1)} - 1/n$). Using this, the transformed data (given by $t_i - \hat{\tau}$) is viewed as data generated from a two-parameter Weibull distribution. The scale and shape parameters are obtained using the procedure outlined in Section 4.5.2.

O'Connor (1997) suggests an alternate procedure to estimate the location parameter and is given by

$$\hat{\tau} = t_{(2)} - \frac{(t_{(3)} - t_{(2)})(t_{(2)} - t_{(1)})}{(t_{(3)} - t_{(2)}) - (t_{(2)} - t_{(1)})} \tag{4.47}$$

Kececioglu (1991) discusses two other methods to obtain estimates of τ. Another method was proposed by Li (1994). It is a two-step iterative procedure. In the first step, the location parameter is assumed known, and the scale and shape parameters are estimated using the traditional graphical method. In the second step, the shape parameter is assumed known, and the scale and location parameters are estimated by transforming the data using the power law transformation so that the transformed data can be modeled by a two-parameter exponential distribution. The process is iterated until the estimates converge.

Yet another method was proposed by Jiang and Murthy (1997a). It is a modification of the method proposed by Li (1994). We omit the details and interested readers can find them in the references cited.

In general, it is possible to fit the empirical distribution to a model by minimizing the sum-of-square deviation. Soman and Misra (1992) and Ahmad (1994) discuss some related issues for the three-parameter Weibull distribution.

4.7.2 Method of Moments

The parameters can be estimated using the equations for the mean [given by (3.29)], the variance [given by (3.18)], and the third central moment given by

$$\mu_3 = \alpha^3\{\Gamma(1+3/\beta) - 3\Gamma(1+1/\beta)\Gamma(1+2/\beta) + 2[\Gamma(1+1/\beta)]^3\} \tag{4.48}$$

Solving the three equations simultaneously yields the parameter estimates.

Note that this method is applicable only for the case of complete data. Also, the parameters cannot be expressed explicitly as functions of the moments.

An alternate approach to estimation is as follows. Note that the coefficient of skewness is a function of the shape parameter alone and is given by

$$\gamma_1^2 = \frac{\{\Gamma(1+3/\beta) - 3\Gamma(1+1/\beta)\Gamma(1+2/\beta) + 2[\Gamma(1+1/\beta)]^3\}^2}{\{\Gamma(1+2/\beta) - [\Gamma(1+1/\beta)]^2\}^3} \tag{4.49}$$

The sample coefficient of skewness can be computed based on the sample moments. Using this in (4.51) and solving for β yields an estimate of β. The remaining two parameters are then obtained using the first two moments.

Menon (1963) proposes a simple formula for estimating the shape parameter. It is given by

$$\tilde{\beta}^{-1} = \frac{\sqrt{6}}{\pi}\sqrt{\frac{1}{n-1}\sum_{i=1}^{n}\left(\ln(t_i) - \sum_{j=1}^{n}\ln(t_j/n)\right)} \tag{4.50}$$

and is an unbiased estimator of β^{-1}.

Cran (1988) suggestes a method that involves the first four central moments. The parameter estimates are given by

$$\hat{\beta} = \frac{\ln 2}{\ln(\hat{\mu}_1 - \hat{\mu}_2) - \ln(\hat{\mu}_2 - \hat{\mu}_4)} \tag{4.51}$$

and

$$\hat{\tau} = \frac{\hat{\mu}_1\hat{\mu}_4 + \hat{\mu}_2^2}{\hat{\mu}_1 + \hat{\mu}_4 - 2\hat{\mu}_2} \tag{4.52}$$

$$\hat{\alpha} = \frac{\hat{\mu}_1 - \hat{\tau}}{\Gamma(1 + 1/\hat{\beta})} \tag{4.53}$$

He investigates the statistical properties of the estimator via simulation and proposes a method to decide whether the location parameter is zero.

Modified Moment Method

A modified moment method is given in Cohen et al. (1984). It uses the first and second moments and another equation involving the mean of the first-order statistic $T_{(1)}$ instead of the third moment. This is given by

$$E(T_{(1)}) = \tau + \frac{\alpha\Gamma(1 + 1/\beta)}{n^{1/\beta}} \tag{4.54}$$

They derive some simplified expressions for the estimates.

Probability-Weighted Moment Method

Bartolucci et al. (1999) propose a method involving probability-weighted moments. For a set of real number (r, s, t), the probability-weighted moment function is defined as

$$M_{r,s,t} = E\{T^r[F(T)]^s[1 - F(T)]^t\} = \int_0^1 \left[F^{-1}(x)\right]^r x^r(1 - x)^t\, dx \tag{4.55}$$

For the three-parameter Weibull distribution,

$$F^{-1}(x) = \tau + \alpha[-\ln(1 - x)]^{1/\beta} \tag{4.56}$$

and as a result we have

$$M_r = M_{1,\tau,r} = \frac{\tau}{1 + r} + \alpha\frac{\Gamma(1 + 1/\beta)}{(1 + r)^{1+1/\beta}}. \tag{4.57}$$

The model parameters can be expressed in terms of these probability-weighted moments as indicated below:

$$\tau = \frac{4(M_0 M_3 - M_1^2)}{M_0 + 4M_3 - 4M_1} \tag{4.58}$$

$$\alpha = \frac{M_0 - \tau}{\Gamma\{\ln[(M_0 - 2M_1)/(M_1 - 2M_3)]/\ln 2\}} \tag{4.59}$$

and

$$\beta = \frac{\ln 2}{\ln[(M_0 - 2M_1)/(2M_1 - 4M_3)]} \tag{4.60}$$

The parameter estimates are obtained from (4.58) to (4.60), using empirical probability-weighted moments that are obtained from the data as indicated below:

$$\hat{M}_r = \frac{1}{n} \sum_{j=1}^{n} \frac{(j-1)(j-2)\cdots(j-r)}{(n-1)(n-2)\cdots(n-r)} t_{(j)} \tag{4.61}$$

where $t_{(j)}$ is the jth item in the order sample.

4.7.3 Method of Percentile

Different variants of the percentile method have been proposed for estimating the parameters. We discuss some of these.

A simple percentile estimator for the location parameter is given in Dubey (1967a) and that for the shape parameter in Zanakis (1979). Using a similar method, Schmid (1997) derived a percentile estimator for the scale parameter. These estimators can be used independently or together with other estimation methods such as the method of maximum likelihood.

For a given $u (0 < u < 1)$, the percentile t_u is given by

$$t_u = \tau + \alpha[-\ln(1-u)]^{1/\beta} \tag{4.62}$$

In the case of complete data, an estimate of the percentile is given by

$$s_u = \begin{cases} t_{(\nu)} & \text{if } \nu \text{ is an integer} \\ t_{(\nu+1)} & \text{if } \nu \text{ is not an integer} \end{cases} \tag{4.63}$$

where $t_{(i)}$ is the ith ordered sample and $[v]$ is the largest integer less than ν.

For three different values of $u_i (1 \le i \le 3)$ with $0 < u_1 < u_2 < u_3 < 1$, let $s_i (1 \le i \le 3)$ denote the estimated percentiles. Estimates of the parameters are given by

$$\hat{\tau} = \frac{s_1 s_3 - s_2^2}{s_1 + s_3 - 2s_2} \tag{4.64}$$

$$\hat{\beta} = \frac{\ln\left[\frac{-\ln(1-u_3)}{-\ln(1-u_2)}\right]}{2\ln\left[\frac{s_3 - s_2}{s_2 - s_1}\right]} \tag{4.65}$$

and

$$\hat{\alpha} = \frac{(s_2 - s_1)^{2w_{13}}(s_3 - s_2)^{2(1-w_{13})}}{s_3 - 2s_2 + s_1} \tag{4.66}$$

where

$$w_{13} = 1 - \frac{\ln[-\ln(1-u_1)]}{\ln[-\ln(1-u_1)] - \ln[-\ln(1-u_3)]} \tag{4.67}$$

Hassanein (1971) proposes several percentile estimators for the location and scale parameters assuming that the shape parameter is known. It is based on the best linear unbiased estimates with two-, four-, and six-sample quantiles. The optimum spacings of the sample quantiles, the coefficients to be used in computing the estimates, and their variances and asymptotic efficiencies are discussed.

4.7.4 Method of Maximum Likelihood

The standard maximum-likelihood method for estimating the parameters can have problems in the context of the three-parameter Weibull model since the regularity conditions are not met (Blischke, 1974; Zanakis and Kyparisis, 1986). Numerous approaches have been proposed to overcome the problem and we discuss these briefly.

Complete Data

In this case the likelihood function is given by

$$L(\alpha, \beta, \tau) = \frac{\beta^n}{\alpha^{\beta n}}\left[\prod_{i=1}^{n}(t_i - \tau)^{\beta-1}\right]\exp\left[-\frac{1}{\alpha^{\beta}}\sum_{i=1}^{n}(t_i - \tau)^{\beta}\right] \tag{4.68}$$

The maximum-likelihood estimates are obtained by setting the partial derivatives to zero. This implies solving the following set of equations:

$$\hat{\alpha}^{\hat{\beta}} - \frac{1}{n}\sum_{i=1}^{n}(t_i - \hat{\tau})^{\hat{\beta}} = 0 \tag{4.69}$$

$$\frac{\sum_{i=1}^{rn}(t_i - \hat{\tau})^{\hat{\beta}}\ln(t_i - \hat{\tau})}{\sum_{i=1}^{n}(t_i - \hat{\tau})^{\hat{\beta}}} - \frac{1}{\hat{\beta}} - \frac{1}{n}\sum_{i=1}^{n}\ln(t_i - \hat{\tau}) = 0 \tag{4.70}$$

and

$$(\hat{\beta}-1)\sum_{i=1}^{n}(t_i-\hat{\tau})^{-1}-\hat{\beta}\hat{\alpha}^{-\hat{\beta}}\sum_{i=1}^{n}(t_i-\hat{\tau})^{\hat{\beta}-1}=0 \tag{4.71}$$

subject to $\hat{\tau} \leq t_{(1)}$.

The existence of solutions to the above set of equations has received considerable attention in the literature as there can be more than one solution or none at all. For a discussion of this and some related issues, see Zanakis and Kyparisis (1986).

Gourdin et al. (1994) examine the numerical problems associated with the maximum likelihood method. They obtain the estimates by reformulating it as a global optimization problem and present some algorithms. See also Qiao and Tsokos (1995) for some further discussions.

Censored Data

For censored data, the maximum-likelihood estimates are obtained in a similar manner to the standard Weibull model. In the case of Type II censoring with $t_1 \leq t_2 \leq \cdots \leq t_k$, the log-likelihood function is given by

$$\begin{aligned}L(\alpha,\beta,\tau) &= \ln\frac{n!}{(n-k)!}+k(\ln\beta-\beta\ln(\alpha))+(\beta-1)\sum_{i=1}^{k}(t_i-\tau)\\&\quad-\alpha^{-\gamma}\left[\sum_{i=1}^{k}(t_i-\tau)^{\beta}+(n-k)(t_k-\tau)^{\beta}\right]\end{aligned} \tag{4.72}$$

The maximum-likelihood estimates are given by

$$\hat{\alpha}=\left[\frac{\sum_{i=1}^{k}(t_i-\hat{\tau})^{\beta}+(n-k)(t_k-\hat{\tau})^{\beta}}{k}\right]^{1/\hat{\beta}} \tag{4.73}$$

with $\hat{\beta}$ and $\hat{\tau}$ obtained from the following two equations:

$$\begin{aligned}&\hat{\beta}k\frac{\sum_{i=1}^{k}(t_i-\hat{\tau})^{\hat{\beta}-1}+(n-k)(t_k-\hat{\tau})^{\hat{\beta}-1}}{k}\Bigg/\frac{\sum_{i=1}^{k}(t_i-\hat{\tau})^{\hat{\beta}}+(n-k)(t_k-\tau)^{\hat{\beta}}}{k}\\&\quad-(\hat{\beta}-1)\sum_{i=1}^{k}(t_i-\hat{\tau})^{-1}=0\end{aligned} \tag{4.74}$$

and

$$\begin{aligned}&\left[k/\hat{\beta}+\sum_{i=1}^{k}\ln(t_i-\hat{\tau})\right]\cdot\left[\sum_{i=1}^{k}(t_i-\tau)^{\hat{\beta}}+(n-k)(t_k-\tau)^{\hat{\beta}}\right]\\&\quad-k\left[\sum_{i=1}^{k}(t_i-\hat{\tau})^{\hat{\beta}}\ln(t_i-\hat{\tau})+(n-k)(t_k-\tau)^{\hat{\beta}}\ln(t_k-\hat{\tau})\right]=0\end{aligned} \tag{4.75}$$

Lemon (1975) develops maximum-likelihood estimators based on various left- and right-censored data situations and studies some iterative algorithms.

Grouped Data

The maximimun-likelihood method can be used for grouped data as discussed for the standard Weibull model.

When compared with the case of complete data, there is a loss of information because of the grouping. As a result, the variances of the estimates are larger. Some discussion on this can be found in Hirose and Lai (1997). For the case of irregular interval group failure data, Kabir (1998) presents three methods for estimating the Weibull distribution parameters. Simulation results indicate that a sequential updating of the distribution function method produces good estimates for all types of failure patterns.

Some Modified Maximum-Likelihood Estimators

To overcome the problem resulting due to the nonregularity discussed earlier, several modifications have been proposed by different authors. Cohen and Whitten (1982) suggest several modified maximum-likelihood estimators. One of these is to replace (4.71) by

$$\hat{\tau} + \frac{\hat{\alpha}}{n^{1/\beta}}\Gamma\left(1 + \frac{1}{\hat{\beta}}\right) = t_{(1)} \tag{4.76}$$

Another alternative proposed is

$$-\ln\left(\frac{n}{n+1}\right) = \frac{(t_{(1)} - \hat{\tau})^{\hat{\beta}}}{\hat{\alpha}^{\hat{\beta}}} \tag{4.77}$$

Yet another alternative is

$$\hat{\tau} + \hat{\alpha}\Gamma\left(1 + \frac{1}{\hat{\beta}}\right) = \hat{\mu}_1 \tag{4.78}$$

Wyckoff and Engelhardt (1980) suggest the following method. The minimum value of the observed data yields the initial estimate $\hat{\tau}_0$. Based on this, an initial estimate of the shape parameter [obtained using the estimator proposed by Dubey (1967b)] is given by

$$\hat{\beta}_0 = \frac{2.989}{\ln(T_{(k)} - t_{(1)}) - \ln(T_{(h)} - t_{(1)})} \tag{4.79}$$

where $T_{(k)} = t_{0.94}$ and $T_{(h)} = t_{0.17}$ are the 94th and 17th percentiles, respectively. An initial estimate of the scale parameter is obtained as follows:

$$\hat{\alpha}_0 = \frac{\bar{t} - \hat{\tau}_0}{\Gamma(1 + 1/\hat{\beta}_0)} \tag{4.80}$$

A refined estimate of the location parameter is given by

$$\hat{\tau}_1 = t_{(1)} - \frac{\hat{\alpha}_0}{n^{1/\hat{\beta}_0}}\Gamma\left(1 + \frac{1}{\beta_0}\right) \tag{4.81}$$

and the procedure is repeated.

For the three-parameter Weibull distribution, the estimation of the location parameter poses a challenge. Zanakis and Kyparisis (1986) provide an interesting review of the problems and compare the different methods with the maximum-likelihood method. Kudlayev (1987) discusses the methods for estimating the parameters of a so-called Weibull–Gnedenko distribution and focuses on the case with three unknown parameters.

4.7.5 Bayesian Estimation

In contrast to the standard Weibull model, there is very little literature on the Bayesian estimation for the three-parameter Weibull model. This is mainly due to computational complexity involved. A method using a rather crude numerical algorithm can be found in Reilly (1976).

Smith and Naylor (1987) compare the Bayesian and maximum-likelihood estimators. Kuo and Yiannoutsos (1993) examine some empirical Bayes estimators for the scale parameter of a Weibull distribution with type II censored data. Under a similar setting, Bayes estimator of the reliability function for a hierarchical Weibull failure model and the robustness issues are discussed in Papadopoulos and Hoekstra (1995). Recently, Tsionas (2000) examines Bayesian analysis in the context of samples from three-parameter Weibull distribution.

4.7.6 Hybrid Method

In theory, it is possible to use one method to estimate one or two parameters and another to estimate the remaining parameters. Because of the problems with the method of maximum likelihood, many hybrid estimation methods have been proposed and studied. We discuss some of them.

McCool (1998) considers the problem of estimating the location parameter. Because

$$\frac{d}{d \ln x} \ln \ln \frac{1}{1 - F(x)} = \frac{\beta}{x - \tau} x \tag{4.82}$$

the slope of the Weibull transformation tends to increase when the data is truncated, and it approaches infinity as x approaches τ from above. A test is suggested based on the estimates of the shape parameter for two truncated samples to determine if the location parameter $\tau = 0$ or not. The following equation

$$\frac{\sum_{i=1}^{r} t_i^{\beta} \ln t_i + (n - r)t_r^{\beta} \ln t_r}{\sum_{i=1}^{r} t_i^{\beta} + (n - r)t_r^{\beta}} - \frac{1}{\beta} - \frac{1}{r}\sum_{i=1}^{r} \ln t_i = 0 \tag{4.83}$$

is solved for two values of r to obtain two estimates of the shape parameter. Let $\hat{\beta}_A$ and $\hat{\beta}_L$ denote these two estimates. The hypothesis that $\tau = 0$ is rejected at $100\delta\%$ level if $\hat{\beta}_L/\hat{\beta}_A > w_{1-\delta}$. A table for 50th, 90th, and 95th percentage points of $w_{1-\delta}$ for different n, r, and r_1 are provided in McCool (1998). It is recommended that a truncation number between 5 and 9 gives nearly maximum power over a sample of size 10 to 100.

Gallagher and Moore (1990) compare the maximum-likelihood and minimum-distance estimates based on testing six combinations of these methods. The minimum-distance methods using both Anderson–Daring and Cramér–Von Mises goodness-of-fit statistics are considered. The minimum distance estimate using Anderson–Daring and Cramer–Von Mises goodness-of-fit statistic on the location parameter (based on the maximum-likelihood estimates of the shape and scale parameters) is recommended as it gives the best results in terms of precision and robustness.

EXERCISES

Date Set 4.1 Complete Data: Twenty Items Tested Till Failure

11.24	1.92	12.74	22.48	9.60
11.50	8.86	7.75	5.73	9.37
30.42	9.17	10.20	5.52	5.85
38.14	2.99	16.58	18.92	13.36

Data Set 4.2 Type I Censored Data: Fifty Items Tested with Test Stopped after the 12 h[a]

0.80	1.26	1.29	1.85	2.41
2.47	2.76	3.35	3.68	4.46
4.65	4.83	5.21	5.26	5.36
5.39	5.53	5.64	5.80	6.08
6.38	7.02	7.18	7.60	8.13
8.46	8.69	10.52	11.25	11.90

[a]The data is the failure times.

Data Set 4.3 Type II Censored Data: Thirty Items Tested with Test Stopped after the 20th Failure[a]

2.45	3.74	3.92	4.99	6.73
7.52	7.73	7.85	7.94	8.25
8.37	9.75	10.86	11.17	11.37
11.60	11.96	12.20	13.24	13.50

[a]The data is the failure times.

Data Set 4.4 Random Censored Data: Plus (+) Denoted Right Censoring

9	14	18	25	27
31^+	35	40	48	50^+

Data Set 4.5 Grouped Data: Failures Times Grouped into Different Intervals[a]

j	t_{j-1} (days)	t_j (days)	n_j
1	0	5	7
2	5	10	12
3	10	15	8
4	15	20	3
5	20	25	7
6	25	30	3
7	30	35	5
8	35	40	2
9	40	45	1
10	45	50	1

[a] n_j denotes the number of items with lifetimes in the interval $[t_j, t_{j+1})$ with $t_0 = 0$.

Data Set 4.6 Complete Data: Failure Times of 24 Mechanical Components

30.94	18.51	16.62	51.56	22.85	22.38	19.08	49.56
17.12	10.67	25.43	10.24	27.47	14.70	14.10	29.93
27.98	36.02	19.40	14.97	22.57	12.26	18.14	18.84

4.1. Carry out the WPP plot of Data Set 4.1. Can the data be modeled by a two-parameter Weibull distribution? Estimate the parameters based on the plot assuming that the data can be modeled by a two-parameter Weibull distribution.

4.2. Repeat Exercise 4.1 for Data Set 4.2.

4.3. Repeat Exercise 4.1 for Data Set 4.3.

4.4. Repeat Exercise 4.1 for Data Set 4.4.

4.5. Repeat Exercise 4.1 for Data Set 4.5.

4.6. Carry out a TTT plot of Data Set 4.1. What does this suggest?

4.7. In Section 4.4.1 we derived the parameter estimates using the first two sample moments. Derive expressions for the estimates based on the second and third sample moments.

4.8. Derive (4.49).

4.9. Derive expressions for the mean and variance for the estimators of the first two moments for the two-parameter Weibull distribution.

4.10. Show that the estimators of the moments in Exercise 4.7 are asymptotically unbiased and consistent. From this, show that the moment estimators of the model parameters also have the same properties.

4.11. Suppose that Data Set 4.1 can be adequately modeled by a two-parameter Weibull distribution. Estimate the model parameters using (i) the method of moments, (ii) the method of percentiles, and (iii) the method of maximum likelihood. Compare the three estimates.

4.12. Suppose that Data Set 4.2 can be adequately modeled by a two-parameter Weibull distribution. Estimate the model parameter using (i) the method of percentile and (ii) the method of maximum likelihood. Compare the two estimates.

4.13. Repeat Exercise 4.11 using Data Set 4.3.

4.14. Repeat Exercise 4.11 using Data Set 4.4.

4.15. Derive (4.35).

4.16. Suppose that Data Set 4.5 can be adequately modeled by a two-parameter Weibull distribution. Estimate the model parameter using the method of maximum likelihood.

4.17. The Bayesian approach requires specifying the prior distribution for the parameters to be estimated. Review the studies cited in Section 4.6.4 and comment on the computational aspects for the different prior distributions that have been used.

4.18. Derive (4.41) to (4.43).

4.19. Consider Data Set 4.6. Carry out the WPP plot and discuss whether it can be modeled by a three-parameter Weibull distribution.

4.20. Assume that Data Set 4.6 can be adequately modeled by a three-parameter Weibull distribution. Obtain estimates of the parameter using (i) the method of moments and (ii) the method of maximum likelihood assuming that the location parameter $\tau = 10$.

4.21. Derive (4.72) to (4.75).

CHAPTER 5

Model Selection and Validation

5.1 INTRODUCTION

In this chapter we focus our attention on step 3 (model selection) and step 5 (model validation) of the modeling process discussed in Section 1.3.

Model selection is a difficult task and the choice of a suitable model $[F(t; \theta)]$ is made on the basis of whatever knowledge is available and with the use of well-considered judgment. It is important that the model selected be flexible enough to model the data adequately, taking into account the compromise between ease of analysis and the complexity of the model. Also, due attention must be paid to model behavior for small and large values of the independent variable t.

Graphical methods of model selection involve preparation of probability plots or other graphs and checking visually to determine whether or not the plots appear to show certain characteristics indicative of a particular model. A side benefit of this analysis is that one can often also estimate parameters from the plots as discussed in the previous chapter. Nelson (1982) and O'Connor (1997) discuss this issue in more detail.

The last step in the modeling process involves validating the model. This involves a *goodness-of-fit* test. Goodness-of-fit tests are statistical procedures for testing hypothesised models. A poor fit (either graphical or analytical) may occur for two reasons: (1) the model is incorrect, or (2) the model is correct, but the parameter values specified or estimated may differ from the true values by too great an amount. In general, validation requires further testing, additional data and other information, and careful evaluation of the results.

In both the selection and validation processes, particularly when graphical methods are employed, both science and art are involved. The science aspect will be emphasized in this book while the art aspect is widely accepted and used in engineering where the model selection is done through visual fitting of the data and the model. Furthermore, there are many partly subjective decisions that must be made

Weibull Models, by D.N.P. Murthy, Min Xie, and Renyan Jiang.
ISBN 0-471-36092-9

during the modeling process. This includes selection of the family of models to be considered, the level of significance in statistical testing, the statistical procedures to be used, and so forth. In this chapter we deal with these issues.

The outline of the chapter is as follows. Graphical methods are discussed in Section 5.2. Section 5.3 deals with goodness-of-fit tests. In both sections, complete and censored data are considered. Sections 5.4 and 5.5 deal with model selection and validation for the two- and three-parameter Weibull models. In Sections 5.6 we discuss model discrimination where we look at choosing the best model from a selection of several appropriate models (that include the standard Weibull model) to model a given data set. Finally, we look briefly at model validation in Section 5.7.

5.2 GRAPHICAL METHODS

The main aim of graphical methods is to "fit" data to one or more of the models (or, conversely, to "fit" the model to data) by plotting the data. Visual fits provided by graphical methods are quite useful. The benefits of the graphical approach are the visual insights obtained and (usually) ease of application.

The problems with the graphical approach are (a) it is, to some extent, subjective and (b) there is no well-developed statistical theory for determining the small sample or asymptotic properties of the procedures. The first problem is not serious; it means that different analysts are likely to arrive at somewhat different results if the plotting is done by hand or by using different algorithms, but they usually will arrive at the same conclusions. The second problem is more serious; it means that standard errors and distributions of estimators are not known, even asymptotically, and that test statistics based on these cannot be obtained. As a result, a proper analysis of a set of data cannot be based exclusively on graphical procedures, and one needs to use statistical and analytical methods as well. As such, the graphical approach must be viewed as providing a good starting point for model selection and validation.

5.2.1 Empirical Distribution Function

The empirical distribution function (EDF) is a plot of the cumulative proportion of the data that lie below t, against $t(-\infty < t < \infty)$ and denoted by $\hat{F}$. The plotting of the EDF for complete data was discussed in Section 4.4.2.

The plotting needs to be modified when the data is censored. For further discussion on this see for example, King (1971), Nelson (1982), Lawless (1982), and Meeker and Escobar (1998). Computer packages vary with regard to the type of plot provided, and some have options for various plots.

5.2.2 Transformed Plots

Transformed plots have been developed as an alternative method of plotting data. As opposed to the sample EDF, which is a nonparametric procedure, transformed plots assume a particular underlying distribution. The idea is to transform the data and/or probability scales so that the plot on the transformed scale is linear (within chance fluctuations). These lead to different kinds of plots. Also, the plotting depends on the type of data (complete, censored, grouped, etc.) available.

Probability Plot

The EDF plot, for complete data, involves plotting $\hat{F}(t_{(i)})$ versus $t_{(i)}$ with $\hat{F}(t_{(i)})$ computed as indicated earlier. Without loss of generality, assume that T is a standardized random variable and $F(t)$ a standardized distribution. [If not, one needs to deal with a random variable Z given by $Z = (T - \tau)/\alpha$, where τ is the location parameter and α is the scale parameter. This is a standardized random variable with distribution function $G(z)$ that is easily derived from $F(t)$.]

A probability plot is a plot of $z_i = F^{-1}(\hat{F}(t_{(i)}))$ versus $t_{(i)}$. If F is the true cumulative distribution function (CDF), the probability plot is approximately a straight line.

Weibull Plot

This is the empirical counterpart of the Weibull transformation discussed in Chapter 3. The plotting depends on the type of data. It involves plotting of y_i versus x_i given by

$$y_i = \ln\left\{-\ln[1 - \hat{F}(t_{(i)})]\right\} \quad \text{and} \quad x_i = \ln\left(t_{(i)}\right) \tag{5.1}$$

The plotting is outlined in Part A of Section 4.5.1 for the case of complete data. (The computing of EDF for other kinds of data will be discussed later in the chapter.) If the plotted data is roughly along a straight line, then one can model the data by the standard Weibull model. If the plot is not a straight line, then depending on the shape one or more models derived from the standard Weibull model might adequately model the given data set. This issue is discussed further in later chapters.

P–P Plot

A P–P plot is a plot of percentages of one distribution versus that of a second distribution. For further details, see Wilk and Gnanadesikan (1968) where they state that while the P–P plots are limited in their usefulness, they are useful for detecting discrepancies in the middle of the distribution (about the median).

Q–Q Plot

A Q–Q plot is a plot of the quantiles (or percentiles) of one distribution versus the quantiles of a second distribution. For further details, see Wilk and Gnanadesikan (1968). If one of these distributions is a hypothesized theoretical distribution, then a Q–Q plot is just a probability plot discussed earlier.

5.2.3 Histogram

A histogram is a normalized plot of the numerical frequency distribution and involves grouping of data into distinct categories or classes. This involves determining an appropriate number k of classes with appropriate class boundaries and simply counting to determine the frequency in each class. Usually k is chosen to be between 5 and 20. A rule of thumb is *Sturge's rule*, which gives k as the nearest integer to $1 + 3.322 \ln(n)$, where n is the sample size. Usually a value of k close to this choice will lead to a good representation of the data set. A rule of thumb for class boundaries is to use classes of equal width unless the data appear to be highly skewed, tailing off substantially on one or both ends, in which case open-ended end intervals may be preferable. Division of the previous counts by the sample size n gives the ordinates of the histogram. It gives the proportion of values in a given class and often expressed as percentages.

Based on an examination of the shape of the histogram, one can select a model that has a density function that roughly matches it. This can involve examining whether the shape of the histogram is (i) decreasing, unimodal, bimodal, and so forth (ii) symmetric, skewed to left or right, and (iii) heavy or light tails.

5.2.4 Hazard Plots

Hazard plots are plots of the sample (or empirical) hazard function or cumulative hazard function. Traditionally, cumulative hazard is plotted as the abscissa and time to failure as the ordinate (both transformed as appropriate so that the plot under the assumed distribution is linear), but the scales may be reversed if desired, and the untransformed data may be plotted instead.

The plot should be linear on suitably transformed paper if the assumed distribution provides a good fit to the data. The information obtained from interpretation of a hazard plot is basically the same as for a probability plot. (In fact, hazard plotting paper often includes the probability scale as well.) In particular, from a hazard plot one can estimate the percentiles of the distribution, the failure rate as a function of time, and a number of related quantities. See Nelson (1982) for more details.

Hazard plot for complete data is discussed in part A of Section 4.5.2. The plotting can be done either on ordinary graph papers or special cumulative hazard plot papers. Most computer packages provide several options for plotting.

5.2.5 Incomplete Data

When the data is incomplete (e.g., as in the case of censored or grouped data) the plotting needs to be modified.

Censored Data

In the case of censored data the procedure is as follows:

1. Order the observations from the smallest to the largest.
2. For each uncensored observation, compute I_j and N_j as follows:

$$I_j = \frac{(n+1) - N_p}{1 + c} \tag{5.2}$$

$$N_j = N_p + I_j \tag{5.3}$$

where I_j is the increment for the jth uncensored dataum, N_p is the order of the previous uncensored observation, and c is the number of data points remaining in the data set, including the current data point. $N_p = 0$ for the first uncensored data.

3. For each uncensored observation $(t_{(j)})$ compute the estimator given by

$$\hat{F}(t_{(j)}) = (N_j - 0.3)/(n + 0.4) \tag{5.4}$$

4. Plot $\ln\{-\ln[-\hat{F}(t_{(j)})]\}$ versus $\ln(t_{(j)})$ for each uncensored data.

For further details, see Dodson (1994).

Grouped Data

In this case let the number of observations in interval $j(1 \leq j \leq J)$ be d_j with the interval given by $[a_{j-1}, a_j)$ with $a_0 = 0$ and $a_{J+1} = \infty$. The EDF is estimated as

$$\hat{F}(t_j) = \frac{\sum_{i=1}^{j} d_i}{n} \tag{5.5}$$

with $n = \sum_{j=1}^{J} d_j$. The WPP plot is obtained by plotting $\ln\{-\ln[1 - \hat{F}(t_j)]\}$ versus $\ln(a_j)$ for $(1 \leq j \leq J)$.

5.3 GOODNESS-OF-FIT TESTS

A general test of fit is a test of a null hypothesis:

H_0: the given data comes from a CDF $F(t; \theta)$

with the form of the CDF specified. One needs to consider the following two cases:

Case 1: The parameter θ is fully specified (or known).

Case 2: The parameter θ is not fully specified (or known) and the unspecified parameters need to be estimated from the given data.

For case 1 the theory is well developed and we consider it first. We confine our attention to the case where the given data set is complete and give appropriate references where interested readers can find the tests for incomplete data. Later we consider case 2.

The basic idea in testing H_0 is to look at the data obtained and evaluate the likelihood of occurrence of this sample given that H_0 is true. If the conclusion is that this is a highly unlikely sample under H_0, H_0 is rejected. It follows that there are two types of errors that can be made: rejecting H_0 when it is true, and failing to reject it when it is not true. These are called type I and type II errors. The level of significance (denoted by α) is the maximum probability of making a type I error and the power of the test is the probability of not making a type II error. The test statistic is a summary of the given data, and the test is usually based on some such statistics. The test involves comparing the computed values of the test statistic with some critical values (which depend on the level of significance), which are usually available in tabulated form. The hypothesis is rejected if the statistics exceeds the critical value.

A test is said to be nonparametric if the distribution of the test statistic does not depend on the CDF $F(t)$. For case 1 many different nonparametric tests have been developed, and we consider some of them in this section. Modifications of these tests when parameters are not specified have been developed for a few distributions. In this case the nonparametric property no longer holds.

In applying goodness-of-fit tests one encounters two types of problems:

1. A specific distribution is suggested by theoretical considerations, past experience, and so forth, and we wish to determine whether or not a set of data is from this distribution.
2. We have only a vague idea of the correct distribution (for example, that it is skewed right) and want to do some screening of possible candidate distributions.

In the former case, it is usually appropriate to test at a low level of significance, say $\alpha = .01$, the idea being that we do not want to reject the theoretical distribution unless there is strong evidence against it. In the second case, it may be more appropriate to test at a much higher level of significance, say $\alpha = .1$ or $.2$, in order to narrow the list of candidates somewhat. The choice is to a great extent subjective and depends on the particular application.

5.3.1 Case 1: Parameters of CDF Specified (Complete Data)

We consider four tests. The first test is the well-known chi-square test, and it involves grouping the observed data into intervals. The next three are tests based on EDF and hence are often referred to as EDF tests.

Chi-Square Test

The classic chi-square test, due to Karl Pearson, may be applied to test the fit of data to any specified distribution (discrete or continuous). We confine our attention to the continuous case.

The data is grouped into k intervals; $[a_{i-1}, a_i)$ is the ith interval, with $a_0 = -\infty$ and $a_k = \infty$. In the goodness-of-fit application of the test, one assumes a sample of size n, with each observation falling into one of k possible classes. The observed frequency in interval i is O_i and is calculated as in the plotting of the histogram; e_i is the frequency that would be expected if the specified distribution were the correct one and is calculated as $e_i = np_i$, where

$$p_i = \int_{a_{i-1}}^{a_i} f(x)\,dx = F(a_i) - F(a_{i-1}) \tag{5.6}$$

and f is the density function associated with the specified distribution function F.

The test statistic is

$$\chi^2 = \sum_{i=1}^{k} \frac{(O_i - e_i)^2}{e_i} \tag{5.7}$$

Under the null hypothesis of a completely specified distribution (including parameter values), the test statistic has approximately a chi-square distribution with $(k-1)$ degrees of freedom df). If r parameters are estimated from the data, the distribution of the test statistic (given by 5.7) is approximately a chi-square with $(k-r-1)$ df. The test is an upper-tail test—large values of χ^2 result when observed and expected frequencies differ significantly, so these lead to rejection of the hypothesized distribution.

In calculating χ^2, it is necessary to assure that expected frequencies are not too small; if they are, this may greatly distort the result. A rule of thumb is that if $e_i < 1$ for some i, combine that class with either the previous or the succeeding class, repeating the process until $e_i \geq 1$ for all i. A more conservative rule is to combine classes if $e_i < 5$.

The χ^2 test has the advantages of being easy to apply and being applicable even when parameters are unknown. However, it is not a very powerful test and is not of much use in small or sometimes even modest size samples.

Kolmogorov–Smirnov (K–S) Test

The test statistic D_n is simply the maximum distance between the hypothesized CDF and EDF. This is most easily calculated as

$$D_n = \max\{D_n^+, D_n^-\} \tag{5.8}$$

where

$$D_n^+ = \max_{i=1,\ldots,n}\left[\frac{i}{n} - F(t_{(i)})\right] \qquad \mathrm{D}_n^- = \max_{i=1,\ldots,n}\left[F(t_{(i)}) - \frac{i-1}{n}\right] \tag{5.9}$$

Percentiles of the distribution of D_n were given in Massey (1951). A very close approximation to these percentiles that is a function only of n and tabulated constants d_α is given in Stephens (1974). The critical value of D_n is calculated as $d_\alpha/(n^{1/2} + 0.11n^{-1/2} + 0.12)$. Values of D_n in excess of the critical value lead to rejection of the hypothesized distribution.

Anderson–Darling (A–D) Test

The Anderson–Darling test is also based on the difference between the hypothesized CDF and EDF. The test statistic is

$$A^2 = A_n^2 = \frac{-1}{n}\sum_{i=1}^{n}(2i-1)\{\ln(F(t_{(i)}) + \ln[1 - F(t_{(n-i+1)})]\} - n \tag{5.10}$$

Percentile w_α of the distribution of A_n^2 can be found in Stephens (1974) and Pearson and Hartley (1972). If the calculated value of A_n^2 exceeds a_α, the hypothesized distribution is rejected at a level of significance α.

Cramér–von Mises Test

The Cramér–von Mises test is also based on the difference between the hypothesized CDF and EDF. The test statistic is

$$W^2 = W_n^2 = \sum_{i=1}^{n} \left[F(t_{(i)}) - \frac{i - 0.5}{n} \right]^2 + \frac{1}{12n} \tag{5.11}$$

As in the earlier case, percentiles w_α of the distribution of W_n^2 can be found in Stephens (1974) and Pearson and Hartley (1972). If the calculated value of W_n^2 exceeds w_α, the hypothesized distribution is rejected at a level of significance α.

Relative Comparison

According to D'Agostino and Stephens (1986):

1. The EDF tests are more powerful than the chi-square test.
2. The Kolmogrov–Smirnov test is the most well-known EDF test, but it is often much less powerful than the other EDF tests (Anderson–Darling and Cramér–von Mises tests).

Further discussion can be found in D'Agostino and Stephens (1986).

5.3.2 Case 1: Parameters of CDF Specified (Incomplete Data)

Censored Data

In this case the EDF tests discussed earlier can still be used for all kinds of censoring (left, right, random, type I, type II, etc.) and the EDF statistics need to be adapted accordingly. For further discussion, see Lawless (1982) and D'Agostino and Stephens (1986).

Grouped Data

In the case of grouped data, the chi-square test is the most appropriate test.

5.3.3 Case 2: Parameters Not Fully Specified (Complete Data)

The goodness-of-fit problem is much more difficult when the parameters of the distribution are not specified in the null hypothesis, unless the sample size is large and the chi-square test is used. In this case, incidentally, *minimum chi-square* estimators, that is, those that minimize the joint distribution expressed in terms of the p_i, given in (5.2), should be used rather than the moment or maximum-likelihood estimators. If the MLEs are used, the resulting statistic is not distributed as χ^2. For further details, see Li and Doss (1993).

A natural approach when H_0 specifies only the form of the distribution would be to use some method of estimating the parameters (e.g., maximum likelihood) and then apply the test statistics of the previous section (or others given in the references). The problem is that the tests are no longer nonparametric in this case, and the distribution of the test statistic is very complicated, depending on the statistic, the estimation procedures used, the hypothesized distribution, and the true distribution.

As indicated in Blischke and Murthy (2000), three basic approaches to the problem are as follows:

1. Embed the distribution in question in a larger family of distributions and test the null hypothesis that the parameter values are those that reduce the distribution to the one in question. For example, the Weibull distribution reduces to the exponential when the shape parameter $\beta = 1$; so a test for the exponential can be done by estimating the Weibull parameters and testing for the null hypothesis H_0: $\beta = 1$. A difficulty that has been encountered in using this approach is that the test may be powerful only against alternatives in the same family (e.g., other Weibull distributions).
2. Determine the asymptotic distribution of the test statistic and use these results to obtain critical points. This has been accomplished for only a few distributions and has the added disadvantage that the results may not be appropriate for small samples.
3. Use the tests discussed in the previous section, but use estimates of the unknown parameters and modify the critical values accordingly. This leads to some very difficult analytical problems, and the modified critical values are usually determined by means of simulation studies. A number of published tables of the results are given or cited in the references previously cited. Quite simple modifications are provided for a number of distributions, and these have been found in many cases to be very good approximations, even for quite small samples.

See Lawless (1982), Liao and Shimokawa (1979) and D'Agostino and Stephens (1986) for additional details and references.

5.4 MODEL DISCRIMINATION

In this section we look at the problem of choosing a model from a set of two or more models. It can be viewed as deciding on which model(s) fits the data better than the others. A special case is one where all the models belong to a particular family (so that the models can be viewed as a submodel of one model).

The simplest case is where the choice is between two models. In the general case the two come from two separate families so that one is not a submodel of the other. This can also be viewed as an hypotheses testing problem.

Let T denote a random variable. The basic problem is to choose between the following:

H_1: T has probability density function $f_1(t;\theta_1)$, $\theta_1 \in \Omega_1$
H_2 : T has probability density function $f_2(t;\theta_2)$, $\theta_2 \in \Omega_2$

One criterion for choosing between the two is the likelihood ratio given by

$$R_{12} = \frac{\sup_{\theta_1 \in \Omega_1} L_1(\theta_1)}{\sup_{\theta_2 \in \Omega_2} L_2(\theta_2)} \tag{5.12}$$

where $L_1(\theta_1)$ and $L_2(\theta_2)$ are the likelihood functions under models H_1 and H_2, respectively. For complete data set $t_1, t_2, \ldots, t_n$,

$$L_i(\theta_i) = \prod_{j=1}^{n} f_i(t_j;\theta_i) \qquad i = 1, 2 \tag{5.13}$$

The decision rule based on the statistics R_{12} is to choose H_1 if $R_{12} \geq c$ and choose H_2 if $R_{12} < c$. The parameter c is selected to make the probability of type I error small and hopefully the power of the test as large as possible.

The problem of choosing between a two-parameter Weibull model and an alternate model has received some attention in the literature. Quesenberry and Kent (1982) and Kappenman (1982, 1989) deal with the choice between Weibull and lognormal based on sample data.

5.5 MODEL VALIDATION

A basic principle in validating a model is through an assessment of the predictive power of the model as it provides a basis for generalization. Assessment of the predictive power of a model is basically a statistical problem. The idea is to use an estimated model to predict outcomes of other observations, which may be data set aside for this purpose or future observations, and to evaluate the closeness of the predictions to observed values. A key requirement for credibility is that data used for validation not be a part of the data set used in model selection and estimation.

The goodness-of-fit test discussed in Section 5.3 can be used for model validation. The approach is as follows:

Step 1: Divide the data into two sets (S_1 and S_2).

Step 2: Carry out model selection (as indicated earlier in this chapter) and estimate the model parameter (using methods discussed in Chapter 4) based on data set S_1.

Step 3: Use K–S or A–D or some other procedure with the selected model as the null hypothesized distribution. The test is applied to data set S_2. Since the hypothesized distribution is completely specified, conditional on the independently estimated parameters, the test is distribution free. This avoids the use of the special tables for each distribution required when the parameters are estimated from the same data that are used in fitting. Another advantage is that the procedure is easily applied if the validation data are incomplete.

5.6 TWO-PARAMETER WEIBULL MODEL

5.6.1 Graphical Approaches

Weibull Probability Plot

The Weibull probability plot involves transforming the data using the Weibull transformation discussed in Chapter 3. The plot involves computing y_i and x_i using (5.1). This depends on the type of data available and the method used to estimate $\hat{F}(t_{(i)})$.

Complete Data

The plotting procedure is as indicated in part A of Section 4.5.2. If the plotted data is scattered close to the fitted straight line, then the two-parameter Weibull model can be accepted (or more correctly, not rejected). If not, the model is rejected. (Note: The parameter estimates can be obtained using the slope of the line and its intercept with the vertical axis as discussed in Chapter 4.)

Since y_i can be viewed as a nonlinear transformation of the estimates of the model parameter, confidence limits for it can be obtained from the confidence limits for the parameter estimates. However, some care needs to be taken in plotting this—see Dodson (1994) for further details. Most computer packages for WPP plots provide these confidence limits.

Censored Data

The procedure is as indicated in Section 5.2.5. As before, if the plotted data is scattered close to the fitted straight line, then the two-parameter Weibull model can be accepted. If not, the model is rejected. The judgment can be subjective or objective based on some goodness-of-fit tests.

Grouped Data

The procedure is similar to the censored case discussed earlier.

TTT Plot

This is the empirical counterpart of the scale TTT transformation discussed in Chapter 3. Let $\{t_{(i)}; i = 1, 2, ..., n\}$ be an ordered sample from F with finite mean; the total time on test to the ith failure is defined as

$$x_i = \sum_{j=1}^{i} t_{(j)} + (n - i)t_{(i)} \qquad i = 1, 2, ..., n \tag{5.14}$$

Plotting $u_i = x_i/x_n$, against i/n and connecting the plotted points by straight lines, we have the so-called empirical TTT plot. If this plot is close to the graph of the scaled TTT transform based on the standard Weibull model, then the model can be accepted as an adequate model.

Hazard Plot

For the case of complete data, the plotting procedure is as indicated in part A of Section 4.5.2. If the plotted data is close to the fitted straight line, then the two-parameter model is accepted (or more correctly, not rejected). If not, it is rejected.

5.6.2 Goodness-of-Fit Tests

Some of the tests are based on the close relationship between the two-parameter Weibull distribution and the exponential and extreme value distribution in the sense that the associated random variables are related. Let T be distributed according to a two-parameter Weibull distribution given by (1.3).

1. Power Transformation Let $Z = T^{\beta}$. Then the CDF for Z is an exponential distribution given by

$$G(z) = 1 - e^{-z/\mu} \tag{5.15}$$

with $\mu = \alpha^{\beta}$.

2. Log Transformation Let $Z = \ln T$. Then the CDF for Z is an extreme value distribution given by

$$G(z) = 1 - \exp\{-\exp[-(z - \eta)/\phi]\} \tag{5.16}$$

with $\eta = -\ln(\alpha)$ (the location parameter) and $\phi = 1/\beta$ (the scale parameter).

Most tests specific to the two-parameter Weibull distribution are based on these two transformations.

As before, we need to consider two cases.

Case 1: Parameters Completely Specified

In this one case use the tests discussed in Section 5.3.1 to test for the goodness of fit. An alternate approach is to transform the data using power (log) transformation and then carrying out the goodness-of-fit test for an exponential (extreme value) distribution using the transformed data. We discuss three specific tests and they are as follows:

S Test (Mann et al., 1974)

This test is based on the log transformation. The test statistic is based on the normalized sample spacings (calculated as differences between successive order statistics) and is given by

$$M = \frac{\left(\frac{n}{2}\right) \sum_{i=[n/2]+1}^{n-1} L_i}{\left(\frac{n-1}{2}\right) \sum_{i=1}^{[n/2]} L_i} \tag{5.17}$$

where $[x]$ is the smallest integer less than or equal to x, and

$$L_i = \frac{\ln(t_{(i+1)}) - \ln(t_{(i)})}{M_{i+1} - M_i} \tag{5.18}$$

The M_i are expected sample spacings and have been extensively tabulated by Mann et al. (1974). An approximation, good for $n \geq 10$, is given by

$$M_i \approx \ln\left(\ln\left(\frac{1}{1 + \frac{i - 0.5}{n + 0.25}}\right)\right) \tag{5.19}$$

[See also Lawless (1982).] For small n and for censored samples, critical values are given in Mann et al. (1974). For $n \geq 20$, the distribution of M is approximately an F distribution with the df for the numerator and denominator given by $\nu_1 = 2[(n-1)/2]$ and $\nu_2 = 2[n/2]$, respectively (where [ ] indicates integer part, as before).

Ratio-Type Test (Thiagrajan and Harris, 1976)
This test is based on the power transformation. Let $z_i = t_i^\beta$ and let $(z_{(1)}, z_{(2)}, \ldots, z_{(n)})$ be the ordered sample. The statistics (with $1 < r < n$) is given by

$$Q(r, n-r) = \left(\frac{n-r}{r}\right) \frac{\sum_{i=1}^{r} S_i}{\sum_{i=r+1}^{n} S_i} \tag{5.20}$$

where S_i is the ith normalized spacing given by

$$S_i = (n - i + 1)(z_{(i)} - z_{(i-1)}) \tag{5.21}$$

where $Q(r, n-r)$ is distributed as an F distribution with $2r$ and $2(n-r)$ degrees of freedom. The goodness-of-fit test can be based on it falling within an appropriate rejection region based on the F distribution.

Test Based on Stabilized Plot (Coles, 1989)
This test is for a CDF of the form $G[(z - \phi)/\eta]$. The log transformation results in this being satisfied so that one is dealing with the transformed data set $\{z_1, z_2, \ldots, z_n\}$ with $z_i = \ln(t_i)$ and testing the CDF given by (1.3).

The stabilized plot is a plot of s_i versus r_i (for $1 \le i \le n$) computed as follows:

$$s_i = (2/\pi)\sin^{-1}[G^{1/2}(z_{(i)})] \tag{5.22}$$

and

$$r_i = (2/\pi)\sin^{-1}([(i - 0.5)/n]^{1/2}) \tag{5.23}$$

The properties of the plot are:

1. Deviations from the line joining the points (0,0) and (1,1) indicates departures from the extreme value distribution.
2. The plotted points have approximately equal variances due to asymptotic properties of the s_i.

Michael (1983) suggests the following statistics for goodness-of-fit test:

$$D_{\text{sp}} = \max|r_i - s_i| \tag{5.24}$$

The critical points for D can be found in Michael (1983) and for further details, see Coles (1989).

Case 2: Parameters Not Completely Specified
Here one needs to consider the following threes subcases:

1. β known and α unknown
2. β unknown and α known
3. Both β and α unknown

For the first subcase, under the power law transformation, goodness-of-fit test for the two-parameter Weibull gets reduced to a goodness-of-fit test for an exponential distribution using the transformed data (Yang et al., 2002).

Littell et al. (1979) discuss the Kolmogorov–Smirnov, Cramer–von Mises, and Anderson–Darling statistics for testing the goodness of fit of the two-parameter Weibull distribution, and they consider the case when the MLEs are used as the estimators of the model parameters. Chandra et al. (1981) consider Kolmogorov statistics for tests of fit for Weibull distributions. Through simulation, Evans et al. (1997) obtained the critical values for the two- and three-parameter Weibull distributions for Kolmogorov–Smirnov, Anderson–Darling, and Shapiro–Walk types of correlation statistics and presented some approximate formulas. See also Shapiro and Brain (1987). D'Agostino and Stephens (1986) deal with EDF tests for the above three subcases by using the log transformation and carrying out the tests for the extreme value distribution. Mann et al. (1974) deal with the S test and give the percentile values for the S statistics for the extreme value distribution with unknown parameters. Thiagarajan and Harris (1979) examine the ratio-type test for the exponential distribution where the power transformation uses the MLE for the shape parameter. Kimber (1985) deals with tests based on stabilized probability plot and presents critical values of the test statistics D_{sp} using estimates that are simple approximations to the best linear unbiased estimates. Coles (1989) extends this by using more refined linear estimation procedure. Shimokawa and Liao (1999) deal with various EDF tests with the parameters estimated using graphical plotting techniques and least-squares method for estimation of the unknown parameters. Many of the above works carry out a study of the power of various tests with different specified distributions as the alternate hypothesis.

5.7 THREE-PARAMETER WEIBULL MODEL

5.7.1 Graphical Approaches

Weibull Probability Plot

The plotting of data depends on the type of data and is the same as for the two-parameter Weibull model. The procedure (involving five steps) is identical to the two-parameter model for the first three steps. The remaining two steps are modified as follows:

4. Fit a smooth curve to the plotted data.
5. If the smooth curve has a shape similar to that in Figure 3.3, then the two-parameter model is accepted (or more correctly, not rejected). If not, it is rejected.

5.7.2 Goodness-of-Fit Tests

We need to consider two cases.

Case 1: Parameters Completely Specified

In this case, by a linear transformation of data ($z_i = t_i - \tau$) the goodness-of-fit test for the three-parameter Weibull model is reduced to that for the two-parameter Weibull model using the transformed data.

Case 2: Parameters Not Completely Specified

Here one needs to consider the several subcases: If τ is known and either β and/or α unknown, then the goodness-of-fit test for the three-parameter Weibull model is reduced to that for the two-parameter Weibull model under the linear transformation discussed above. Khamis (1997a) suggests a delta-corrected Kolmogorov–Smirnov statistic, which is a simple modification of the Kolmogorov–Smirnov statistic, for the three-parameter Weibull distribution assuming a known location parameter and unknown shape and scale parameters.

If τ is unknown, then the problem is more difficult. Lockhart and Stephens (1994) discuss this topic and give critical values for three different EDF tests.

EXERCISES

Data Set 5.1 Complete Data: Failure Times of 20 Electric Bulbs

1.32	12.37	6.56	5.05	11.58
10.56	21.82	3.60	1.33	12.62
5.36	7.71	3.53	19.61	36.63
0.39	21.35	7.22	12.42	8.92

Data Set 5.2 Complete Data: Failure Times of 20 Identical Components

15.32	8.29	8.09	11.89	11.03	10.54	4.51	1.79	7.93	6.29
5.46	2.87	11.12	11.23	3.58	9.74	8.45	2.99	3.14	1.80

Data Set 5.3 Distance Between Cracks in a Pipe

30.94	18.51	16.62	51.56	22.85	22.38	19.08	49.56
17.12	10.67	25.43	10.24	27.47	14.70	14.10	29.93
27.98	36.02	19.40	14.97	22.57	12.26	18.14	18.84

Data Set 5.4 Complete Data: Lifetimes of 20 Electronic Components

0.03	0.12	0.22	0.35	0.73	0.79	1.25	1.41	1.52	1.79
1.80	1.94	2.38	2.40	2.87	2.99	3.14	3.17	4.72	5.09

5.1. It is hypothesized that the lifetimes of bulbs (given by Data Set 5.1) are given by a two-parameter Weibull distribution with $\alpha = 20$ and $\beta = 1.5$. Use P–P and Q–Q plots to determine whether the hypothesis is true or not.

5.2. Use the K–S test to test the hypothesis of Exercise 5.1 at the following three levels of significance: 0.05, 0.1, and 0.2.

5.3. Repeat Exercise 5.2 using the A–D test.

5.4. Repeat Exercise 5.3 using the Cramer–von Mises test.

5.5. Repeat Exercise 5.1 with the hypothesis $\alpha = 10$ and $\beta = 1.5$.

5.6. Repeat Exercises 5.2 to 5.4 based on the hypothesis in Exercise 5.5.

5.7. Suppose that the data in Data Set 5.2 can be modeled by a two-parameter Weibull model. How would you test whether the hypothesis $\beta = 1$ is true or not.

5.8. Use A–D test to decide whether the distance between cracks of Data Set 5.3 can be modeled by a two-parameter Weibull model or not?

5.9. Generate a total time on test plot for the failure data in Data Set 5.4. Discuss the aging property of the electronic component based on the plot.

PART C

Types I and II Models

CHAPTER 6

Type I Weibull Models

6.1 INTRODUCTION

Type I models are derived from the transformation of the standard Weibull variable. The transformation can be either linear or nonlinear. The linear transformation results in four models [Model I(a)1–4]. Model 1(a)-1 is a special case of the standard Weibull model and hence will not be discussed. Model I(a)-2 is the three-parameter Weibull model and is discussed in Chapter 4. The remaining two models [Model I(a)-3 and -4] are discussed in this chapter. The nonlinear transformation results in three models [Model I(b)1–3] and these are also discussed in this chapter.

Let T denote the random variable from the standard Weibull model and Z denote the transformed variable. For some models, it can take negative values. The models and their underlying transformation are as follows:

1. **Model I(a)-3:** Linear transformation

$$Z = -T + \tau \tag{6.1}$$

2. **Model I(a)-4:** Linear transformation

$$Z = \begin{cases} T/2 + \tau & \text{for } Z \geq 0 \\ -T/2 + \tau & \text{for } Z < 0 \end{cases} \tag{6.2}$$

3. **Model I(b)-1:** Power transformation

$$\frac{Z - \tau}{\eta} = \left(\frac{T}{\alpha}\right)^{\beta} \tag{6.3}$$

4. **Model I(b)-2:** Log transformation

$$\frac{Z - \tau}{\eta} = \beta \ln\left(\frac{T}{\alpha}\right) \tag{6.4}$$

Weibull Models, by D.N.P. Murthy, Min Xie, and Renyan Jiang.
ISBN 0-471-36092-9

5. **Model I(b)-3:** Inverse transformation

$$Z = \frac{\alpha^2}{T} \tag{6.5}$$

For the above models, let $G(t;\theta)$ denote the distribution function for Z; $G(t;\theta)$ is obtained easily from the distribution function for the standard Weibull model given by (1.3). For each model, we first discuss the form of $G(t;\theta)$ and its properties. Following this, we present the results of model analysis and then discuss the estimation of model parameters.

The outline of the chapter is as follows. Section 6.2 deals with Model I(a)-3 (reflected Weibull model), Section 6.3 with Model I(a)-4 (double Weibull model), Section 6.4 with Model I(b)-1 (two-parameter exponential model), Section 6.5 with Model I(b)-2 (log Weibull model), and Section 6.6 with Model I(b)-3 (inverse Weibull model).

6.2 MODEL I(a)-3: REFLECTED WEIBULL DISTRIBUTION

The random variables Z and T are related by (6.1).

6.2.1 Model Structure

Distribution Function

The distribution function for Z is given by

$$G(t) = \exp\left[-\left(\frac{\tau - t}{\alpha}\right)^{\beta}\right] \qquad -\infty < t < \tau \tag{6.6}$$

Note that the support for $G(t)$ is different from that for $F(t)$ and the two are disjoint.

The model was first proposed in Cohen (1973). It is obtained by a reflection of the standard three-parameter Weibull model about a vertical axis at $t = \tau$ and hence called the *reflected Weibull distribution*.

Density Function

The density function is given by

$$g(t) = \left(\frac{\beta}{\alpha}\right)\left(\frac{\tau - t}{\alpha}\right)^{\beta-1} \exp\left[-\left(\frac{\tau - t}{\alpha}\right)^{\beta}\right] \tag{6.7}$$

Hazard Function

The hazard function is given by

$$h(t) = \left(\frac{\beta}{\alpha}\right)\left(\frac{\tau - t}{\alpha}\right)^{\beta-1} \frac{\exp\left[-\left(\frac{\tau-t}{\alpha}\right)^{\beta}\right]}{1 - \exp\left[-\left(\frac{\tau-t}{\alpha}\right)^{\beta}\right]} \tag{6.8}$$

The reflected Weibull distribution can also be viewed as the third asymptotic distribution of largest values, or the Fisher–Trippet Type III distribution of largest values (Cohen, 1973).

6.2.2 Model Properties

Moments

From (6.1) the jth moment of Z is given by

$$M_j(\theta) = E(Z^j) = E[(\tau - T)^j] \tag{6.9}$$

As a result, the moments can be expressed in terms of the moments of the three-parameter Weibull model. The mean and variance are given by

$$\mu = \tau - \alpha\Gamma(1 + 1/\beta) \tag{6.10}$$

and

$$\sigma^2 = \mu_2 = \alpha^2\{\Gamma(1 + 2/\beta) - [\Gamma(1 + 1/\beta)]^2\} \tag{6.11}$$

respectively.

The coefficient of skewness and the coefficient of kurtosis are given by

$$\gamma_1^2 = \frac{(\Gamma_3 - 3\Gamma_2\Gamma_1 + 2\Gamma_1^3)^2}{(\Gamma_2 - \Gamma_1^2)^3} \tag{6.12}$$

and

$$\gamma_2 = \frac{\Gamma_4 - 4\Gamma_3\Gamma_1 + 6\Gamma_2\Gamma_1 - 2\Gamma_1^4}{(\Gamma_2 - \Gamma_1^2)^2} \tag{6.13}$$

with

$$\Gamma_k = \Gamma\left(1 + \frac{k}{\beta}\right) = \int_0^\infty s^{k/\beta}e^{-s}\, ds \tag{6.14}$$

Both γ_1 and γ_2 are functions of the shape parameter only. In the interval $\beta < \beta_0 = 3.60235$, γ_1 is a decreasing function of β, and it is an increasing function after that. In the interval $\beta < \beta_1 = 3.35$, γ_2 is a decreasing function of β, whereas it becomes an increasing function after that. The minimum values of γ_1 and γ_2 are $\gamma_1(\beta_0) = 0$ and $\gamma_2(\beta_1) \approx 2.72$.

The reflected Weibull distribution is positively skewed for $\beta < \beta_0$ and negatively skewed for $\beta > \beta_0$.

Median and Mode
The median is given by

$$t_{\text{median}} = \tau - \alpha(\ln 2)^{1/\beta} \tag{6.15}$$

There is a single mode when $\beta > 1$ and is given by

$$t_{\text{mode}} = \tau - \alpha(1 - 1/\beta)^{1/\beta} \tag{6.16}$$

6.2.3 Parameter Estimation

By reversing the sign of the data, it can be viewed as data from the three-parameter Weibull model. The parameters can be estimated using the methods discussed in Section 4.7.

Method of Moments
Cohen (1973) discusses this. The first three sample moments are used to obtain the estimates.

6.3 MODEL I(a)-4: DOUBLE WEIBULL DISTRIBUTION

The random variables Z and T are related by (6.2).

6.3.1 Model Structure

Distribution Function
The distribution function is given by

$$G(t) = \begin{cases} 0.5\exp\{-[(\tau - t)/\alpha]^{\beta}\} & \text{for } t < \tau \\ 1 - 0.5\exp\{-[(t - \tau)/\alpha]^{\beta}\} & \text{for } t \geq \tau \end{cases} \tag{6.17}$$

Note that the support is $-\infty < t < \infty$. For $t \geq \tau$ the distribution is similar to the three-parameter Weibull distribution, and for $t < \tau$ it is similar to the reflected Weibull distribution.

This model is a combination of the standard three-parameter Weibull model and its reflection about the vertical line through $t = \tau$ and hence called the double Weibull distribution. When $\beta = 1$, it is called the double exponential distribution.

Special Case
Balakrishnan and Kocherlakota (1985) studied a special case where $\alpha = 1$ and $\tau = 0$. In this case, the density function is given by

$$g(t) = \frac{\beta}{2}|t|^{\beta-1}\exp(-|t|^{\beta}) \qquad -\infty < t < \infty, \beta > 0 \tag{6.18}$$

6.3.2 Model Analysis

Moments

For the special case, the rth moment is zero when r is odd, and when r is even it is given by

$$E(Z^r) = \Gamma\left(1 + \frac{r}{\beta}\right) \tag{6.19}$$

The moment generating function of $\ln|x|$ is given by

$$\Psi(s) = \Gamma\left(1 + \frac{s}{\beta}\right) \tag{6.20}$$

which is the same as that for the log transformation of the standard Weibull (Balakrishnan and Kocherlakota, 1985). From this, we have

$$E[|Z|^r] = \Gamma\left(1 + \frac{r}{\beta}\right) \tag{6.21}$$

6.3.3 Parameter Estimation

Balakrishnan and Kocherlakota (1985) also considered the estimation problem. In particular, the best linear unbiased estimators are provided for the more general class of double Weibull distribution (with location and scale parameter).

Define $\mathbf{X}' = (X_{(1)}, X_{(2)}, \ldots, X_{(n)})$. The best linear unbiased estimators of (τ, α) are given by

$$\tau^* = \frac{\mathbf{1}'\Omega\mathbf{X}}{\mathbf{1}'\Omega\mathbf{1}} \quad \text{and} \quad \alpha^* = \frac{\boldsymbol{\alpha}'\Omega\mathbf{X}}{\boldsymbol{\alpha}'\Omega\boldsymbol{\alpha}} \tag{6.22}$$

where $\mathbf{1}' = (1, \ldots, n)$, $\boldsymbol{\alpha} = (\alpha_{1:n}, \alpha_{2:n}, \ldots, \alpha_{n:n})$, and Ω is the inverse of the covariance matrix of the vector $\mathbf{X}$. Tables for the coefficient of the best linear unbiased estimator of (τ, α) are provided in Balakrishnan and Kocherlakota (1985).

Rao et al. (1991) studied an optimum unbiased estimation of the scale parameter with the shape parameter known. The new estimator, which makes use of the expected values, variance, and covariance of order statistics from Balakrishnan and Kocherlakota (1985), is generally more efficient than the corresponding best linear unbiased estimator.

6.4 MODEL I(b)-1: POWER LAW TRANSFORMATION

The random variables Z and T are related by (6.3).

6.4.1 Model Structure

The distribution function for Z is given by

$$G(t) = 1 - \exp[-(t - \tau)/\alpha] \qquad t \geq \tau \tag{6.23}$$

This is the two-parameter exponential distribution and can be viewed as a three-parameter Weibull distribution with $\beta = 1$. The density function is given by

$$g(t) = \frac{1}{\alpha}\exp\left(-\frac{t - \tau}{\alpha}\right) \qquad t \geq \tau \tag{6.24}$$

The hazard function is

$$h(t) = \frac{1}{\alpha} \qquad t \geq \tau \tag{6.25}$$

Note that the support for a two-parameter exponential distribution is smaller than that for the standard Weibull model when $\tau > 0$. When $\tau = 0$, $G(t)$ reduces to the standard (one-parameter) exponential distribution. Finally, $G(t)$ can be viewed as a special case of (1.2) with the shape parameter $\beta = 1$.

6.4.2 Model Analysis

Since the two-parameter exponential model is a special case of the three-parameter Weibull model, the model analysis is given by the results of Section 3.4 with $\beta = 1$. Additional results can be found in Balakrishnan and Basu (1996), which deals with this model in great detail.

Graphical Plots

Under the Weibull transformation [given by (1.7)], (6.23) gets transformed into (3.33) with $\beta = 1$. When $\tau = 0$, this is a straight line with a slope of 45°.

Define

$$y = \ln[1 - G(t)] \quad \text{and} \quad x = t \tag{6.26}$$

This is different from the Weibull transformation. Under this transformation, (6.23) reduces to

$$y = (x - \tau)/\alpha \tag{6.27}$$

This is a linear relationship. A plot of y versus x is called the exponential probability plot.

6.4.3 Parameter Estimation

The estimation of model parameters has also received considerable attention. Chapter 3 of Bain and Engelhardt (1991) deals with different methods of estimation, for different data structures, in great detail.

6.4.4 Modeling Data Set

Given a set of data (complete, censored, or grouped), one can plot the transformed data using the transformation given by (6.26) in a manner similar to the WPP plotting discussed in Section 4.5.1. This yields the exponential probability plot of the data. If the data is scattered along a straight line, then the two-parameter exponential distribution is an appropriate model to model the data set. In this case, the slope yields an estimate of α and the intercept with the y axis yields an estimate of τ.

More refined statistical tests for model selection can be found in Bain and Engelhardt (1991) and Balakrishnan and Basu (1996).

6.5 MODEL I(b)-2: LOG WEIBULL TRANSFORMATION

The variables Z and T are related by (6.4).

6.5.1 Model Structure

The distribution function for Z is given by

$$G(t) = 1 - \exp\left[-\exp\left(\frac{t-a}{b}\right)\right] \qquad -\infty < t < \infty \tag{6.28}$$

with $a = \ln \alpha$ and $b = 1/\beta$. This is an extreme value distribution and also called as the log Weibull distribution.

The extreme value distribution has received a lot of attention in statistics because of the location-scale interpretation of the model parameters. In fact, many statistical results for the standard Weibull model are given in terms of the results for the extreme value distribution using the transformation given by (6.4). Fisher and Tippett (1928) carried out a study of the limiting distributions for the extreme values, and they showed that they can only be one of three types, with (6.28) being one of them.

The density and hazard functions are given by

$$g(t) = \frac{1}{b}\exp\left(\frac{t-a}{b}\right)\exp\left\{-\exp\left(\frac{t-a}{b}\right)\right\} \tag{6.29}$$

and

$$h(t) = \frac{g(t)}{1-G(t)} = \frac{1}{b}\exp\left(\frac{t-a}{b}\right) \tag{6.30}$$

respectively.

Special Case
Let T be a random variable from the Weibull distribution. Consider the transformation

$$Z = \beta(\ln(T) - \ln(\alpha)) \tag{6.31}$$

Then the distribution for Z is given by

$$G(z) = 1 - \exp(-e^{z}) \tag{6.32}$$

This is a special case of (6.28) with $a = 0$ and $b = 1$. White (1969) calls this the reduced log Weibull distribution.

6.5.2 Model Analysis

Moments
The moment generating function for the reduced log Weibull model is given by

$$\Psi(s) = \int_{-\infty}^{\infty} e^{ts} \exp(t - e^{t})\, dt = \Gamma(1 + s) \qquad s > -1 \tag{6.33}$$

From this the mean and variance are given by

$$\mu = -\gamma \quad \text{and} \quad \sigma^2 = \pi^2/6 \tag{6.34}$$

where $\gamma = 0.5772\ldots$ is the Euler constant.

The moments for the log Weibull model can be expressed in terms of the moments for the reduced log Weibull model as the random variables from the two distributions are related to each other through a simple linear relationship. As a result, the mean and variance for log Weibull model are given by

$$\mu = a - \gamma b \quad \text{and} \quad \sigma^2 = \pi^2 b^2/6 \tag{6.35}$$

respectively.

Percentiles
The pth percentile for the log Weibull model is given by

$$z_p = a + b\ \ln[-\ln(1 - p)] \tag{6.36}$$

Order Statistics
White (1969) deals with moments of ordered statistics for the log Weibull and reduced log Weibull model. In particular, the distribution function of the first-order

statistic in a sample of size n from the reduced log Weibull distribution is

$$P(X_{(1)} < t) = 1 - [1 - F(t)]^n = 1 - \exp(-ne^t) = F(t + \ln(n)) \tag{6.37}$$

which implies that $X_{(1)}$ has the same distribution as $X - \ln n$, and hence

$$E(X_{(1)}) = E(X) - \ln(n) = -\gamma - \ln(n) \tag{6.38}$$

$$\text{Var}(X_{(1)}) = \text{Var}(X) = \frac{\pi^2}{6} \tag{6.39}$$

and

$$E(X^2_{(1)}) = \frac{\pi^2}{6} + (\gamma + \ln(n))^2 \tag{6.40}$$

The moments of $X_{(i)}$ may be expressed in terms of the moments of the first-order statistic of sample size j, for $j \le n$, so that

$$E(X_{(i)}) = -\gamma - \sum_{j=0}^{i-1} (-1)^j \binom{n}{j} \Delta^j \ln(n-j) \tag{6.41}$$

where Δ is the difference operator.

For the second moment, it can be shown (White, 1969) that

$$E(X^2_{(i)}) = \frac{\pi^2}{6} + \gamma^2 + 2\gamma \sum_{j=0}^{i-1} (-1)^j \binom{n}{j} \Delta^j \ln(n-j) + \sum_{j=0}^{i-1} (-1)^j \binom{n}{j} \Delta^j [\ln(n-j)]^2 \tag{6.42}$$

and

$$\text{Var}(X_{(i)}) = \frac{\pi^2}{6} + \sum_{j=0}^{i-1} (-1)^j \binom{n}{j} \Delta^j [\ln(n-j)]^2 - \left[\sum_{j=0}^{i-1} (-1)^j \binom{n}{j} \Delta^j \ln(n-j)\right]^2 \tag{6.43}$$

Graphical Plots

Define

$$y = \ln\{-\ln[1 - G(t)]\} \quad \text{and} \quad x = t \tag{6.44}$$

Then under this transformation (6.28) gets reduced to

$$y = (x - a)/b \tag{6.45}$$

which is a straight-line relationship. Note that (6.44) is identical to the WPP transformation due to the log transformation (6.4).

Characterization

Bain and Engelhardt (1991) summarize a number of characterization results for the extreme value distribution. Many of them are in fact a modified version of that for the Weibull distribution because of the close relationship between the two.

6.5.3 Parameter Estimation

The estimation of parameters for the log Weibull distribution can be viewed as a special case of the estimation for extreme value distributions. Maximum-likelihood estimation can be carried out in the usual way. The moment estimation has been studied in Dekkers (1989) and Christopeit (1994). See Harter (1978), Lawless (1982), and Johnson et al. (1995) for more information about statistical inference on extreme value distribution. Kotz and Nadarajah (2000) provide an excellent coverage of this important statistical distribution.

6.5.4 Modeling Data Sets

Given a data set (complete, censored, or grouped), one can plot the transformed data using the transformation given by (6.44) in a manner similar to the WPP plotting discussed in 4.5.1. If the data is scattered along a straight line, then the log Weibull model is an appropriate model to model the data set. In this case, the slope yields an estimate of b and the intercept with the y axis yields an estimate of a.

6.6 MODEL I(b)-3: INVERSE WEIBULL DISTRIBUTION

The variables Z and T are related by (6.5).

6.6.1 Model Structure

The distribution function for Z is given by

$$G(t) = \exp[-(t/\alpha)^{-\beta}] \qquad t \geq 0 \tag{6.46}$$

The inverse Weibull is also a limiting distribution of the largest order statistics (Type II asymptotic distribution of the largest extreme). Drapella (1993) calls it the *complementary Weibull distribution*, and Mudholkar and Kollia (1994) call it the *reciprocal Weibull distribution*.

The density function is given by

$$g(t) = \beta\alpha^{\beta}t^{-\beta-1}\exp[-(t/\alpha)^{-\beta}] \qquad t \geq 0 \tag{6.47}$$

The hazard function is given by

$$h(t) = \frac{\beta\alpha^{\beta}t^{-\beta-1}\exp[-(t/\alpha)^{-\beta}]}{1 - \exp[-(t/\alpha)^{-\beta}]} \tag{6.48}$$

6.6.2 Model Analysis

Moments

The kth moment of Z is given by

$$\begin{aligned} M_k &= \int_0^{\infty} t^k g(t)\,dt = \int_0^{\infty} \eta\beta t^{(k-\beta-1)} e^{-z(t)}\,dt \\ &= \int_0^{\infty} \left(\frac{z}{\eta}\right)^{-k/\beta} e^{-z}\,dz = \alpha^k \Gamma\left(1 - \frac{k}{\beta}\right) \end{aligned} \tag{6.49}$$

The integral does not have a finite value for $k \geq \beta$. As such, when $\beta \leq 2$ the variance is not finite.

The inverse Weibull distribution generally exhibits a long right tail, and its hazard function is similar to that of the log normal and inverse Gaussian distributions. Johnson et al. (1995) provide some further information on the inverse Weibull distribution.

Parametric Study of Density Function

Jiang et al. (2001a) show that the density function is unimodal (Type 2) with mode at t_{mode} given by

$$t_{\text{mode}} = \alpha(1 + 1/\beta)^{-1/\beta} \tag{6.50}$$

This is in contrast to the Weibull model for which the density function is either monotonically decreasing (for $\beta \leq 1$) or unimodal (for $\beta > 1$).

Parametric Study of Hazard Function

Jiang et al. (2001a) show that the hazard function is unimodal (Type 5) with the mode at $t = t_M$ given by the solution of the following equation:

$$\frac{z(t_M)}{1 - e^{-z(t_M)}} = 1 + \frac{1}{\beta} \tag{6.51}$$

where $z(t) = (\alpha/t)^{\beta}$ and

$$\lim_{t\to 0} h(t) = \lim_{t\to\infty} h(t) = 0 \tag{6.52}$$

This is in contrast to the standard Weibull model for which the failure rate is decreasing (for $\beta < 1$), constant (for $\beta = 1$), or increasing (for $\beta > 1$).

Graphical Plots

WPP Plot The WPP plot is given by

$$y = \ln[-\ln(1 - e^{-z})] \tag{6.53}$$

where

$$z = (\alpha/t)^{\beta} = \exp(\beta \ln(\alpha) - \beta x) \tag{6.54}$$

Note that y is a highly nonlinear function of x. Jiang et al. (2001a) show that the WPP plot is concave and the two asymptotes are as follows.

- Left asymptote: For small t (or large z) we have

$$y = \ln[-\ln(1 - e^{-z})] \approx \ln(e^{-z}) = -z = -e^{-\beta(x - \ln(\alpha))} \tag{6.55}$$

- Right asymptote: For large t (or small z) we have

$$y = \ln[-\ln(1 - e^{-z})] \approx \ln(-\ln(z)) = \ln[\beta(x - \ln(\alpha))] \tag{6.56}$$

The WPP plot and the two asymptotes are shown in Figure 6.1. Note that the left (right) asymptote matches the WPP plot very closely for small (large) t.

Inverse Weibull Transform Drapella (1993) proposed the following transformation:

$$y = -\ln[-\ln F(t)] \quad \text{and} \quad x = \ln(t) \tag{6.57}$$

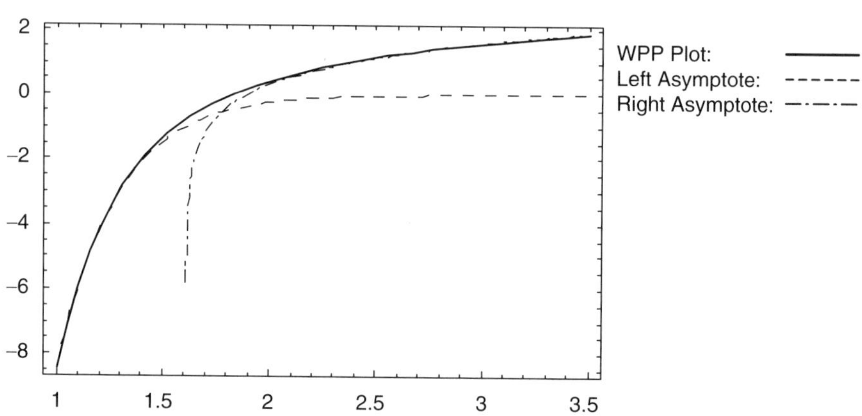

Figure 6.1 WPP plot for the inverse Weibull model ($\alpha = 5.0$, $\beta = 3.5$).

which he defined as the inverse Weibull transform. Applying this transform to (6.46) yields the following relationship:

$$y = \beta(x - \ln(\alpha)) \tag{6.58}$$

The plot y vs. x is called the IWPP (inverse Weibull probability paper) plot. The interesting feature is that the IWWP plot for the inverse Weibull model is a straight line as opposed to the WPP plot [given by (6.53)] is highly nonlinear.

Comparison of WPP and IWPP Plots

The standard Weibull model gets transformed into

$$y = -\ln[-\ln(1 - e^{-1/z})] \tag{6.59}$$

under the inverse Weibull transform. This is convex in shape and the two asymptotes are as follows:

- Left asymptote: For small t (or small z) we have

$$y = -\ln[-\ln(1 - e^{-z})] \approx -\ln(-\ln(z)) = -\ln[-\beta(x - \ln(\alpha))] \tag{6.60}$$

- Right asymptote: For large t (or large z) we have

$$y = -\ln[-\ln(1 - e^{-z})] \approx -\ln(e^{-z}) = z = \exp[\beta(x - \ln(\alpha))] \tag{6.61}$$

This is in contrast to the Weibull transform, which results in a linear relationship for the standard Weibull model.

Figure 6.2 shows the IWPP plot for the two-parameter Weibull model. The WPP plot for the inverse Weibull model and the IWPP plot for the Weibull model are mirror images of each other.

6.6.3 Parameter Estimation

Graphical Estimation

The graphical method for estimation based on the IWPP plot is the same as that for the standard Weibull model based on the WPP plot discussed in Section 4.5.1.

Maximum-Likelihood Estimation

Calabria and Pulcini (1989, 1994) carry out a study of the maximum-likelihood method. With a change of variables $x = 1/t$, the inverse Weibull distribution can be transformed to standard Weibull distribution with shape parameter $c = \beta$ and scale parameter $b = \alpha$. The maximum-likelihood estimates of the parameters of the inverse Weibull distribution can be obtained by using the same two equations derived for the Weibull distribution. Hence, the MLE $(\hat{\beta}, \hat{\alpha})$ have the same statistical properties of the corresponding estimator $(\hat{c}, \hat{b})$. In particular, we have that

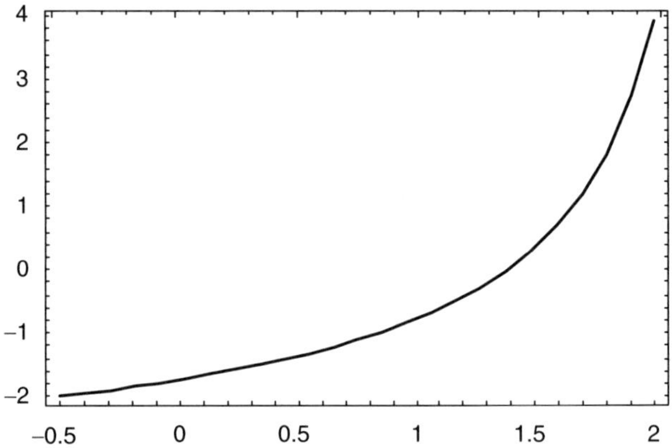

Figure 6.2 IWPP plots for the two-parameter Weibull model ($\alpha = 5.0, \beta = 3.5$).

1. $\hat{\beta}_s = \hat{\beta}/\beta$ and $\hat{\alpha}_s = \hat{\alpha}/\alpha$ are pivotal quantities, as well as $\hat{c}_s = \hat{c}/c$ and $\hat{b}_s = \hat{b}/b$; $\hat{\beta}_s$ and $\hat{\alpha}_s$ have the same distributions as $\hat{c}_s$ and $\hat{b}_s$, respectively.
2. The elements from the asymptotic covariance matrix of $(\hat{\beta}, \hat{\alpha})$ are equal to those from the matrix of $(\hat{c}, \hat{b})$. As a result,

$$\text{Var}[\hat{\beta}] = 0.608\beta^2/n \qquad \text{Var}[\hat{\alpha}] = 1.109\alpha^2/(\beta^2 n) \qquad \text{Cov}(\hat{\alpha}, \hat{\beta}) = 0.257\alpha/n$$

Bayesian Estimation

Calabria and Pulcini (1994) studied a Bayesian approach of predicting the ordered lifetimes in a future sample when samples are assumed to follow the inverse Weibull distribution. The Bayesian approach allows both Type I and Type II data to be analyzed in the same manner. Under Type I or Type II censoring, denote by $\bar{T}$ the time the testing is stopped, and m is the number of failures, the likelihood function is

$$L(\text{data} \mid \alpha, \beta) \propto \beta^m \alpha^{-m\beta} \left(\sum_{i=1}^{m} t_i \right)^{-\beta-1} \exp\left(-\alpha^{-\beta} \sum_{i=1}^{m} t_i^{-\beta} \right) \left\{ 1 - \exp[-(\alpha\bar{T})^{-\beta}] \right\}^{n-m} \tag{6.62}$$

Let $g(\alpha, \beta)$ be the joint prior density on the distribution parameters, measuring the uncertainty about the true values. The joint posterior density for (α, β) is

$$h(\alpha, \beta \mid \text{data}) = \frac{L(\text{data} \mid \alpha, \beta) \cdot g(\alpha, \beta)}{\int\int_{\alpha,\beta} L(\text{data} \mid \alpha, \beta) \cdot g(\alpha, \beta)\, d\alpha\, d\beta} \tag{6.63}$$

and the denominator is independent of (α, β).

For a second ordered sample of future observation from the same distribution, the conditional joint density of jth ordered observation, y_j, of (α, β) is obtained from

$$h(y_j, \alpha, \beta \mid \text{data}) = h(y_j \mid \alpha, \beta) \cdot h(\alpha, \beta \mid \text{data}) \tag{6.64}$$

and the statistical inference can be carried out. Calabria and Pulcini (1994) studied two procedures: (i) no prior information is available and (ii) prior information on the unreliability level at a fixed time is introduced in the predictive procedure. They demonstrate via simulation that under Type II censoring, the informative procedure outperforms the noninformative one, not only when the prior information is correct but also when a wrong choice of the prior information is made.

6.6.4 Modeling Data Set

The IWPP plot of a given data set can be used to determine if it can be modeled by an inverse Weibull distribution or not. The procedure is identical to that for the standard Weibull model discussed in Section 5.2.

6.6.5 Applications

The inverse Weibull distribution has been derived as a suitable model for describing degradation phenomena of mechanical components such as the dynamic components of diesel engines. The physical failure process given by Erto and Rapone (1984) also leads to this model. They show that the inverse Weibull distribution provides a good fit to several data sets such as the times to breakdown of an insulating fluid subject to the action of a constant tension (Nelson, 1982). Calabria and Pulcini (1994) provide an interpretation of the inverse Weibull in the context of the load–strength relationship for a component.

EXERCISES

Data Set 6.1 Complete Data: Failure Times of 20 Components

0.481	1.196	1.438	1.797	1.811
1.831	1.885	2.104	2.133	2.144
2.282	2.322	2.334	2.341	2.428
2.447	2.511	2.593	2.715	3.218

Data Set 6.2 Complete Data: Failure Times of 20 Components

0.067	0.068	0.076	0.081	0.084
0.085	0.085	0.086	0.089	0.098
0.098	0.114	0.114	0.115	0.121
0.125	0.131	0.149	0.160	0.485

Data Set 6.3 Censored Data: 50 Items Tested and the Test Stopped after the 40th Failure

0.602	0.603	0.603	0.615	0.652
0.663	0.688	0.705	0.761	0.770
0.868	0.884	0.898	0.901	0.911
0.918	0.935	0.953	0.983	1.009
1.040	1.097	1.097	1.148	1.296
1.343	1.422	1.540	1.555	1.653
1.752	1.885	2.015	2.015	2.030
2.040	2.123	2.175	2.443	2.548

6.1. Derive (6.10) to (6.15).

6.2. Consider Data Set 6.1. Plot the data using the transformations given by (1.7), (6.44), and (6.57). What can you infer from these three plots?

6.3. Assume that Data Set 6.1 can be modeled by a log Weibull distribution. Estimate the model parameters using (i) the method of moments and (ii) the method of maximum likelihood. Compare the two estimates.

6.4. Show that the WPP plot for the inverse Weibull model is concave.

6.5. Repeat Exercise 6.2 with Data Set 6.2.

6.6. Assume that Data Set 6.2 can be adequately modeled by an inverse Weibull distribution. Estimate the parameters based on the IWPP plot.

6.7. Assume that Data Set 6.2 can be modeled by an inverse Weibull distribution. Estimate the model parameters using (i) the method of moments and (ii) the method of maximum likelihood. Compare these estimates with the estimates obtained in Exercise 6.6.

6.8. Repeat Exercise 6.2 with Data Set 6.3.

6.9. Assume that Data Set 6.3 can be adequately modeled by an inverse Weibull distribution. Estimate the parameters based on the IWPP plot.

6.10. Assume that Data Set 6.3 can be modeled by an inverse Weibull distribution. Estimate the model parameters using (i) the method of moments and (ii) the method of maximum likelihood. Compare these estimates with the estimates obtained in Exercise 6.9

CHAPTER 7

Type II Weibull Models

7.1 INTRODUCTION

Type II models are derived from the standard Weibull model. One can classify Type II models into two categories [Type II(a) and II(b)]. Type II(a) models involve no additional parameters, whereas Type II(b) models involve one or more additional parameters. Some of the Type I models can also be viewed as Type II models, for example, the three-parameter Weibull model viewed as a Type II(b) model.

Let $G(t;\theta)$ denote the distribution function for the derived model, and it is related to the distribution function $F(t;\theta)$ for the standard Weibull model given by (2.1). The Type II(a) and Type II(b) models that we discuss are as follows:

Model II(a)-1 Pseudo-Weibull distribution (Voda, 1989)

Model II(a)-2 Stacy and Mihram (1965) model

Model II(b)-1 Extended Weibull distribution (Marshall and Olkin, 1997)

Model II(b)-2 Exponentiated Weibull distribution (Mudholkar and Srivastava, 1993)

Model II(b)-3 Modified Weibull distribution (Lai et al., 2003)

Model II(b)-4–6 Generalized Weibull family (Mudholkar et al., 1996)

Model II(b)-7–8 Generalized gamma distribution

Model II(b)-9–10 Four-parameter generalized Weibull distribution (Kies, 1958; Phani, 1987)

Model II(b)-11 Truncated Weibull distributions

Model II(b)-12 Slymen–Lachenbruch (1984) modified Weibull distribution

Model II(b)-13 Modified Weibull extension (Xie et al., 2002b)

For each of these we discuss model analysis, parameter estimation, and model applications. As in Chapter 6, we review the relevant literature and present the main results for each model.

Weibull Models, by D.N.P. Murthy, Min Xie, and Renyan Jiang.
ISBN 0-471-36092-9

The outline of the chapter is as follows. Sections 7.2 and 7.3 deal with Model II(a)-1 (pseudo-Weibull distribution) and Model II(a)-2 (Stacy and Mihram Weibull model). Sections 7.4 to 7.6 deal with Model II(b)-1 (extended Weibull distribution), Model II(b)-2 (modified Weibull model), and Model II(b)-3 (exponentiated Weibull model), respectively. Section 7.7 deals with Model II(b)-4–6 (generalized Weibull family), and they are combined because of their close relationship via quartile function. Sections 7.8 and 7.9 deal with Model II(b)-7 (three-parameter generalized gamma) and Model II(b)-8 (four-parameter generalized gamma), respectively. Section 7.10 considers the four-parameter Weibull distribution [Model II(b)-9], and we also briefly discuss a five-parameter Weibull [Model II(b)-10]. Section 7.11 deals with Model II(b)-11 (general truncated Weibull) and some special cases. In Section 7.12 we discuss Model II(b)-12 (Slymen–Lachenbruch modified Weibull distribution). Finally, in Section 7.13 we consider Model II(b)-13.

7.2 MODEL II(a)-1: PSEUDO-WEIBULL DISTRIBUTION

7.2.1 Model Structure

Density Function

The density function is given by

$$g(t) = kt^k[\theta^{1+1/k}\Gamma(1+1/k)]^{-1}\exp(-t^k/\theta) \qquad \theta, k > 0 \tag{7.1}$$

The relationship between this model and the standard Weibull model is as follows. Let

$$g_1(t) = \frac{t}{\mu} f(t) \qquad t > 0 \tag{7.2}$$

where $f(t)$ is the density function for the standard Weibull [given by (3.11)], and μ is the mean of the standard Weibull random variable and given by (3.16). As a result, (7.2) can be written as

$$g_1(t) = \frac{\beta t^\beta}{\alpha^{\beta+1}\Gamma(1+1/\beta)} \exp\left\{-\left(\frac{t}{\alpha}\right)^\beta\right\} \tag{7.3}$$

With $\theta = \alpha^\beta$ and $k = \beta$, (7.3) is the same as (7.1).

Distribution Function

The distribution function associated with (7.1) has the following form:

$$G(t; \theta, k) = \frac{\Gamma_{\sqrt{\theta t}}(1+1/k)}{\Gamma(1+1/k)} \qquad t \geq 0 \tag{7.4}$$

where $\Gamma_*(\cdot)$ is the incomplete gamma function (Pearson, 1965).

Special Case

When $k = 1$, the model reduces to the gamma distribution with parameter $(\alpha, 2)$, and the density function given by

$$g(t) = \alpha^{-2} t \exp(-t/\alpha) \qquad t > 0 \tag{7.5}$$

The distribution and hazard functions are given by

$$F(t) = 1 - \left(1 + \frac{t}{\alpha}\right)\exp\left(-\frac{t}{\alpha}\right) \qquad t > 0 \tag{7.6}$$

and

$$h(t) = \alpha^{-1}\left(1 - \frac{1}{1 + t/\alpha}\right) \tag{7.7}$$

respectively.

The pseudo-Weibull distribution is neither a generalization of the gamma distribution nor of the Weibull distribution.

7.2.2 Model Analysis

Moments

The jth moment is given by

$$M_j = \frac{\alpha^{j/k}\Gamma[1 + (1 + j)/k]}{\Gamma(1 + 1/k)} \tag{7.8}$$

As a result, the mean and variance are given by

$$\mu = \frac{\alpha^{1/k}\Gamma(1 + 2/k)}{\Gamma(1 + 1/k)} \tag{7.9}$$

and

$$\sigma^2 = \alpha^{2/k}\left[\frac{\Gamma(1 + 3/k)}{\Gamma(1 + 1/k)} - \frac{\Gamma^2(1 + 2/k)}{\Gamma^2(1 + 1/k)}\right] \tag{7.10}$$

respectively. The coefficient of variation is given by

$$\rho = \frac{[\Gamma(1 + 3/k) \cdot \Gamma(1 + 1/k) - \Gamma^2(1 + 2/k)]^{1/2}}{\Gamma(1 + 2/k)} \tag{7.11}$$

Finally, Voda (1989) proves the following result. If Y is a pseudo-Weibull variable, then the variable Y^k belongs to the class of gamma distribution, $G(x; \alpha, 1 + 1/k)$.

Special Case

When $k = 1$, we have $\mu = 2\alpha$ and $\sigma^2 = 2/\alpha^2$. As a result, the coefficient of variation is given by $\rho = 1/\sqrt{2}$ and is independent of the model parameters.

7.2.3 Parameter Estimation

Method of Moments

For complete data, estimates of the model parameters (k, α) can be obtained from (7.9) and (7.10) using the first two sample moments.

Method of Maximum Likelihood

Voda (1989) considers the estimation of α with k known and complete data given by $(t_1, t_2, \ldots, t_n)$. The likelihood function is given by

$$L(\alpha) = \frac{k^n \prod_{i=1}^n t_i^k}{\alpha^{n+n/k}\Gamma(1 + 1/k)} \exp\left[-\frac{1}{\alpha}\sum_{i=1}^n t_i^k\right] \tag{7.12}$$

and the maximum-likelihood estimate is given by

$$\hat{\alpha} = \left[n\left(1 + \frac{1}{k}\right)\right]^{-1} \sum_{i=1}^n t_i^k \tag{7.13}$$

This is an unbiased estimator of α.

7.3 MODEL II(a)-2: STACY–MIHRAM MODEL

The standard Weibull model requires that the shape parameter β be positive. In this model, it can be either positive or negative. The density function is given by

$$g(t) = \frac{|\beta| t^{\beta-1}}{\alpha^\beta} \exp[-(t/\alpha)^\beta] \qquad t > 0 \tag{7.14}$$

The distribution function is given by

$$G(t) = \begin{cases} 1 - \exp[-(t/\alpha)^\beta] & \beta > 0 \\ \exp[-(t/\alpha)^\beta] & \beta < 0 \end{cases} \tag{7.15}$$

The model reduces to the standard Weibull model when $\beta > 0$ and to the inverse Weibull distribution (see Section 6.6) when $\beta < 0$.

This model is referred to in Fisher and Tippet (1928) and Gumbel (1958). It was studied by Stacy and Mihram (1965) and they called it a *generalized gamma*.

7.4 MODEL II(b)-1: EXTENDED WEIBULL DISTRIBUTION

7.4.1 Model Structure

Marshall and Olkin (1997) proposed a modification to the standard Weibull model through the introduction of an additional parameter $\nu(0 < \nu < \infty)$. The model is given by

$$\bar{G}(t) = \frac{\nu\bar{F}(t)}{1-(1-\nu)\bar{F}(t)} = \frac{\nu\bar{F}(t)}{F(t)+\nu\bar{F}(t)} \qquad t \geq 0 \tag{7.16}$$

where $F(t)$ is the two-parameter Weibull distribution given by (1.3) and $\bar{G}(t) = 1 - G(t)$.

Distribution Function

When $\nu = 1$, $\bar{G}(t) = \bar{F}(t)$ the model reduces to the standard Weibull model. Using (3.16) in (7.16), the distribution function is given by

$$G(t) = 1 - \frac{v \exp[-(t/\alpha)^{\beta}]}{1-(1-v)\exp[-(t/\alpha)^{\beta}]} \tag{7.17}$$

Marshall and Olkin call this the *extended Weibull distribution*.

Density Function

The density associated with (7.17) is given by

$$g(t) = \frac{(v\beta/\alpha)(t/\alpha)^{\beta-1}\exp[-(t/\alpha)^{\beta}]}{\left\{1-(1-v)\exp[-(t/\alpha)^{\beta}]\right\}^2} \tag{7.18}$$

Hazard Function

The corresponding hazard function is

$$h(t) = \frac{(\beta/\alpha)(t/\alpha)^{\beta-1}}{1-(1-v)\exp[-(t/\alpha)^{\beta}]} \tag{7.19}$$

7.4.2 Model Analysis

The family of distributions given by (7.16) has the following interesting property. It is geometric-minimum (maximum) stable. That is, if $X_i; i \leq N$ is a sequence of independent identically distributed random variables with distribution from this family, and N is geometrically distributed, then the minimum (maximum) of $X_i; i \leq N$ also has a distribution in the family.

Moments

For $|1 - v| \leq 1$, the kth moment of the random variable from this distribution is given by

$$M_k = \frac{k}{\alpha\beta} \sum_{j=0}^{\infty} \frac{(1-v)^j}{(j+1)^{k/\beta}} \Gamma\left(\frac{k}{\beta}\right) \tag{7.20}$$

When $|1 - v| > 1$, the moments cannot be obtained in closed form and have to be computed numerically.

It can be shown that

$$\lim_{\beta \to \infty} E(X^k) = \alpha^k \qquad s > 0 \tag{7.21}$$

The first moment for various combinations of β and ν are tabulated in Marshall and Olkin (1997). For the special case $\beta = 1$, we have the mean given by

$$\mu = -\frac{\alpha\nu \ln(\nu)}{1 - \nu} \tag{7.22}$$

When $|1 - \nu| \leq 1$, the moments can also be expressed as infinite series given by

$$M_k = r\alpha^r \nu \sum_{j=1}^{\infty} \frac{(1-\nu)^j \Gamma(k)}{(j+1)^k} \tag{7.23}$$

Hazard Function

Marshall and Olkin (1997) carried out a partial study of the hazard function. It is increasing when $\nu \geq 1, \beta \geq 1$ and decreasing when $\nu \leq 1, \beta \leq 1$. If $\beta > 1$, then the hazard function is initially increasing and eventually increasing, but there may be an interval where it is decreasing. Similarly, when $\beta < 1$, the hazard function is initially decreasing and eventually decreasing, but there may be an interval where it is increasing. In summary, the results are as follows:

- $\nu \geq 1$ and $\beta \geq 1$: Type 3 (increasing)
- $\nu \leq 1$ and $\beta \leq 1$: Type 1 (decreasing)
- $\beta > 1$: Type 6
- $\beta < 1$: Type 7

WPP Plot

There has been no study of the WPP plot for the extended Weibull distribution. Simulation results indicate that the plot is concave for $\nu < 1$, straight line for

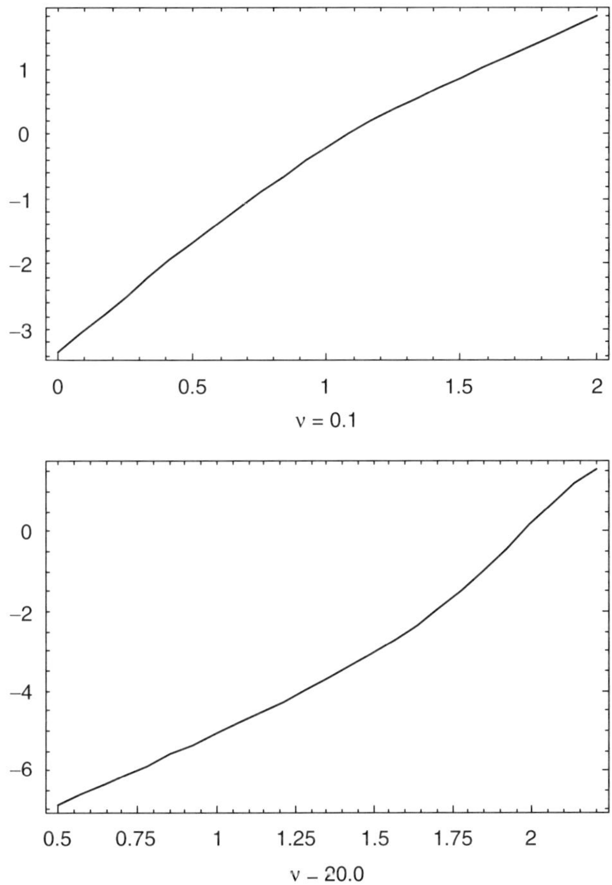

Figure 7.1 WPP plot for the extended Weibull distribution ($\alpha = 5.0, \beta = 3.5$).

$\nu = 1$ and convex for $\nu > 1$. Figure 7.1 shows the WPP plots for two different values of ν.

7.5 MODEL II(b)-2: EXPONENTIATED WEIBULL DISTRIBUTION

Mudholkar and Srivastava (1993) proposed a modification to the standard Weibull model through the introduction of an additional parameter ν $(0 < \nu < \infty)$.

7.5.1 Model Structure

Distribution Function

The distribution function is given by

$$G(t) = [F(t)]^{\nu} = \{1 - \exp[-(t/\alpha)^{\beta}]\}^{\nu} \qquad t \geq 0 \tag{7.24}$$

where $F(t)$ is the standard two-parameter Weibull distribution given by (3.10). The support for $G(\ )$ is $[0, \infty)$.

When $\nu = 1$, the model reduces to the standard two-parameter Weibull model. When ν is an integer, the model is a special case of the multiplicative model discussed in Chapter 10.

Density Function
The density function is given by

$$g(t) = \nu\{F(t)\}^{\nu-1}f(t) \tag{7.25}$$

where $f(t)$ is the density function of the standard two-parameter Weibull distribution and given by (3.11).

Hazard Function
The hazard function, after some simplification, is given by

$$h(t) = \nu\left[\frac{\{F(t)\}^{\nu} - \{F(t)\}^{\nu+1}}{F(t) - \{F(t)\}^{\nu+1}}\right]\left[\left(\frac{\beta}{\alpha}\right)\left(\frac{t}{\alpha}\right)^{\beta-1}\right] \tag{7.26}$$

7.5.2 Model Analysis

From (7.24), we see that $G(t) > F(t)$ when $\nu < 1$, $G(t) = F(t)$ when $\nu = 1$ and $G(t) < F(t)$ when $\nu > 1$. This implies that $\bar{G}(t) = 1 - G(t)$ increases with ν.

Moments
In general, the moments for this model are intractable. If ν is a positive integer, then

$$\mu_k = \alpha^k \nu \Gamma\left(1 + \frac{k}{\beta}\right) \sum_{j=0}^{\nu-1} (-1)^j \binom{\nu - 1}{j} \frac{1}{(j+i)^{1+k/\beta}} \tag{7.27}$$

If k/β is a positive integer (m), then the moments can be expressed in terms of the derivatives of the beta function $B(s, \nu)$ and are given by

$$\mu_k = \alpha^k \nu(-1)^m \left[\frac{d^m}{ds^m} B(s, \nu)\right]_{s=1}$$

which can be simplified by expressing the derivatives of the beta function in terms of the di-gamma function and its derivatives [see Abromowitz and Stegun (1964), pp. 258–260].

Mudholkar and Hutson (1996) carry out a detailed numerical study of the coefficient of skewness (γ_1) and coefficient of kurtosis (γ_2). They give surface and contour plots of these on the two-dimensional (β, ν) plane and discuss the exponentiated Weibull aliases of some common distributions.

Percentiles
The percentile, or quantile function, $Q(u)$, is given by

$$Q(u) = \alpha[-\ln(1 - u^{1/\nu})]^{1/\beta}$$

Order Statistics
Let $Y_{(i)}$, $i = 1, 2, \ldots, n$, denote the sample order statistics from the exponentiated model. Let $S_n = Y_{(n)} - Y_{(n-1)}$ and $S_2 = Y_{(2)} - Y_{(1)}$. The following results are from Mudholkar and Hutson (1996).

Theorem 7.1. Let Z be the standard exponential random variable with distribution function $F(z) = 1 - e^{-z}$, $z \geq 0$. Then as $n \to \infty$,

$$n^{1/(\beta\nu)} Y_{(1)} \xrightarrow{L} Z^{1/(\beta\nu)}$$

and

$$\beta[\ln(n)]^{1-1/\beta} Y_{(n)} - \beta \ln(n) - \ln(\nu) \xrightarrow{L} -\ln(Z)$$

where $\xrightarrow{L}$ indicates the convergence in distribution.

Theorem 7.2. For a random sample of size n from the exponentiated Weibull distribution and (Z, X) distributed with joint density function

$$f_{Z,X} = \begin{cases} e^{-z}, & \text{if } 0 \leq x \leq z \\ 0, & \text{otherwise} \end{cases}$$

As $n \to \infty$,

$$n^{1/(\beta\nu)} S_{(1)} \xrightarrow{L} Z^{1/(\beta\nu)} - X^{1/(\beta\nu)}$$

and

$$[\ln(n)]^{1-1/\beta} S_{(n)} \xrightarrow{L} \frac{1}{\beta}[\ln(Z) - \ln(X)]$$

Density Function
Mudholkar and Hutson (1996) and Jiang and Murthy (1999a) have studied the shapes of $g(t)$ and its characterization in the parameter space. The shape of $g(t)$ does not depend on α but changes with β and ν. The parametric characterization on the (β, ν) plane is as follows:

- $\beta\nu \leq 1$: Type 1 (monotonically decreasing)
- $\beta\nu > 1$: Type 2 (unimodal)

Also note that $g(0) = \infty$ when $\beta\nu < 1$, $g(0) = 1/\alpha$ when $\beta\nu = 1$, and $g(0) = 0$ when $\beta\nu > 1$.

When $\beta\nu > 1$, $g(t)$ is unimodal and let t_{mode} denote the mode of $g(t)$. It is not possible to obtain an analytical expression for t_{mode}. Mudholkar and Hutson (1996) propose an analytical approximation that is given by

$$t_{\text{mode}} = \alpha\left\{\frac{1}{2}\left[\frac{\sqrt{\beta(\beta - 8\nu + 2\beta\nu + 9\beta\nu^2)}}{\beta\nu} - 1 - \frac{1}{\nu}\right]\right\}^{\nu} \tag{7.28}$$

It is interesting to note that the effect of the product $(\beta\nu)$ on the shape of the density function for the exponentiated Weibull distribution is similar to the influence of the shape parameter β in the case of two-parameter Weibull distribution.

Mudholkar and Hutson (1996) classify the density function into three major classes based on the shapes and tails of the density function and the results of the extreme value distributions. Their classification is as follows:

- *Class I* ($\beta\nu < 1$) Monotone decreasing densities, becoming unbounded in the left tail and having medium right tail.
- *Class II* ($\beta\nu = 1$) Monotone decreasing densities with left tail ordinate equal to 1 and medium right tail.
- *Class III* ($\beta\nu > 1$) Unimodal densities with short left tail and medium right tail.

They subdivide each of these into three subclasses as indicated below:

- *Subclass a* ($\beta > 1$) Right tail is classified as medium short.
- *Subclass b* ($\beta = 1$) Right tail is classified as medium medium.
- *Subclass c* ($\beta < 1$) Right tail is classified as medium long.

For the precise details of light, medium, and heavy tails, see Mudholkar and Hutson (1996).

Hazard Function

The hazard function is given by

$$h(t) = \frac{\beta\nu(t/\alpha)^{\beta-1}[1 - \exp(-(t/\alpha)^{\beta})]^{\nu-1}\exp(-(t/\alpha)^{\beta})}{1 - [1 - \exp(-(t/\alpha)^{\beta})]^{\nu}} \tag{7.29}$$

It can be shown that for small t, we have

$$h(t) \approx \left(\frac{\beta\nu}{\alpha}\right)\left(\frac{t}{\alpha}\right)^{\beta\nu-1} \tag{7.30}$$

In other words, for small t, $h(t)$ can be approximated by the failure rate of a two-parameter Weibull distribution with shape parameter ($\beta\nu$) and scale parameter α. For large t (limiting case $t \to \infty$), the first square bracket in (7.29) $\to$ $(1/\nu)$. Using this in (7.29) yields

$$h(t) \approx \left(\frac{\beta}{\alpha}\right)\left(\frac{t}{\alpha}\right)^{\beta-1} \tag{7.31}$$

In other words, for large t, $h(t)$ can be approximated by the failure rate of a two-parameter Weibull distribution with shape parameter β and scale parameter α.

Mudholkar et al. (1996), Mudholkar and Hutson (1996), and Jiang and Murthy (1999a) deal with the shapes of $h(t)$ and its characterization in the parameter space. The shape of $h(t)$ does not depend on α but changes with β and ν. The characterization on the (β, ν) plane is as follows:

- $\beta \leq 1$ and $\beta\nu \leq 1$: Type 1 (monotonically decreasing)
- $\beta \geq 1$ and $\beta\nu \geq 1$: Type 3 (monotonically increasing)
- $\beta < 1$ and $\beta\nu > 1$: Type 5 (unimodal)
- $\beta > 1$ and $\beta\nu < 1$: Type 4 (bathtub shape)

The monotonicities are strict except for the case $\beta = \nu = 1$.

WPP Plot

Under the Weibull transformations [given by (3.13)], (7.24) gets transformed into

$$y = \ln[-\ln(1 - \exp\{-[\exp(x)/\alpha]^{\beta}\})^{\nu}] \tag{7.32}$$

Jiang and Murthy (1999a) carry out a detailed study of the WPP plot and the results are as follows:

- The WPP plot is concave for $\nu > 1$ and convex for $\nu < 1$.
- Right asymptote: As $x \to \infty$ (or $t \to \infty$) the asymptote is a straight line (L_w) given by

$$y \approx \beta(x - \ln(\alpha)) \tag{7.33}$$

- Left asymptote: As $x \to -\infty$ (or $t \to 0$) the asymptote is a straight line (L_a) given by

$$y \approx \beta\nu(x - \ln(\alpha)) \tag{7.34}$$

The left and right asymptotes intersect the x axis at $(\ln(\alpha))$, but the slopes are different unless $\nu = 1$ in which case they collapse into a single line, which is the WPP plot for the two-parameter Weibull distribution.

Typical WPP plots are shown in Figure 7.2 (for $\nu > 1$) and in Figure 7.3 (for $\nu < 1$). Also shown are the left and right asymptotes.

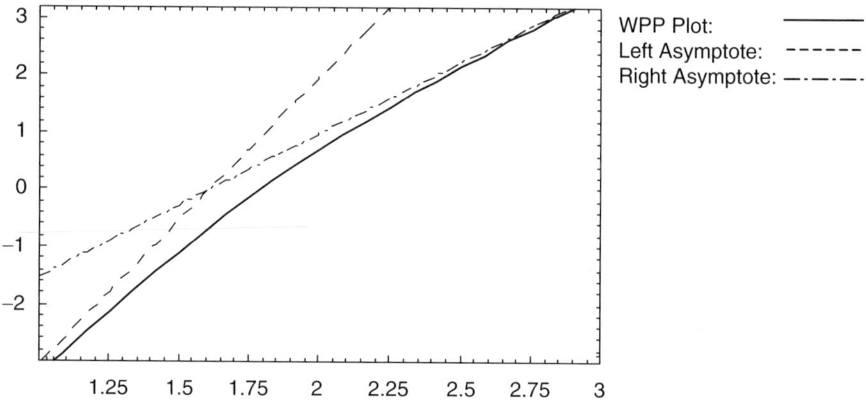

Figure 7.2 WPP plot for the exponentiated Weibull model ($\alpha = 5.0, \beta = 2.5, \nu = 2$).

TTT Plot

Mudholkar and Srivastava (1993) discuss the TTT transform plot. Its shape depends on the shape of the failure rate function and is convex (for decreasing failure rate), concave (for increasing failure rate), and S-shaped (for bathtub-shaped failure rate).

7.5.3 Parameter Estimation

Based on WPP Plot

This is discussed in Section 7.5.4.

Method of Percentile

Analytical expressions for the percentiles as a function of the model parameters are easily derived using (7.24). The method yields parameter estimates based on three sample percentiles.

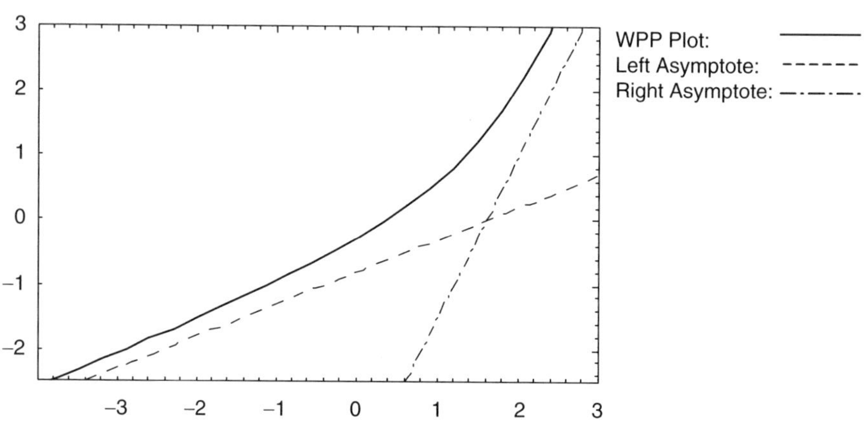

Figure 7.3 WPP plot for the exponentiated Weibull model ($\alpha = 5.0, \beta = 2.5, \nu = 0.2$).

Method of Maximum Likelihood
Mudholkar and Srivastava (1993) deal with estimating the model parameters with complete data. Mudholkar et al. (1995) deal with estimating the model parameters with grouped data.

Bayesian Approach
Cancho et al. (1999) deal with a Bayesian analysis for parameter estimation.

7.5.4 Modeling Data Set

The modeling based on WPP plot involves two stages.

Stage 1: Plotting Data
This depends on the type of data and is done as indicated in Section 5.5.2. If a smooth fit to the plotted data has a shape similar to that in Figure 7.1 or 7.2, the data can be modeled by an exponentiated Weibull family. If not, then the exponentiated Weibull family is not an appropriate model.

Stage 2: Parameter Estimation
If stage 1 indicates that the exponentiated Weibull family is an appropriate model, then the model parameters (β, ν, and α) can be estimated using the following steps:

Step 1: The asymptotic fit to the right side of the plot yields L_w. The slope yields $\hat{\beta}$ and the intercept yields $\hat{\alpha}$.

Step 2: The asymptotic fit to the left side of the plot yields L_a. The slope yields an estimate of ($\beta\nu$). Using $\hat{\beta}$ from step 1 yields $\hat{\nu}$.

(Note: In drawing the asymptotes, it is important to ensure that the two asymptotes intersect the x axis at ln (α). This might require an iterative approach based on trial and error.)

Goodness-of-Fit Test
Mudholkar and Srivastava (1993) and Mudholkar et al. (1995) use the likelihood ratio goodness-of-fit test where the null hypothesis is the exponentiated Weibull model (with three parameters), and the alternate hypotheses are submodels obtained by setting either $\nu = 1$ (so that the model reduces to the standard two-parameter Weibull distribution) or $\nu = 1$ and $\beta = 1$ (so that the model reduces to an exponential distribution).

7.5.5 Applications

Mudholkar and Srivastava (1993) consider the modeling of failure data obtained from Aarset (1987) and estimate the parameters using the method of maximum likelihood. They compare the fit between the model and the data using TTT transform plot.

Mudholkar and Srivastava (1993) dealt with the modeling of two data sets. The first is the bus-motor failure data from Davis (1952), and the second deals with survival time data (of head-and-neck cancer patients) from Efron (2000). The model parameters are estimated using the method of maximum likelihood and goodness-of-fit based on likelihood ratio test is studied. They also compare the exponentiated Weibull model against two submodels (by setting either $\nu = 1$ or $\beta = 1$). Jiang and Murthy (1999a) used the same two data sets but carried out the modeling based on the WPP plot.

Mudholkar and Hutson (1996) model flood data and estimate the model parameters using the method of maximum likelihood. They compare the fit between the model and the data using TTT transform plot.

7.6 MODEL II(b)-3: MODIFIED WEIBULL DISTRIBUTION

Lai et al. (2003) proposed a new modification to the standard Weibull model involving an additional parameter.

7.6.1 Model Structure

Distribution Function

The distribution function is given by

$$G(t) = 1 - \exp(-\lambda t^{\beta} e^{\nu t}) \qquad t \geq 0 \tag{7.35}$$

where the parameters $\lambda > 0$, $\beta \geq 0$, and $\nu \geq 0$. The model reduces to the standard two-parameter model when $\nu = 0$.

Density Function

The density function is given by

$$g(t) = \lambda(\beta + \nu t)t^{\beta-1} e^{\nu t} \exp(-\lambda t^{\beta} e^{\nu t}) \tag{7.36}$$

Hazard Function

The hazard function is

$$h(t) = \lambda(\beta + \nu t)t^{\beta-1} e^{\nu t} \tag{7.37}$$

Relationship to Other Models

The distribution given by (7.35) is related to some well-known distributions as indicated below.

1. When $\nu = 0$, the model gets reduced to the standard Weibull model given by (1.3).

2. When $\beta = 0$, the model gets reduced to a log Weibull model (Type I extreme-value distribution) Weibull discussed in Section 6.5.
3. The beta-integrated model (Lai et al., 1998), specified in terms of the cumulative failure rate function, $H(t) = \int_0^t h(x)\,dx$, is given by

$$H(t) = at^b(1 - dt)^c \qquad 0 < t < 1/d \tag{7.38}$$

and a, b, $d > 0$, and $c < 0$. The distribution function is given by

$$G(t) = 1 - \exp[-H(t)] \tag{7.39}$$

Set $d = 1/n$, $b = \beta$, and $c = \nu n$. In the limit $n \to \infty$, we have $(1 - t/n)^{\nu n} \to e^{\nu t}$ and this yields

$$H(t) = \lambda t^{\beta} e^{\nu t} \tag{7.40}$$

The derivative of $H(t)$ is (7.37). Hence, this model can be viewed as the limiting case of the beta-integrated model.

7.6.2 Model Analysis

Hazard Function

The shape of the hazard function depends only on β because of the factor $t^{\beta-1}$, and the remaining two parameters have no effect. For $\beta \geq 1$:

1. $h(t)$ is increasing in t, implying an increasing failure rate (type 3).
2. $h(0) = 0$ if $\beta > 1$ and $h(0) = \lambda\beta$ if $\beta = 1$.
3. $h(t) \to \infty$ as $t \to \infty$.

For $0 < \beta < 1$:

1. $h(t)$ initially decreases and then increases in t, implying a bathtub shape (type 4).
2. $h(t) \to \infty$ as $t \to 0$ and $h(t) \to \infty$ as $t \to \infty$.
3. The change point t^* (where the derivative of the hazard function changes sign) is given by

$$t^* = \frac{\sqrt{\beta} - \beta}{\nu} \tag{7.41}$$

The interesting feature is that t^* increases as ν decreases. The limiting case $\nu = 0$ reduces the model to the standard Weibull model for which the hazard function can be increasing, decreasing, or constant.

WPP Plot

Under the Weibull transformations [given by (1.7)], (7.35) gets transformed into

$$y = \ln(\lambda) + \beta x + \nu e^x \tag{7.42}$$

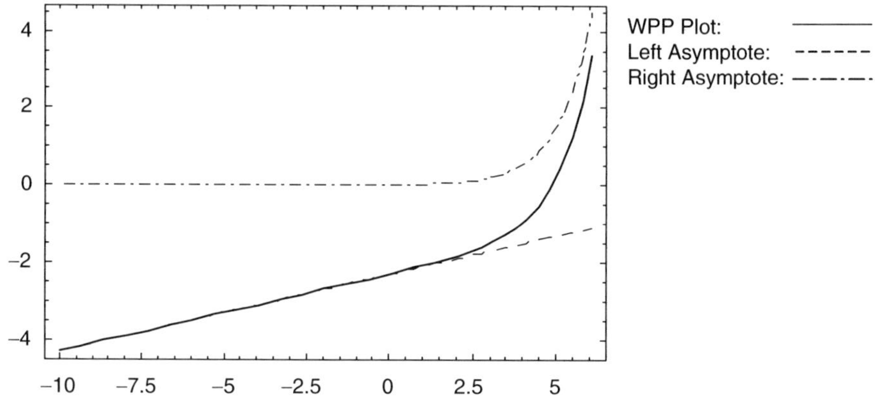

Figure 7.4 WPP plot for the modified Weibull distribution ($\lambda = 0.1, \beta = 0.2, \nu = 0.1$).

The WPP plot is convex and the two asymptotes are as follows:

- Left asymptote: As $x \to -\infty$ (or $t \to 0$), it is given by $y \approx \beta x$.
- Right asymptote: As $x \to \infty$ (or $t \to \infty$), it is given by $y \approx \nu e^x$.

Let x_0 and y_0 denote the coordinates of the WPP plot intersecting the x (corresponding to $y = 0$) and y (corresponding to $x = 0$) axes. That is, x_0 and y_0 are the x intercept and y intercept, respectively. Then, we have

$$\ln(\lambda) + \beta x_0 + \nu e^{x_0} = 0 \quad \text{and} \quad y_0 = \ln(\lambda) + \nu \tag{7.43}$$

It is easily seen that either $x_0 < 0$ and $y_0 > 0$ or vice versa and that $\beta|x_0| < |y_0|$.

Figure 7.4 shows a typical WPP plot along with the left and right asymptotes.

An Alternate Plot

An alternate plot is the following:

$$y = \ln[-\ln(\bar{F}(t))] \quad \text{and} \quad x = t \tag{7.44}$$

Using (7.44) in (7.35) yields the following:

$$y = \ln(\lambda) + \beta \ \ln(x) + \nu x \tag{7.45}$$

The plot is concave and the two asymptotes are as follows:

- Left asymptote: As $x \to 0$ (or $t \to 0$), it is given by $y \approx \beta \ \ln(x)$.
- Right asymptote: As $x \to \infty$ (or $t \to \infty$), it is given by $y \approx \nu x$.

7.6.3 Parameter Estimation

Based on WPP Plot

The procedure involves the following steps:

Step 1: Fit a smooth curve to the transformed data.

Step 2: Estimate ν from the slope of the straight-line asymptote.

Step 3: Estimate the remaining two parameters from the two intercepts given by (7.43).

A possible approach is simple regression analysis using (7.42) and estimating the parameters by least-squares fit. An alternate is a multiple linear regression using $x_1 = x$ and $x_2 = e^x$ so that the WPP plot can be represented as a linear equation:

$$y = \ln(\lambda) + \beta x_1 + \nu x_2 \tag{7.46}$$

noting that x_1 and x_2 are not independent. The parameter estimates are obtained by solving a system of linear equations.

Method of Percentiles

Expressions for the percentiles are analytically tractable, and, using the estimates of the two quartiles and median, one can obtain the parameter estimates.

Method of Maximum Likelihood

For the case of complete data, the log likelihood is given by

$$\begin{aligned} L(\lambda, \beta, \nu) = n \ \ln(\lambda) + \sum_{i=1}^{n} \ln(\beta + \nu t_i) + (\beta - 1) \ln\left(\sum_{i=1}^{n} t_i\right) \\ + \nu\left(\sum_{i=1}^{n} t_i\right) - \lambda \sum_{i=1}^{n} t_i^{\beta} e^{\nu t_i} \end{aligned} \tag{7.47}$$

The estimates $\hat{\beta}$ and $\hat{\nu}$ are obtained by solving the following three equations:

$$\frac{\partial L}{\partial \beta} = \sum_{i=1}^{n} \frac{1}{\hat{\beta} + \hat{\nu} t_i} + \ln\left(\sum_{i=1}^{n} t_i\right) - \hat{\lambda}\hat{\beta}\left(\sum_{i=1}^{n} t_i^{\hat{\beta}-1} e^{\hat{\nu} t_i}\right) = 0 \tag{7.48}$$

$$\frac{\partial L}{\partial \nu} = \sum_{i=1}^{n} \frac{t_i}{\hat{\beta} + \hat{\nu} t_i} + \sum_{i=1}^{n} t_i - n\hat{\nu} = 0 \tag{7.49}$$

and

$$\hat{\lambda} = \frac{n}{\sum_{i=1}^{n} t_i^{\hat{\beta}} e^{\hat{\nu} t_i}} \tag{7.50}$$

7.6.4 Modeling Data Set

The procedure based on the WPP plot is similar to that discussed in Section 7.5.4 with the parameters estimated as indicated in Section 7.6.3.

7.7 MODELS II(b)4–6: GENERALIZED WEIBULL FAMILY

Mudholkar and associates (Mudholkar and Srivastava, 1993; Mudholkar et al., 1995) developed three different models that we discuss together in this section. The link between these models and the standard Weibull is best seen through the quantile function. For the standard Weibull model, the quantile function is given by

$$Q(U) = \alpha[-\ln(1-U)]^{1/\beta} \tag{7.51}$$

and this forms the starting point.

7.7.1 Model Structure

Model II(b)-4: Generalized Weibull Family

Quantile Function
The quantile function for this model is given by (Mudholkar et al., 1996)

$$Q(U) = \alpha\left[\frac{1-(1-U)^{\nu}}{\nu}\right]^{1/\beta} \tag{7.52}$$

where the new parameter ν is unconstrained so that $-\infty < \nu < \infty$.

Distribution Function
The distribution function is

$$G(t) = 1 - \left[1 - \nu\left(\frac{t}{\alpha}\right)^{\beta}\right]^{1/\nu} \qquad t \geq 0 \tag{7.53}$$

An interesting feature of the model is the support for $G(t)$ is $(0, \infty)$ for $\nu \leq 0$ and $(0, \alpha/\nu^{1/\beta})$ for $\nu > 0$.

Density Function
The density function is given by

$$g(t) = \frac{(t/\alpha)^{\beta-1}}{(\alpha/\beta)[1-\nu(t/\alpha)^{\beta}]}\left[1 - \nu\left(\frac{t}{\alpha}\right)^{\beta}\right]^{1/\nu} \tag{7.54}$$

Hazard Function
The hazard function is given by

$$h(t) = \frac{(t/\alpha)^{\beta-1}}{(\alpha/\beta)[1 - \nu(t/\alpha)^{\beta}]} \tag{7.55}$$

Note that the model reduces to the basic two-parameter Weibull when $\nu \to 0$.

Model II(b)-5: More Generalized Weibull Family

Quantile Function
A different version of the Weibull family can be obtained with $\alpha = 0$, and it is given by (Mudholkar and Kollia, 1994)

$$Q(U) = \beta\left[\frac{1-(1-U)^{\nu}}{\nu}\right]^{1/\beta} - \beta \tag{7.56}$$

with ν unconstrained so that $-\infty < \nu < \infty$. Here the parameter β can take negative values so that the inverse Weibull is a special case of this distribution.

Distribution Function
The distribution function is given by

$$G(t) = 1 - [1 - \nu(1 + t/\beta)^{\beta}]^{1/\nu} \tag{7.57}$$

Note that the support for $G(t)$ depends on the model parameters and is as follows. For $\beta < 0$: if $\nu < 0$, $(-\infty, -\beta)$ and if $\nu > 0$, $[-\infty, (\beta/\nu^{1/\beta} - \beta]$, while for $\beta > 0$ it is if $\nu < 0$, $(-\beta, \infty)$ and if $\nu > 0$, $(-\beta, \beta/\nu^{1/\beta} - \beta)$. This model was proposed in Mudholkar and Kolia (1994) and called a *more generalized Weibull family.*

Model II(b)-6: Extended Weibull Distribution

Quantile Function
The quantile function for the model is related to (7.51) and is given by

$$Q(U) = \begin{cases} \beta[(-\ln((1-U)^{1/\beta})) - 1] & \text{for } \beta \neq 0 \\ \ln[-\ln(1-U)] & \text{for } \beta = 0 \end{cases} \tag{7.58}$$

Distribution Function
The distribution function for this model is given by

$$G(t) = \begin{cases} 1 - \exp[-(1 + t/\beta)^{\beta}] & \text{for } \beta \neq 0 \\ 1 - \exp(-\exp\ (t)) & \text{for } \beta = 0 \end{cases} \tag{7.59}$$

The support for $G(t)$ depends on the model parameters. It is $(-\infty, \beta)$ for $\beta < 0$, $(-\beta, \infty)$ for $\beta > 0$, and $(-\infty, \infty)$ for $\beta = 0$. This model was proposed in Freimer et al. (1989) and called the *extended Weibull distribution*.

7.7.2 Model Analysis

Model II(b)-4

Moments

The kth moment can be shown to be (Mudholkar and Kollia, 1994)

$$E[X^k] = \frac{B(1/\nu, k/\beta + 1)}{\nu^{k/\beta+1}} \tag{7.60}$$

where $B(\ )$ is the beta function.

Characterization of Hazard Function

The following result is from Mudholkar et al. (1996). The shape of the hazard function depends on model parameters and is as follows:

- $\beta < 1$ and $\nu > 0$: Type 4
- $\beta \leq 1$ and $\nu \leq 0$: Type 1
- $\beta > 1$ and $\nu < 0$: Type 5
- $\beta \geq 1$ and $\nu \geq 0$: Type 3
- $\beta = 1$ and $\nu = 0$: Type 2

Model II(b)-5

Moment

Mudholkar and Kollia (1994) discuss the moments, skewness, kurtosis, and extremes of the distribution. The kth moment, for $\nu > 0$, is given by

$$E[Y^k] = \sum_{i=0}^{k} \binom{k}{i} (-1)^{k-i} \beta^k \frac{B(1/\nu, k/\beta + 1)}{\nu^{k/\beta+1}} \tag{7.61}$$

Model II(b)-6

Freimer et al. (1989) give a characterization of the extremes and extreme spacings.

7.8 MODEL II(b)-7: THREE-PARAMETER GENERALIZED GAMMA

Stacy (1962) first introduced a generalization of the gamma distribution, which is also a generalization of the Weibull distribution. In this section we discuss this model.

7.8.1 Model Structure

Density Function

The density function is given by

$$g(t) = \frac{\alpha^{-\beta k}\beta t^{\beta k-1}}{\Gamma(k)}\exp[-(t/\alpha)^{\beta}] \qquad \alpha, \beta > 0,\ t \geq 0 \tag{7.62}$$

and the new additional parameter $k > 0$. When $k = 1$, it reduces to the standard Weibull model, and when $\beta = 1$ it reduces to gamma distribution.

Distribution Function

The distribution function does not have a closed analytical form. Define

$$\Gamma_z(k) = \int_0^z s^{k-1}e^{-s}\, ds \tag{7.63}$$

Then the distribution function is given by

$$G(t) = \frac{\Gamma_{(t/\alpha)^{\beta}}(k)}{\Gamma(k)} \tag{7.64}$$

Hazard Function

The hazard function is given by

$$h(t) = \frac{\alpha^{-\beta k} b t^{\beta k-1}\exp[-(t/\alpha)^{\beta}]}{\Gamma(k) - \Gamma_z(k)} \tag{7.65}$$

with $z = (t/\alpha)^{\beta}$.

The generalized gamma distribution is considered in a number of other contexts in the statistics and engineering literature.

The three-parameter generalized gamma distribution contains no location parameter. A straightforward extension of this model is to introduce a location parameter and the model can be viewed as a generalization of the three-parameter Weibull distribution as well. Harter (1967) first mentioned this and the model is given by

$$f(t; \alpha, \beta, \tau, k) = \frac{\alpha^{-\beta k}\beta(t-\tau)^{\beta k-1}}{\Gamma(k)}\exp\left[-\left(\frac{t-\tau}{\alpha}\right)^{\beta}\right] \qquad t > \tau \tag{7.66}$$

7.8.2 Model Analysis

Moment Generating Function
The moment generating function is given by

$$\psi(s) = \sum_{r=0}^{\infty} \frac{(s\alpha)^r}{r!} \left[\Gamma\left(\frac{\beta k + r}{\beta}\right) \Big/ \Gamma(k) \right] \qquad s \geq 0 \tag{7.67}$$

The mean, variance, and the rth moment are given by

$$\mu = \frac{\alpha\Gamma(k + 1/\beta)}{\Gamma(k)} \tag{7.68}$$

$$\sigma^2 = \alpha^2 \left\{ \frac{\Gamma(k + 2/\beta)}{\Gamma(k)} - \left[\frac{\Gamma(k + 1/\beta)}{\Gamma(k)} \right]^2 \right\} \tag{7.69}$$

and

$$E[X^r] = \frac{\alpha^k \Gamma(k + r/\beta)}{\Gamma(k)}$$

Shapes of Density and Hazard Functions
Pham and Almhana (1995) studied the shapes of the density and hazard function for three-parameter generalized gamma distribution and presented some characterization results. They show that if $(1 - k\beta)/[\beta(\beta - 1)] > 0$, then the hazard function is

- Type 4: (bathtub) if $\beta > 1$.
- Type 5: (upside-down bathtub) if $0 < \beta < 1$.

Otherwise, the hazard function is increasing for $\beta > 1$ and decreasing for $0 < \beta < 1$ and constant when $\beta = 1$.

7.8.3 Parameter Estimation

Maximum-Likelihood Estimation
For the case of complete data, the estimates are obtained by solving the following equations:

$$\sum_{i=1}^{n} \left(\frac{t_i}{\hat{\alpha}}\right)^{\hat{\beta}} - n\hat{k} = 0 \tag{7.70}$$

$$\frac{n}{\hat{\beta}} + \hat{k} \sum_{i=1}^{n} \ln\left(\frac{t_i}{\hat{\alpha}}\right) - \sum_{i=1}^{n} \left(\frac{t_i}{\hat{\alpha}}\right)^{\hat{b}} \ln\left(\frac{t_i}{\hat{\alpha}}\right) = 0 \tag{7.71}$$

$$\hat{\beta} \sum_{i=1}^{n} \ln\left(\frac{t_i}{\hat{\alpha}}\right) - n\phi(\hat{k}) = 0 \tag{7.72}$$

where $\phi(\hat{k}) = d \ln \Gamma(y)/dy$. Hager and Bain (1970) showed some independence properties of the estimators and discuss numerical algorithms for obtaining the estimates of the parameters. Wingo (1987) and Hirose (2000) also discuss some numerical problems associated with the maximum-likelihood estimation.

In an earlier work, Stancy and Mihram (1965) studied some estimation procedures with the focus on estimating the scale parameter. Bain and Weeks (1965) developed one-sided tolerance limits and confidence limits with one parameter at time being unknown. Parr and Webster (1965) presented the asymptotic density of the maximum likelihood estimators.

Bell (1988) considers the problem of parameter estimation via reparameterization. Agarwal and Kalla (1996) present some results related to reliability applications and a modification to the generalized gamma. Rao et al. (1991) considered the linear estimation of location and scale parameters, and Wong (1993) discusses the problems of simultaneous estimation of all three parameters. Hirose (2000) discusses the MLE for a generalized four-parameter gamma distribution.

7.9 MODEL II(b)-8: EXTENDED GENERALIZED GAMMA

A number of modified or generalized versions of the generalized gamma distribution have been studied. These include Ghitany (1998), Kalla et al. (2001), and others. In this section we consider the model proposed by Ghitany (1998).

7.9.1 Model Structure

Density Function

The density function is given by

$$f(t; \alpha, \beta, \lambda, n) = \frac{\alpha^{(m-1)/\beta+1-\lambda}\beta t^{m-\beta-2}(n + t^{\beta})^{-\lambda}}{\Gamma_{\lambda}[(m-1)/\beta, \alpha n]} \exp\left(-\left(\frac{t}{\alpha}\right)^{\beta}\right) \tag{7.73}$$

defined for $t \geq 0$, and the model parameters are positive valued and $\Gamma_{\lambda}(\cdot, \cdot)$ is given by (7.63).

Hazard Function

The hazard function is given by

$$h(t) = \frac{\alpha^{m-\kappa}}{\Gamma_{\lambda}(m, \alpha n)} \frac{t^{m-1}e^{-\alpha t}}{(t+n)^{\lambda}} \qquad t \geq 0 \tag{7.74}$$

Here $\Gamma(\cdot, \cdot, \cdot)$ is given by

$$\Gamma_{\lambda}(m, \alpha n) = \int_{\alpha t}^{\infty} \frac{y^{m-1}e^{-y}}{(y + \alpha n)^{\lambda}} dy \tag{7.75}$$

7.9.2 Model Analysis

Moment Generating Function

The moment generating function is given by

$$\psi(s) = \left(1 - \frac{s}{\alpha}\right)^{\lambda - m} \frac{\Gamma_\lambda[m, \alpha n(1 - s/\alpha)]}{\Gamma_\lambda(m, \alpha n)} \tag{7.76}$$

The rth derivative with respect to t is

$$\psi^{(r)}(s) = \alpha^{-r}\left(1 - \frac{s}{\alpha}\right)^{\lambda - m - r} \frac{\Gamma_\lambda[m + r, \alpha n(1 - s/\alpha)]}{\Gamma_\lambda(m, \alpha n)} \tag{7.77}$$

and as a result, the rth moment about the origin is given by

$$\psi^{(r)}(0) = \alpha^{-r} \frac{\Gamma_\lambda(m + r, \alpha n)}{\Gamma_\lambda(m, \alpha n)} \tag{7.78}$$

Hazard Function

It can be shown by the L'Hopital rule (Ghitany,1998) that $h(\infty) = \alpha$. Ghitany (1998) shows that the shapes for the hazard function can be one of four types, and this depends only on the values of λ and m. They are as follows:

- Type 1: if $\lambda = 0$ and $m < 1$, or $\lambda > 0$ and $m \leq 1$.
- Type 2: if and only if $\lambda = 0$ and $m = 1$.
- Type 3: if $\lambda = 0$ and $m > 1$, or $0 < \lambda \leq m$.
- Type 5 (upside-down bathtub shaped) with a unique change point $x_h \in (0, x_0)$ where

$$x_0 = \frac{m - 1 + \sqrt{(m - 1)\lambda}}{\lambda - (m - 1)} n \tag{7.79}$$

  if $\lambda > m - 1 > 0$.

If $\lambda = 0$, the above results reduce to that of a gamma distribution. That is,

$$h(t) \begin{cases} \text{decreasing} & \text{if } m < 1 \\ \text{constant} & \text{if } m = 1 \\ \text{increasing} & \text{if } m > 1 \end{cases} \tag{7.80}$$

Mean Residual Life

The mean residual life is given by

$$m(t) = E(T - t | T \geq t) = \frac{1}{1 - F(t)} \int_t^\infty (y - t) f(y)\, dy \tag{7.81}$$

Increasing (decreasing) mean residual life functions represent beneficial (adverse) aging. Mean residual life functions that first decrease (increase) and then increase (decrease) are usually termed bathtub (upside-down bathtub) shaped.

For the generalized gamma distribution, we have that

$$m(t) = \alpha^{-1}\frac{\Gamma_\lambda(m+1,\alpha n)}{\Gamma_\lambda(m,\alpha n)} - t \qquad t \geq 0 \tag{7.82}$$

where $\Gamma(\cdot,\cdot,\cdot)$ is given by (7.75). From this, we have

1. $m(t) = 1/\alpha$ if and only if $\lambda = 0$ *and* $m = 1$.
2. $m(t)$ is increasing if $\lambda = 0$ *and* $m < 1$, or $\lambda > 0$ *and* ≤ 1.
3. $m(t)$ is decreasing if $\lambda = 0$ *and* $m > 1$, or $0 < \lambda \leq m$.
4. $m(t)$ is bathtub shaped with a unique change point if $\lambda > m - 1 > 0$.

If $\lambda = 0$, the above results reduce to that of a gamma distribution. That is, for $\lambda = 0$,

$$m(t) \text{ is } \begin{cases} \text{increasing} & \text{if } m < 1 \\ \text{constant} & \text{if } m = 1 \\ \text{decreasing} & \text{if } m > 1 \end{cases} \tag{7.83}$$

7.10 MODELS II(b)9–10: FOUR- AND FIVE-PARAMETER WEIBULLS

Kies (1958) proposed a four-parameter Weibull distribution, and it was further studied by Phani (1987). The results are briefly discussed in this section.

7.10.1 Model II(b)–9: Four-Parameter Weibull

Distribution Function

The *four-parameter Weibull distribution* has the following distribution function:

$$G(t) = 1 - \exp\{-\lambda[(t-a)/(b-t)]^\beta\} \qquad 0 \leq a \leq t \leq b < \infty \tag{7.84}$$

This implies that the support is a finite interval.

WPP Plot

Under the Weibull transformation [given by (3.13)], (7.84) gets transformed into

$$y = \ln(\lambda) - \beta[\ln(e^x - a) - \ln(b - e^x)] \qquad \ln(a) \leq x \leq \ln(b) \tag{7.85}$$

The WPP plot is S-shaped with the left asymptote a vertical line through $x = \ln(a)$ and the right asymptote a vertical line through $x = \ln(b)$. Figure 7.5 shows a typical WPP plot along with the two asymptotes.

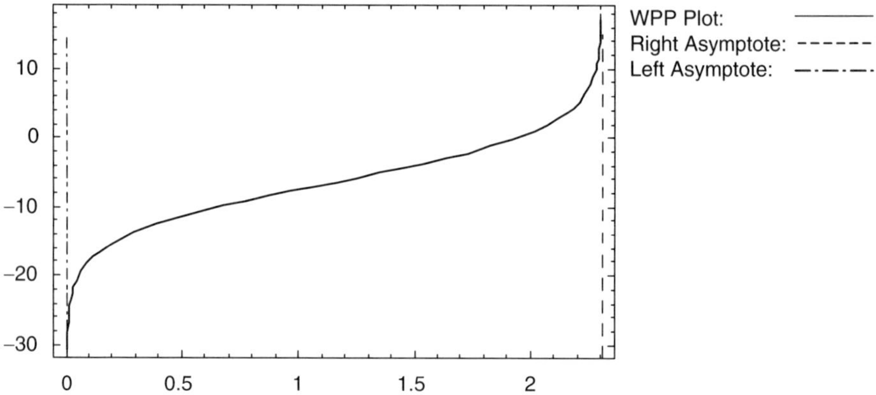

Figure 7.5 WPP plot for the four-parameter Weibull distribution ($\lambda = 0.1, \beta = 0.2, \nu = 0.1$).

For further details, see Smith and Hoeppner (1990) and Phani (1987), who also discusses the modeling based on the WPP plot.

7.10.2 Model II(b)-10: Five-Parameter Weibull

Distribution Function

The four-parameter Weibull model has been extended by Kies (1958), and the new model is given by

$$G(t) = 1 - \exp\{-\lambda[(t-a)^{\beta 1}/(b-t)^{\beta 2}]\} \tag{7.86}$$

and called it a *modified Weibull distribution*. This has also been called a *five-parameter Weibull distribution*.

7.11 MODEL II(b)-11: TRUNCATED WEIBULL DISTRIBUTION

By restricting the support of a two- or three-parameter Weibull distribution to a smaller range, truncated Weibull distributions are obtained. Since a truncation point can be viewed as an additional parameter, this model is classified under this group. The general formulation includes two truncation points, one to the left and the other to the right, so that it can be viewed as a four- or five-parameter model.

7.11.1 Model Structure

Distribution Function

The general truncated Weibull model is given by the distribution function

$$G(t) = \frac{F(t) - F(a)}{F(b) - F(a)} \qquad 0 \le a \le t \le b < \infty \tag{7.87}$$

where $F(t)$ is given by (1.3). This is also referred to as the *doubly truncated Weibull distribution.*

Density Function
The density function is given by

$$g(t) = \frac{f(t)}{F(b) - F(a)} \tag{7.88}$$

Using (1.3), the density function is given by

$$g(t) = \frac{(\beta/\alpha)(t/\alpha)^{\beta-1}\exp\{-(t/\alpha)^{\beta}\}}{\exp\{-(b/\alpha)^{\beta}\} - \exp\{-(a/\alpha)^{\beta}\}} \tag{7.89}$$

Special Cases

Case (i): When $a = 0$ and $b < \infty$, the model is called the *right truncated Weibull distribution.*

Case (ii): When $a > 0$ and $b \to \infty$, the model is called the *left truncated Weibull distribution.*

7.11.2 Model Analysis

Moments
McEwen and Parresol (1991) derive expressions for the moments. For the general case, the kth moment is given by

$$E(X^k) = \frac{\exp\{[(b-\tau)/\alpha]^{\beta}\}}{1 - \exp\{-[(a-\tau)/\alpha]^{\beta}\}} \sum_{n=0}^{k} \binom{k}{n} \alpha^{k-n}\tau^{n} \left[\Gamma\left(\frac{k-n}{\beta} + 1, \left(\frac{b-\tau}{\alpha}\right)^{\beta}\right)\right.$$
$$\left. - \Gamma\left(\frac{k-n}{\beta} + 1, \left(\frac{a-\tau}{\alpha}\right)^{\beta}\right)\right] \tag{7.90}$$

The mean, variance, skewness, and kurtosis can be obtained from this. See also Sugiura and Gomi (1995) for related discussions.

Density Function
The shape of $g(t)$ is determined by the shape of $f(t)$ over the interval $a \le t \le b$. As a result we have the following results:

- $\beta < 1$ or $\beta > 1$ and $a > t_{\text{mode}}$: Type 1 (decreasing)
- $\beta > 1$ and $b < t_{\text{mode}}$: (increasing)
- $\beta > 1$ and $a < t_{\text{mode}} < b$: Type 3 (unimodal)

where t_{mode} is the mode for the standard Weibull model and given by (3.24).

Hazard Function
The hazard function is given by

$$h(t) = \frac{f(t)}{F(b) - F(t)} = \left[\frac{f(t)}{1 - F(t)}\right]\left[\frac{1 - F(t)}{F(b) - F(t)}\right] \tag{7.91}$$

The first bracket is the failure rate for $F(t)$. The second bracket is a monotonically increasing function, approaching infinity as $t \to b$. This implies that the failure rate can be either type 1 [decreasing], type 3 [increasing], or bathtub shaped [type 4].

Finally, Wingo (1989b) discusses some other issues and Wingo (1988a) deals with model fitting.

7.11.3 Parameter Estimation

Mittal and Dahiya (1989) discuss the maximum-likelihood estimators for the truncated Weibull distribution. They propose a modified likelihood estimator and compare it with the maximum-likelihood estimator.

Further results concerning parameter estimation for this model can be found in Martinez and Quintana (1991), Shalaby (1993), Shalaby and Elyousef (1993), and Hwang (1996), among others.

7.12 MODEL II(b)-12: SLYMEN–LACHENBRUCH DISTRIBUTIONS

One can derive a family of distributions such that

$$\ln[-\ln(1 - G(t)] = \alpha + \beta w(t) \tag{7.92}$$

which can be viewed as a kind of generalization of Weibull transformation, where $w(t)$ satisfies some regularity conditions. Slymen and Lachenbruch (1984) introduced two families of distributions. The families were derived from a general linear form by specifying a form for the survival function with certain restrictions. They study a specific distribution and called it *modified Weibull distribution*.

7.12.1 Model Structure

The Slymen–Lachenbruch modified Weibull has the following distribution function:

$$G(t) = 1 - \exp\left\{-\exp\left[\alpha + \frac{\beta(t^k - t^{-k})}{2k}\right]\right\} \qquad t \geq 0 \tag{7.93}$$

The model is based on the transformation

$$\psi[\bar{F}(t)] = \alpha + \beta w(t) \tag{7.94}$$

where $w(t)$ is a monotonically increasing function, which may contain one or more parameters, that satisfies the following condition:

$$\lim_{t\to 0} w(t) = -\infty \quad \text{and} \quad \lim_{t\to\infty} w(t) = \infty$$

For the Slymen–Lachenbruch model

$$\psi[\bar{F}(t)] = \ln\{-\ln[\bar{F}(t)]\} \tag{7.95}$$

and

$$w(t) = \frac{t^k - t^{-k}}{2k} \qquad t, k \geq 0 \tag{7.96}$$

In fact, one can recognise ψ ( ) in (7.94) as the Weibull transformation.

The hazard function is given by

$$h(t) = \frac{1}{2}\beta(t^{k-1} - t^{-k-1})\exp\left\{\left[\alpha + \frac{\beta(t^k - t^{-k})}{2k}\right]\right\} \tag{7.97}$$

7.12.2 Model Analysis

Moments

It is not possible to obtain closed-form expressions for the moments. On the other hand, expressions using the probability integral transformation can be derived.

Let $U = S(T)$ where S denotes the survival function, U is distributed uniformly over (0,1), so that $T = S^{-1}(U)$. The mean of $T, E[T]$ can be written as

$$\mu = E(T) = \int_0^1 S^{-1}(U)\, dU \tag{7.98}$$

and the rth moment about the mean is given by

$$E[(T - \mu)^r] = \int_0^1 [S^{-1}(U) - \mu]^r\, dU \tag{7.99}$$

Percentile

The $(1 - p)$th percentile is given by

$$t_{1-p} = (C + \sqrt{C^2 + 1})^{1/k} \tag{7.100}$$

where

$$C = \frac{k\{\ln[-\ln(1-p)] - \alpha\}}{\beta} \tag{7.101}$$

Hazard Function

When $k > 0$, it can be seen that as t approaches infinity, the hazard function, $h(t)$, also approaches infinity. This implies that the hazard function can never be a monotonically decreasing function. However, over a limited range of t, the hazard function could be decreasing, and, in fact, it may have this property except in the extreme right tail of the distribution. Slymen and Lachenbruch (1984) derive the following result.

The hazard function $h(t)$ is monotonically increasing when

$$\beta > \frac{2[(k+1)t^{-k-2} - (k-1)t^{k-2}]}{(t^{k-1} + t^{-k-1})^2} = \beta_1(t) \qquad \text{for} \quad 0 < t < \infty \tag{7.102}$$

It is of interest to find the maximum value of β_1 if it exists. By taking the derivative of the above expression and making it equal to zero, this value can be found.

The hazard function is increasing for small values of t. If $\beta_1 < \infty$, the hazard function is decreasing followed by increasing. This implies that the model can have a general bathtub or the roller-coast-shaped hazard function.

7.12.3 Parameter Estimation

Least-Squares Estimation

Slymen and Lachenbruch (1984) discuss the least-squares estimation method for estimating the model parameters. The sum-of-square errors of the empirical survival function and the model survival function are minimized with respect to the parameters. The product limit estimate is used for obtaining the empirical survival function.

Since 7.95 is the Weibull transformation, Slymen and Lachenbruch (1984) use this to simplify the procedure. As a result, the estimates are obtained by solving the following equations:

$$\hat{\beta} = \frac{\sum_{i=1}^{n} (g_i - \hat{g})(w_i - \hat{w})}{\sum_{i=1}^{n} (w_i - \hat{w})^2} \qquad \hat{\alpha} = \hat{g} - \hat{\beta}\hat{w} \tag{7.103}$$

where

$$\hat{g} = \sum_{i=1}^{n} \frac{g_i}{n} \quad \text{and} \quad \hat{w} = \sum_{i=1}^{n} \frac{w_i}{n} \tag{7.104}$$

This is easily solved using a simple linear regression program. However, this is only possible when k is specified. When k is an unknown parameter, a simple expression for k is not available, and it has to be solved numerically.

Maximum-Likelihood Estimation
General results for randomly censored data are presented in Slymen and Lachenbruch (1984).

7.12.4 Model Validation

Slymen and Lachenbruch (1984) discuss testing the null hypothesis $k = 0$ versus the alternative hypothesis $k > 0$ to choose between these two models.

7.13 MODEL II(b)-13: WEIBULL EXTENSION

This model was proposed by Xie et al. (2002b). It is an extension of a two-parameter model proposed by Chen (2000).

7.13.1 Model Structure

Distribution Function
The distribution function is given by

$$G(t) = 1 - \exp\left[\lambda\alpha\left(1 - e^{(t/\alpha)^{\beta}}\right)\right] \qquad t \geq 0 \tag{7.105}$$

with $\lambda, \alpha, \beta > 0$.

Density Function
The density function is given by

$$g(t) = \lambda\beta(t/\alpha)^{\beta-1}\exp\left[(t/\alpha)^{\beta}\right]\exp\left[\lambda\alpha\left(1 - e^{(t/\alpha)^{\beta}}\right)\right] \tag{7.106}$$

Hazard Function
The hazard function is given by

$$h(t) = \lambda\beta(t/\alpha)^{\beta-1}\exp\left[(t/\alpha)^{\beta}\right] \tag{7.107}$$

Relation to Other Models
The model reduces to the model of Chen (2000) when $\alpha = 1$. When the scale parameter α is large, then $1 - \exp(t/\alpha)^{\beta} \approx (t/\alpha)^{\beta}$, and in this case the distribution can

be approximated by a Weibull distribution. When $\lambda\alpha = 1$, the model is related to the exponential power model proposed by Smith and Bain (1975). Finally the model can be related to the extreme value distribution. Let T be a random variable from (7.105). Let Z be a random variable defined as $Z = t^{\beta}$ and let $\eta = \lambda\alpha$. Then T follows the extreme value distribution with scale parameter α^{β} and location parameter $[-\alpha^{\beta}\ln(\eta)]$.

7.13.2 Model Analysis

Hazard Function

The shape of the hazard function depends only on the shape parameter β. For $\beta \geq 1$:

1. $h(t)$ is an increasing function (type 3).
2. $h(0) = 0$ if $\beta > 1$ and $h(0) = \lambda$, if $\beta = 1$.
3. $h(t) \to +\infty$ as $t \to +\infty$.

For $0 < \beta < 1$:

1. $h(t)$ is decreasing for $t < t^*$ and increasing for $t > t^*$ where

$$t^* = \alpha(1/\beta - 1)^{1/\beta} \tag{7.108}$$

   This implies that the hazard function has a bathtub shape (type 4).
2. $h(t) \to +\infty$ for $t \to 0$ or $t \to +\infty$.
3. t^* increases as the shape parameter β decreases.

WPP Plot

Under the Weibull transformation given by (1.7), (7.105) gets transformed into

$$y = \ln(\lambda\alpha) + \ln\left[\exp(t/\alpha)^{\beta} - 1\right] \tag{7.109}$$

with $t = e^x$. For small t, $\exp(t/\alpha)^{\beta} - 1 \approx (t/\alpha)^{\beta}$ so that (7.109) can be approximated by a straight line given by

$$y = \beta x + \ln(\lambda\alpha^{1-\beta}) \tag{7.110}$$

For large t, the second term in the right-hand side of (7.109) can be approximated as

$$\ln\left\{\exp[(t/\alpha)^{\beta}] - 1\right\} \approx (t/\alpha)^{\beta} \tag{7.111}$$

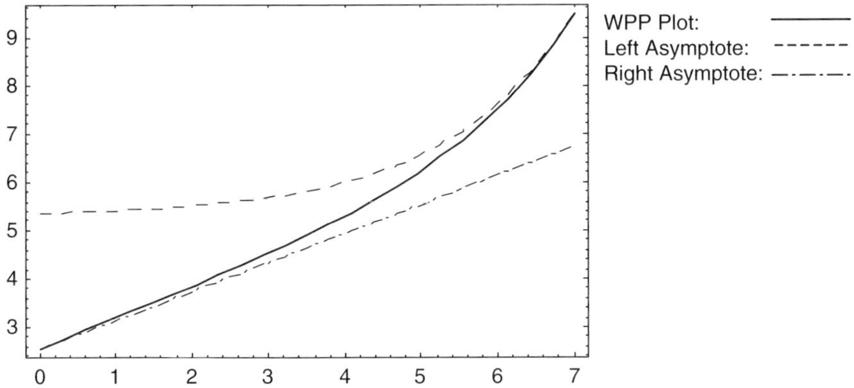

Figure 7.6 WPP plot for the Weibull extension ($\alpha = 100, \beta = 0.6, \lambda = 2$).

As a result, y (as a function of t) can be approximated by the curve $(t/\alpha)^{\beta}$ so that $\ln(y)$ is asymptotically a linear function of t, that is, for large t,

$$\ln(y) = \beta \ln(t) - \beta \ln(\alpha) \tag{7.112}$$

A typical plot of WPP along with the above approximations is as shown in Figure 7.6.

7.13.3 Parameter Estimation

Based on WPP Plot

The procedure is the same as discussed in Section 7.5.4. If the plot indicates that the Weibull extension is an appropriate model, then the parameters are obtained as follows:

Step 1: The asymptotic fit to the right side yields the line given by (7.112). The slope yields $\hat{\beta}$ and from the intercept we obtain $\hat{\alpha}$.

Step 2: From the left side of the plot, $\hat{\lambda}$ can be obtained from the intercept using (7.110).

Method of Maximum Likelihood

For the Type II censored case let $t_{(1)}, t_{(2)}, \ldots, t_{(k)}$ be the ordered data. The likelihood function is given by

$$L(\lambda, \alpha, \beta) = \lambda^k \beta^k \prod_{i=1}^{k} \left(\frac{t_{(i)}}{\alpha}\right)^{\beta-1} \exp\left\{\sum_{i=1}^{k} \left(\frac{t_{(i)}}{\alpha}\right)^{\beta} + \sum_{i=1}^{k} \lambda\alpha \left[1 - \exp\left(\frac{t_{(i)}}{\alpha}\right)^{\beta}\right]\right.$$
$$\left. + (n-k)\lambda\alpha\left[1 - \exp\left(\frac{t_{(k)}}{\alpha}\right)^{\beta}\right]\right\} \tag{7.113}$$

The estimates are obtained by solving the following three equations simultaneously:

$$\hat{\lambda} = \frac{k}{\hat{\alpha}\sum_{i=1}^{k}\exp\left(t_{(i)}/\hat{\alpha}\right)^{\hat{\beta}} + (n-k)\hat{\alpha}\exp\left(t_{(k)}/\hat{\alpha}\right)^{\hat{\beta}} - n\hat{\alpha}} \tag{7.114}$$

$$\frac{k}{\hat{\beta}} + \sum_{i=1}^{k}\ln\frac{t_{(i)}}{\hat{\alpha}} + \sum_{i=1}^{k}\left[\left(\frac{t_{(i)}}{\hat{\alpha}}\right)^{\hat{\beta}}\ln\frac{t_{(i)}}{\hat{\alpha}}\right] - \lambda\alpha\sum_{i=1}^{k}\left[\exp\left(\frac{t_{(i)}}{\hat{\alpha}}\right)^{\beta}\left(\frac{t_{(i)}}{\hat{\alpha}}\right)^{\hat{\beta}}\ln\left(\frac{t_{(i)}}{\hat{\alpha}}\right)\right]$$
$$- (n-k)\hat{\lambda}\hat{\alpha}e^{\left(t_{(k)}/\alpha\right)^{\hat{\beta}}}\left(\frac{t_{(k)}}{\hat{\alpha}}\right)^{\hat{\beta}}\ln\left(\frac{t_{(k)}}{\hat{\alpha}}\right) = 0 \tag{7.115}$$

and

$$-\frac{k(\hat{\beta}-1)}{\hat{\alpha}} + n\lambda - \frac{1}{\hat{\alpha}}\sum_{i=1}^{k}\left(\frac{t_{(i)}}{\hat{\alpha}}\right)^{\hat{\beta}} - \lambda\sum_{i=1}^{k}\left\{e^{\left(t_{(i)}/\hat{\alpha}\right)^{\hat{\beta}}}\left[1-\left(\frac{t_{(i)}}{\hat{\alpha}}\right)^{\hat{\beta}}\right]\right\}$$
$$- (n-k)\lambda e^{\left(t_{(k)}/\hat{\alpha}\right)^{\hat{\beta}}}\left[1-\left(\frac{t_{(k)}}{\hat{\alpha}}\right)^{\hat{\beta}}\right] = 0 \tag{7.116}$$

7.13.4 Hypothesis Testing

Xie et al. (2002) discuss testing the hypothesis $\alpha = 1$ (Chen's model) versus $\alpha \neq 1$ based on the likelihood ratio test.

EXERCISES

Data Set 7.1 Complete Data: Failure Times of 20 Components

0.072	0.477	1.592	2.475	3.597
4.763	5.284	7.709	7.867	8.661
8.663	9.511	10.636	10.729	11.501
12.089	13.036	13.949	16.169	19.809

Data Set 7.2 Censored Data: 30 Items Tested with Test Stopped after 20th Failure[a]

0.0014	0.0623	1.3826	2.0130	2.5274
2.8221	3.1544	4.9835	5.5462	5.8196
5.8714	7.4710	7.5080	7.6667	8.6122
9.0442	9.1153	9.6477	10.1547	10.7582

[a] The data is the failure times.

Data Set 7.3 Complete Data: Failure Times of 20 Components

2.968	4.229	6.560	6.662	7.110
8.608	8.851	9.763	9.773	10.578
19.136	30.112	37.386	48.442	54.145
57.337	57.637	70.175	79.333	85.283

Data Set 7.4 Complete Data: Failure Times of 20 Components

0.0003	0.0298	0.1648	0.3529	0.4044
0.5712	0.5808	0.7607	0.8188	1.1296
1.2228	1.2773	1.9115	2.2333	2.3791
3.0916	3.4999	3.7744	7.4339	13.6866

7.1. Derive (7.8) to (7.11).

7.2. Derive (7.20).

7.3. Derive (7.27).

7.4. Carry out a WPP plot of Data Set 7.1. Based on the WPP plot, determine which of the Type II models discussed in Chapter 7 are not suitable for modeling the data set.

7.5. Assume that Data Set 1 can be modeled by the modified Weibull distribution. Estimate the model parameters based on the WPP plot and using (i) the method of moments (if possible) and (ii) the method of maximum likelihood. Compare the estimates obtained.

7.6. Suppose that the data in Data Set 7.1 can be adequately modeled by the modified Weibull distribution with the following parameter values: $\alpha = 0.1$, $b = 0.5$, and $\lambda = 0.1$. Plot the P–P and Q–Q plots and discuss whether the hypothesis should be accepted or not?

7.7. How would you test the hypothesis of Exercise 7.3 based on A–D and K–S tests for goodness of fit?

7.8. Repeat Exercises 7.4 to 7.7 with Data Set 7.2.

7.9. Repeat Exercise 7.4 with Data Set 7.3.

7.10. Assume that Data Set 7.3 can be modeled by the exponentiated Weibull distribution. Estimate the model parameters based on the WPP plot and using (i) the method of moments (if possible) and (ii) the method of maximum likelihood. Compare the estimates obtained.

7.11. Suppose that the data in Data Set 7.3 can be adequately modeled by the exponentiated Weibull distribution with the following parameter values: $\beta = 5$, $\nu = 0.1$, and $\alpha = 100$. Plot the P–P and Q–Q plots and discuss whether the hypothesis should be accepted or not?

7.12. How would you test the hypothesis of Exercise 7.11 based on A–D and K–S tests for goodness of fit.

7.13. Repeat Exercise 7.1 with Data Set 7.4.

7.14. Assume that Data Set 7.4 can be modeled by the Weibull extension model. Estimate the model parameters based on the WPP plot and using (i) the method of moments (if possible) and (ii) the method of maximum likelihood. Compare the estimates obtained.

7.15. Suppose that the data in Data Set 7.4 can be adequately modeled by the Weibull extension model with the following parameter values: $\alpha = 10$, $\beta = 0.5$, and $\lambda = 0.2$. Plot the P–P and Q–Q plots and discuss whether the hypothesis should be accepted or not?

7.16. How would you test the hypothesis of Exercise 7.15 based on A–D and, K–S tests for goodness of fit.

7.17. Show that the hazard function for Model II(b)-1 is increasing for $\nu \geq 1$ and $\beta \geq 1$.

7.18. Derive (7.31).

7.19. Show that the hazard function for Model II(b)-2 is of bathtub shape for $\beta > 1$ and $\beta\nu < 1$.

7.20. The mean residual life is defined as $E[T - t|T > t]$. Compute the mean residual life for the various Type II models discussed in this chapter.

PART D

Type III Models

CHAPTER 8

Type III(a) Weibull Models

8.1 INTRODUCTION

Type III models involve more than one distribution with one or more of them being the Weibull or the inverse Weibull model. As indicated in Chapter 2, these models can be categorised into four groups [Type III(a) to (d)]. Type III(a) models are referred to as mixture models and are discussed in this chapter.

A general n-fold (or finite) mixture model is given by (2.36). Mixture models also arise in a different context. Let $F(t;\theta)$ be a distribution function with a random parameter θ. Let $H(a)$ denote the distribution function for θ. Then, the distribution function $G(t)$ given by

$$G(t) = \int F(t;a)\, dH(a) \tag{8.1}$$

is also a mixture distribution. When $F(t;\theta)$ is a Weibull distribution, then this is the Type IV(d) model and is discussed in Chapter 12.

Titterington et al. (1985) mention that mixture distribution models have been used for a long time and give a comprehensive reference list of the applications of such models. The earliest reference (dating to 1886) involves a normal mixture. Earliest Weibull mixture models can be traced to the late 1950s [see, Mendenhall and Hader (1958) and Kao (1959)]. Since then, the literature on Weibull mixture models has grown at an increasing pace and with many different applications of the model.

As indicated in Chapter 2, the three mixture models are the following:

Model III(a)-1: Weibull Mixture

Here the n subpopulations are either the standard two- or the three-parameter Weibull distributions. As a result, subpopulation i, $1 \le i \le n$, is given by either

$$F_i(t) = 1 - \exp\left[-\left(\frac{t}{\alpha_i}\right)^{\beta_i}\right] \qquad t > 0 \tag{8.2}$$

Weibull Models, by D.N.P. Murthy, Min Xie, and Renyan Jiang.
ISBN 0-471-36092-9

or

$$F_i(t) = \begin{cases} 0 & t < \tau_i \\ 1 - \exp\left[-\left(\frac{t-\tau_i}{\alpha_i}\right)^{\beta_i}\right] & t \geq \tau_i \end{cases} \tag{8.3}$$

where τ_i, α_i and β_i are the location (or shift), scale, and shape parameters for the subpopulation i. The Weibull mixture model has been referred to by many other names, such as, additive-mixed Weibull distribution, bimodal-mixed Weibull (for a twofold mixture), mixed-mode Weibull distribution, Weibull distribution of the mixed type, multimodal Weibull distribution, and so forth.

Model III(a)-2: Inverse Weibull Mixture
Here all the subpopulations are inverse Weibull distributions with subpopulation $i, 1 \leq i \leq n$, given by

$$F_i(t) = \exp\left[-\left(\frac{\alpha_i}{t}\right)^{\beta_i}\right] \qquad t \geq 0 \tag{8.4}$$

Model III(a)-3: Hybrid Mixtures
Here only some subpopulations are either two- or three-parameter Weibull distributions and others are non-Weibull distributions.

The outline of the chapter is as follows: Section 8.2 deals with the Weibull mixture model. The results for the general case are limited. In contrast, the special case $n = 2$ has been studied more thoroughly. Section 8.3 deals with the twofold inverse Weibull model. In each of these sections we discuss the model structure, analysis, and estimation of parameters. We conclude with a brief discussion of hybrid Weibull mixture models in Section 8.4.

8.2 MODEL III(a)-1: WEIBULL MIXTURE MODEL

8.2.1 Model Structure

The n subpopulations are given either by (8.2) or by (8.3). When all the subpopulations are given by (8.2), we can assume, without loss of generality, that $\beta_i \leq \beta_j, i < j$, and $\alpha_i > \alpha_j$ when $\beta_i = \beta_j$.

The density function is given by

$$g(t) = \sum_{i=1}^{n} p_i f_i(t) \tag{8.5}$$

where $f_i(t)$ is the density function associated with $F_i(t)$.

The hazard function $h(t)$ is given by

$$h(t) = \frac{g(t)}{1 - G(t)} = \sum_{i=1}^{n} w_i(t) h_i(t) \tag{8.6}$$

where $h_i(t)$ is the hazard function associated with subpopulation i, and

$$w_i(t) = \frac{p_i R_i(t)}{\sum_{i=1}^{n} p_i R_i(t)} \qquad \sum_{i=1}^{n} w_i(t) = 1 \tag{8.7}$$

with

$$R_i(t) = 1 - F_i(t) \tag{8.8}$$

for $1 \leq i \leq n$. From (8.7) we see that the failure rate for the model is a weighted mean of the failure rates for the subpopulations with the weights varying with t.

Special Case: Twofold Weibull Mixture Model
The twofold Weibull mixture model is given by

$$G(t) = pF_1(t) + (1 - p)F_2(t) \tag{8.9}$$

with $F_1(t)$ and $F_2(t)$ given by either (8.2) or (8.3). As a result, when the two subpopulations are given by (8.2), the model is characterized by five parameters—the shape and scale parameters for the two subpopulations and the mixing parameter $p(0 < p < 1)$.

The density and hazard functions are given by

$$g(t) = pf_1(t) + (1 - p)f_2(t) \tag{8.10}$$

and

$$h(t) = \frac{pR_1(t)}{pR_1(t) + (1 - p)R_2(t)} h_1(t) + \frac{(1 - p)R_2(t)}{pR_1(t) + (1 - p)R_2(t)} h_2(t) \tag{8.11}$$

Mixture with Negative Weights
A basic feature of the finite mixture models discussed so far is that the mixing weights are all positive. However, Titterington et al. (1985) suggest that it is possible to have mixture models with negative weights as long as the density function for the model is a valid density function. Jiang, et al. (1999a) discuss a multicomponent system (with series, parallel, or k-out-of-n structure) where the component failures are given by either the Weibull or inverse Weibull distributions with common shape parameters, and the system failure is given by a Weibull or inverse Weibull mixture model allowing negative weights.

8.2.2 Model Analysis

Moments
The jth moment about the origin for the model is given by

$$M_j = \sum_{i=1}^{n} p_i M_{ji} \tag{8.12}$$

where M_{ji} is jth moment associated with the distribution $F_i(t)$. If $F_i(t)$ is given by (8.2), then from (3.15) we have

$$M_{ji} = \alpha_i^j \Gamma\left(\frac{j}{\beta_i} + 1\right) \tag{8.13}$$

When $F_i(t)$ is given by (8.3), the expressions are slightly more complicated.

Approximations to Distribution Function

Denote $z_i = (t/\alpha_i)^{\beta_i}$. Then it can be shown [see Jiang and Murthy (1995a)] that

$$\lim_{t\to 0}\left(\frac{z_i}{z_1}\right) = \begin{cases} 0 & \text{if } \beta_i > \beta_1 \\ (\alpha_1/\alpha_i)^{\beta_1} & \text{if } \beta_i = \beta_1 \end{cases} \tag{8.14}$$

and

$$\lim_{t\to \infty}\left(\frac{z_i}{z_1}\right) = \begin{cases} \infty & \text{if } \beta_i > \beta_1 \\ (\alpha_1/\alpha_i)^{\beta_1} > 1 & \text{if } \beta_i = \beta_1 \end{cases} \tag{8.15}$$

From this we have the following result:

1. For small t (very close to zero) $G(t)$ can be approximated by

$$G(t) \approx cF_1(t) \tag{8.16}$$

where

$$c = \sum_{i=1}^{k} p_i \left(\frac{\alpha_1}{\alpha_i}\right)^{\beta_1} \tag{8.17}$$

and k is the number of the subpopulations with the common shape parameter β_1. When $k = 1$, then $c = p_1$.

2. For large t, $G(t)$ can be approximated by

$$G(t) \approx 1 - p_1[1 - F_1(t)] \tag{8.18}$$

These will be used to derive the asymptotes of the WPP plot and to discuss the behavior of the hazard and density functions for very small and large t.

Density Function

From (8.14) and (8.15) the density function can be approximated by

$$g(t) \approx cf_1(t) \tag{8.19}$$

for small t, and by

$$g(t) \approx p_1 f_1(t) \tag{8.20}$$

for large t. This implies that the density function is increasing (decreasing) for small t if $\beta_1 > 1$ ($\beta_1 \leq 1$).

The shape of the density function depends on the model parameters, and the different possible shapes are as follows:

- Type $(2k - 1)$: Decreasing followed by $k - 1$ modal $(k = 1, \ldots, n - 1)$
- Type $(2k)$: k Modal $(k = 1, 2, \ldots, n)$

Special Case: Twofold Weibull Mixture Model
The shape of the density function depends on the parameter values. The possible shapes (see Fig. 3.1) are as follows:

- Type 1: Decreasing
- Type 2: Unimodal
- Type 3: Decreasing followed by unimodal
- Type 4: Bimodal

Although the twofold mixture model is characterized by five parameters, the shape of the density function is only a function of the two shape parameters, the ratio of the two scale parameters, and the mixing parameter. Jiang and Murthy (1998) give a parametric characterization of the density function in this four-dimensional parameter space. The boundaries separating the different shapes are fairly complex.

Hazard Function
From (8.14) and (8.15) the hazard function can be approximated by

$$h(t) \approx ch_1(t) \tag{8.21}$$

for small t, and by

$$h(t) \approx h_1(t) \tag{8.22}$$

for large t, where c is given by (8.17). This implies that the hazard function is increasing (decreasing) for small t if $\beta_1 > 1$ ($\beta_1 < 1$).

The shape of the hazard function depends on the model parameters and different possible shapes are as follows:

- Type 1: Decreasing
- Type 3: Increasing
- Type $4k + 3$: Decreasing followed by k modal $(k = 1, \ldots, n - 1)$
- Type $4k + 2$: k Modal followed by increasing $(k = 1, 2, \ldots, n)$

Special Case: Twofold Weibull Mixture Model
The hazard function is given by (8.11). As indicated earlier, we assume $\beta_1 \leq \beta_2$. Jiang and Murthy (1998) prove the following asymptotic result:

- For large t, $h(t) \to h_1(t)$.
- For small t, $h(t) \to ch_1(t)$ with $c = p$ for $\beta_1 < \beta_2$ and $c = p + (1-p)/k$ for $\beta_1 = \beta_2$.

The implication of this is that for both small and large t the shape of $h(t)$ is similar to that for $h_1(t)$. As a result, if $h_1(t)$ has an increasing (decreasing) failure rate, then $h(t)$ is increasing (decreasing) for small and large t. This implies that $h(t)$ can never have a bathtub shape.

Some authors (e.g., Krohn, 1969) suggest that when one of the shape parameters is less than one and the other greater than one, the hazard function has a bathtub shape (type 4 in Fig. 3.2). This is not possible since the shape of the hazard function $h(t)$ is the same for small and large values of t. The results of Glaser (1980) also show that a twofold Weibull mixture can never have a bathtub-shaped failure rate.

The shape of the hazard function depends on the parameter values. The possible shapes (see, Figure 3.2) include, but not limited to the following:

- Type 1: Decreasing
- Type3: Increasing
- Type 6: Uni-modal followed by increasing
- Type 7: Decreasing followed by uni-modal
- Type 10: Bi-modal followed by increasing

Jiang and Murthy (1998) give a complete parametric characterization of the different shapes in the four dimensional parameter space. The boundaries separating the different regions have complex shapes.

WPP Plot
Under the Weibull transform (see Section 3.2.5) the WPP plot for the general model is a complex curve that is difficult to study analytically. However, from (8.16) and (8.18) we have the following asymptotic result:

Left asymptote: The asymptote, as $x \to -\infty$, is given by the straight line

$$y = \beta_1[x - \ln(\alpha_1)] + \ln(c) \tag{8.23}$$

Right asymptote: The asymptote, as $x \to \infty$, is given by the straight line

$$y = \beta_1[x - \ln(\alpha_1)] \tag{8.24}$$

Note that the two asymptotes are parallel to each other, and the vertical separation is $\ln(c)$ where c is given by (8.17).

Special Case: Twofold Weibull Mixture Model

The WPP plot is a smooth curve in the x–y plane. The shape of the plot is significantly influenced by the two shape parameters of the subpopulations. We need to consider the following two cases separately:

- Case 1: Different shape parameters ($\beta_1 < \beta_2$)
- Case 2: Same shape parameters ($\beta_1 = \beta_2$)

The WPP plot for subpopulation 1 is a straight line L_1 given by (8.24), and for the subpopulation 2 is another straight line L_2 given by

$$y = \beta_2[x - \ln(\alpha_2)] \tag{8.25}$$

When $\beta_1 \neq \beta_2$, L_1 and L_2 intersect and the coordinates of this point (denoted as I) in the x–y plane are given by

$$x_I = \frac{\ln(\alpha_1^{\beta_1}/\alpha_2^{\beta_2})}{\beta_1 - \beta_2} \qquad y_I = \frac{\beta_1\beta_2 \ln(\alpha_1/\alpha_2)}{\beta_1 - \beta_2} \tag{8.26}$$

When $\beta_1 = \beta_2$, the two lines are parallel to each other.

Case 1: $\boldsymbol{\beta_1 < \beta_2}$ A typical plot of the curve for the case when $0 < p < 1$ is shown in Figure 8.1 along with the asymptotes. Jiang and Murthy (1995a) show that the curve has the following properties:

- *Left asymptote*: As $x \to -\infty$, the asymptote is the straight line given by (8.23) with $c = p$.
- *Right asymptote*: As $x \to -\infty$, the asymptote is the straight line L_1 given by (8.24).

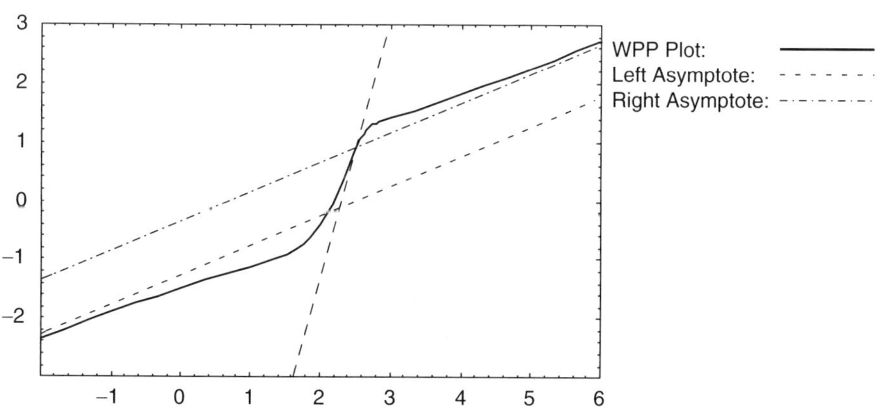

Figure 8.1 WPP plot for Weibull mixture model ($n = 2, \alpha_1 = 2, \beta_1 = 0.5, \alpha_2 = 10, \beta_2 = 4.3, p = 0.4$).

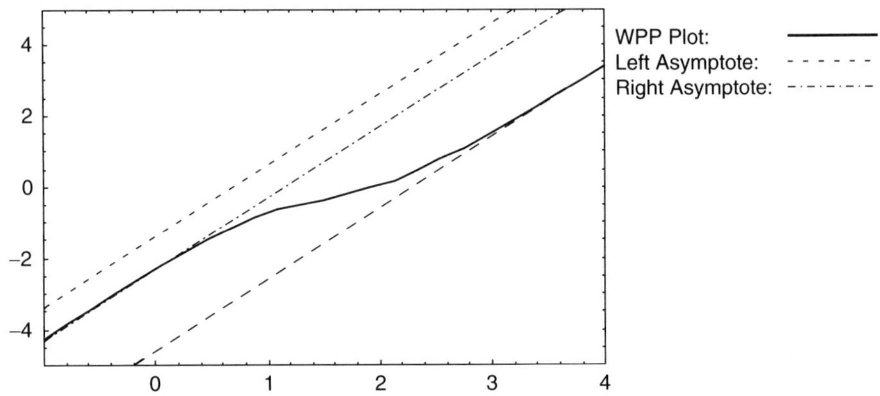

Figure 8.2 WPP plot for Weibull mixture model ($n = 2, \alpha_1 = 2, \alpha_2 = 10, \beta_1 = \beta_2 = 2, p = 0.4$).

- The curve passes through the intersection of L_1 and L_2 and the slope of the curve at this point is given by

$$\bar{\beta} = p\beta_1 + (1 - p)\beta_2 \tag{8.27}$$

- The curve has three inflection points.

Case 2: $\boldsymbol{\beta_1 = \beta_2}$ Let $\beta_0 = \beta_1 = \beta_2$ and define $\xi = (\alpha_2/\alpha_1)^{\beta_0} < 1$. The two straight lines, L_1 and L_2, are parallel to each other. A typical plot of the curve is shown in Figure 8.2. Jiang and Murthy (1995a) show that the curve has the following properties:

- *Left asymptote*: As $x \to -\infty$, the asymptote is the straight line:

$$y = \beta_1[x - \ln(\alpha_1)] + \ln\left(p + \frac{1-p}{\xi}\right) \tag{8.28}$$

- *Right asymptote*: As $x \to \infty$, the asymptote is the straight line L_1.
- The curve has only one inflection point.

Special Case: Threefold Weibull Mixture Model

Jensen and Petersen (1982) and Moltoft (1983) consider a mixture of three Weibull distributions and use the terms *strong*, *freak*, and *infant mortality* for the three subpopulations. They discuss the use of WPP plots for the special case where the three scale parameters and three mixing proportions differ by orders of magnitude.

Jiang and Murthy (1996) study a special case where the shape parameters are different and the scale and mixing parameters differ significantly so that

$$\alpha_{\text{I}} \ll \alpha_{\text{F}} \ll \alpha_{\text{S}} \qquad \text{and} \qquad p_I \ll p_F \ll p_S$$

where the subscripts I, F, and S denote the subpopulations corresponding to infant mortality, freak, and strong. They examine six different combinations resulting from the ranking of the shape parameters for the three subpopulations. The cases studied by Jensen and Petersen (1982) and Moltoft (1983) are two special cases of these six combinations. According to Jiang and Murthy (1996), the WPP plot for the model has the following properties:

- As $x \to -\infty$, the asymptote is a straight line given by (8.23) with $c = p_1$.
- As $x \to \infty$, the asymptote is a straight line given by (8.24).
- The plot has two plateaus that correspond to $y = \ln(p_I)$ and $y = \ln[-\ln(p_S)]$, respectively.

8.2.3 Parameter Estimation

The literature on parameter estimation for the mixture model is vast. Both graphical and analytical approaches have been used. Because the principles and procedure of analytical methods are the same (except some details) as those discussed in Chapter 4, we only present a chronological review of the literature and omit giving any results. However, we will discuss the graphical method based on the WPP plot in more detail.

Graphical Method

Kao (1959) was the first to use the graphical method for parameter estimation in the Weibull mixture model. His analysis is based on the observation that there are two types of failure—sudden catastrophic failures (modeled by a two-parameter Weibull distribution with shape parameter < 1) and wear-out failures (modeled by a three-parameter Weibull distribution with shape parameter > 1). He first estimates the mixing parameter p and then the location parameter τ (for the subpopulation characterizing wear-out failure). He then separates the two subpopulations and estimates the remaining four parameters (scale and shape parameters for the two subpopulations) separately. This approach has been cited by several authors—for example, Everitt and Hand (1981), Mann et al. (1974), and Falls (1970). The shortcomings of the approach are discussed in Jiang (1996).

Cran (1976) basically follows Kao's method with some minor modifications. It does not lead to any significant improvement on the original method proposed by Kao (1959).

Lawless (1982) assumes that the mixing parameter p is known and separates the data into two groups. The first group corresponds to the data from subpopulation 1. It is given by the initial p fraction of the ordered data for the WPP plotting. The remaining data is viewed as coming from subpopulation 2.

Jensen and Petersen (1982) introduce an approximation to graphically estimate the parameters of a twofold mixture. Their approximation is valid only when the scale parameters differ significantly, and the shape parameter for subpopulation 1 is bigger than that for subpopulation 2.

Jensen and Petersen (1982) and Moltoft (1983) deal with a mixture of three Weibull distributions. They discuss the use of WPP plots to estimate the model parameters for the special case where the three scale parameters and three mixing proportions differ by orders of magnitude.

Natesan and Jardine (1986) propose yet another approach to segregate the mixed population into subpopulations and then estimate the parameters for each population separately.

Jiang and Kececioglu (1992a) study the WPP plot for a twofold Weibull mixture model. They give a characterization of the asymptotic behavior of the plots and estimate the parameters based on this. As indicated in Jiang and Murthy (1995a) their characterization is not correct and hence the method is not correct.

Kececioglu and Sun (1994) propose a method to separate subpopulations of a mixture, which is similar to that of Lawless (1982), and the method does not discuss the estimation of the mixing parameter.

Jiang and Murthy (1995a) deal with the special case $n = 2$ and propose a method for estimation. The steps involved are discussed later in the section.

Jiang and Murthy (1996) deal with a special case of the threefold Weibull mixture model where the three scale parameters and the three mixture proportions differ by orders of magnitude. Using the asymptotes and some approximations to these plots, they first obtain estimates of the parameters of one subpopulation and its mixing proportion. Once this is done, the original data is transformed so that it effectively removes the data from this subpopulation. The new data is viewed as coming from a mixture of two Weibull distributions, and the parameters are estimated using the approach outlined in Jiang and Murthy (1995a).

In summary, the graphical method has been used widely to estimate the parameters of the Weibull mixture model. The bulk of the literature deals with the well-separated case (i.e., $\alpha_1 \gg \alpha_2$ or $\alpha_1 \ll \alpha_2$ in the case of the twofold model) and uses various approximations for plotting and the characterization of the asymptotes. Unfortunately, many of them are not valid due to incorrect approximations.

As mentioned earlier, the graphical method has two serious limitations. These are as follows:

1. They yield very crude estimates unless repeated iteration and visional evaluation. As such, they should be used as starting points for more sophisticated statistical methods.
2. The graphical method does not provide any statistical confidence limits for the estimates.

Special Case $(n = 2)$

Jiang and Murthy (1995a) carry out a thorough study of the WPP plot and discuss parameter estimation based on the WPP plot. Of particular significance are the inflection points of the WPP plot and approximations to the coordinates of these points in terms of the model parameters. It involves two parts (parts A and B) as discussed in Section 4.5.1. Part A (involving five steps) deals with plotting the data and is dependent on the type of data available. The steps involved are the same as in Section 4.5.1. If a smooth fit to the plotted data has a shape similar to that in

Figure 8.1 or 8.2, then one can proceed to part B to estimate the model parameters. We outline the steps involved in part B and need to consider the following three cases separately:******

Case 1: Well-mixed case ($\beta_1 \neq \beta_2$ and $\alpha_1 \approx \alpha_2$)
Case 2: Well-separated case ($\beta_1 \neq \beta_2$ and $\alpha_1 \gg \alpha_2$ or $\alpha_1 \ll \alpha_2$)
Case 3: Common shape parameter ($\beta_1 = \beta_2$)

Note: The WPP plot of the data allows one to determine which of the above three cases is appropriate for a given data.

Case 1 The two asymptotes yield estimates of the scale and shape parameters for the two subpopulations, and the intersection coordinates are used to obtain an estimate of the mixing parameter. As a result, the steps for part B are as follows:

Step 6: Draw the asymptotes L_1 and L_a ensuring that they are parallel to each other.

Step 7: Obtain estimates β_1 and α_1 from the slope and intercept of L_1.

Step 8: Obtain an estimate of p from the vertical separation $[= \ln(p)]$ between the two lines.

Step 9: Determine the point I where L_1 and the WPP plot intersect.

Step 10: Obtain the slope to the WPP at this intersection point. This provides an estimate of $\bar{\beta}$. Using this in (8.27) yields an estimate of β_2.

Step 11: Draw line L_2 (with a slope given by the estimate of β_2 and passing through the intersection point I). The intercept of this line with the x or y axis yields an estimate of α_2.

Case 2 The approximation at $t = \alpha_1$ (for $\alpha_1 \ll \alpha_2$) or $t = \alpha_2$ (for $\alpha_1 \gg \alpha_2$) is used to estimate the mixing parameter. The asymptotes and the intersection coordinates are used to obtain estimates of the parameters of the two subpopulations. As a result, the steps for part B are as follows:

When $\alpha_1 \ll \alpha_2$ ($\alpha_1 \gg \alpha_2$) most of the data points are to the left (right) of intersection point I. As a result, the procedure is different for the two cases.

Step 6: Determine visually whether most of the data is scattered along the bottom (or top) of the WPP plot. This determines whether $\alpha_1 \ll \alpha_2$ ($\alpha_1 \gg \alpha_2$).

Step 7: If $\alpha_1 \ll \alpha_2$ ($\alpha_1 \gg \alpha_2$), then locate the inflection and this is to the left (right) of intersection point I. From Jiang and Murthy (1995a) its y coordinate (denoted by y_a) is given by $y_a \cong \ln[-\ln(1-p)]$ or $\{y_a \cong \ln[-\ln(p)]\}$. The estimate of this coordinate yields an estimate for p.

Step 8: If $\alpha_1 \ll \alpha_2$, then estimates of α_1 and α_2 are obtained as follows. Compute $y_1 = \ln[-\ln(1-p+p/e)]$ or $(y_2 = \ln\{-\ln[(1-p)/e]\})$. Draw a horizontal line with y intercept given by y_1 (y_2). Let x_1 (x_2) be the x coordinate of the intersection of this line with the fitted WPP plot. Then an estimate of α_1 (α_2) is obtained from $x_1 = \ln(\alpha_1)$ [$x_2 = \ln(\alpha_2)$].

If $\alpha_1 \gg \alpha_2$, then estimates of α_1 and α_2 are obtained as follows. Compute $y_1 = \ln[-\ln(p/e)]$ (or $y_2 = \ln\{-\ln[p + (1-p)/e]\}$). Draw a horizontal line with y intercept given by y_1 (y_2). Let x_1 (x_2) be the x coordinate of the intersection of this line with the fitted WPP plot. Then an estimate of α_1 (α_2) is obtained from $x_1 = \ln(\alpha_1)$ [$x_2 = \ln(\alpha_2)$].

Step 9: If $\alpha_1 \ll \alpha_2$, then draw the tangent to the end of the fitted WPP plot to approximate L_a and obtain an estimate of β_1 from its slope.

If $\alpha_1 \gg \alpha_2$, then draw the tangent to the right end of the fitted WPP plot to approximate L_1 [ensuring that it intersects the x axis at $\ln(\alpha_1)$] and obtain an estimate of β_1 from its slope.

Step 10: Determine the intersection point I and compute its x coordinate. Using this in (8.26) yields an estimate of β_2.

Case 3 Define $\xi = (\alpha_2/\alpha_1)^{\beta_0}$. We need to consider the following two subcases $\xi \approx 1$ and $\xi \ll 1$ separately. As a result, the steps for part B are as follows:

We first consider the case $\xi \approx 1$.

Step 6: Draw the two asymptotes L_1 and L_a.

Step 7: Obtain estimates of β_0 and α_1 from the slope and intercept of L_1.

Step 8: From the vertical separation between the L_1 and L_a one can obtain an estimate of $p + (1-p)/\xi$.

Step 9: Draw a vertical line through $x = \ln(\alpha_1)$ and let this intersect with the WPP plot with the y coordinate given by y_1. This yields an estimate of $(p/e) + (1-p)/e^{1/\xi}$.

Step 10: From the estimates obtained in steps 8 and 9, one can obtain estimates of p and ξ. This then yields an estimate of α_2 from $\xi = (\alpha_2/\alpha_1)^{\beta_0}$.

When $\xi \ll 1$, the steps are as follows:

Step 6: Locate the inflection point for the fitted WPP plot. Let y_T denote the estimated y coordinate of this point. An estimate of p is obtained from $y_T \cong \ln[-\ln(p)]$.

Step 7: Obtain α_1 and α_2 from step 8 for Case 2.

Step 8: Depending on the plot, fit either the right or the left asymptote to the plot. If it is the right asymptote, one needs to ensure that it intersects the x axis at $\ln(\alpha_1)$. No such restriction applies if it is the left asymptote. The slope of this yields an estimate of β_0.

For further details of the approximation and the application to model real data, see Jiang and Murthy (1995a) and Jiang (1996).

Method of Moments

Rider (1961) discusses the method of moments for estimating the parameters of a twofold Weibull mixture model involving two-parameter Weibull distributions with a known common shape parameter. Estimates of the two scale parameters and mixing parameter are obtained using the first three sample moments.

Maximum-Likelihood Method

Maximum-likelihood estimation can be carried out in the traditional way. Many studies deal with this topic, and we briefly review this literature.

Mendenhall and Hader (1958) consider an n-fold Weibull mixture model. They derive the maximum-likelihood estimates for the scale and mixing parameters assuming that the shape parameters are known. Beetz (1982) estimates the parameters of a mixed Weibull distribution by fitting the mixed probability density to the experimental histogram using the maximum-likelihood method. Ashour (1987) considers the problem of maximum-likelihood estimation with five unknown parameters of the mixed Weibull distribution for a multistage censored type I sample. Chen et al. (1989) discuss a twofold Weibull mixture and derive the maximum likelihood and Bayes estimators for the model parameters. A comparison of these two estimators is done using the Monte Carlo simulation.

Chou and Tang (1992) consider a Weibull mixture model with one shape parameter less than one and the other equal to one (implying an exponential distribution for one of the subpopulations). They use the model to describe two modes of failure—one due to infant mortality and the other represents failure under normal mode. The change point (where the failure rate changes from decreasing to a constant) is estimated using the method of maximum likelihood. They also study the effect of sample size on the estimation error and carry out a sensitivity analysis of the model with respect to parameter variations.

Jiang and Kececioglu (1992b) present an algorithm for estimating the parameters of a Weibull mixture model with right censored data using the method of maximum likelihood. Numerical examples involving two, three, and five subpopulations are presented. Ahmad and Abdelrahman (1994) present a procedure for finding maximum-likelihood estimates of the parameters of a mixture of two Weibull distributions. Estimation of a nonlinear discriminant function on the basis of small sample size is studied via simulation experiments.

Several authors have discussed the problems associated with the MLE method. The main difficulty with the method is the lack of analytical tractability and the need for iterative computational methods.

General Curve Fitting Method

General curve fitting can be used to fit the density function to the histogram, the distribution function to the empirical distribution plot, or the hazard function to the empirical hazard function. In some instances, the fit is done to transformed data (as in the case of the WPP plot). This has been widely used to estimate the parameters of the Weibull mixture models.

Cheng and Fu (1982) deal with estimating the parameters of a twofold Weibull mixture model using the weighted least-squares method. They develop the method in the context of life testing with grouped and censored data. Woodward and Gunst (1987) consider a twofold Weibull mixture model with each subpopulation a three-parameter Weibull distribution. They use a minimum distance estimation method to estimate the mixing parameter. Several simulated data sets are used to evaluate the effectiveness of the estimation method.

Contin et al. (1996) use a Weibull mixture model to fit experimental data for partial discharge phenomena. They use nonparametric kernel regression technique to preprocess the data in order to determine the number of subpopulations and basic characteristics of the mixture distribution. Ling and Pan (1998) consider a twofold mixture model involving three-parameter Weibull distributions. The model parameters are obtained by minimizing the maximum absolute difference between the observed probability of failure and the expected probability of failure.

Schifani and Candela (1999) present a heuristic algorithm for the evaluation of the parameters of additive Weibull distributions having more than two elementary functions with application in partial discharge analysis. The algorithm makes use of a linear combination of three quality functions (representing the distance or error between the fitted CDF and empirical CDF) that work well in three different intervals of partial discharge: for example, low, middle, and high variable values. The model parameters are estimated by minimizing the overall quality function using an iterative evolutionary procedure.

Nagode and Fajdiga (2000) also present an algorithm for parameter estimation of an n-fold Weibull mixture model by minimizing a kind of absolute relative deviation. The algorithm starts with a plot of the histogram using all the data. It assumes that the highest peak of the histogram corresponds to a subpopulation, whose parameters can be approximately estimated from the magnitude and location of the mode. Once the parameters of this subpopulation are numerically determined, a new histogram, corresponding to a $(n-1)$-fold model, can be obtained. The process is repeated till only one subpolpulation is left. As a result, the number and the parameters of subpopulations are estimated successively. Limitations with this approach are (1) it requires a large sample, (2) complete data, and (3) well-separated mixture with the shape parameters >1.

Bayesian Method

When the model builder has some intuitive assessment of the parameters that can be described through an appropriate prior distribution, the Bayesian method is more appropriate than any other method. Due to mathematical intractability, the bulk of the studies on the Bayesian approach deal with exponential mixture models and very few with the Weibull mixture. The Bayesian approach involves obtaining the posterior distribution for the unknown parameters using the prior distribution and the data collected. As such, the prior distribution has a significant impact on the estimator.

Jensen and Petersen (1982) consider a Bayesian approach to estimating the parameters of a Weibull mixture model in the context of burn-in problem. Sinha (1987) obtains Bayes estimators of the parameters and the reliability function for a mixture of two time-censored Weibull distributions. Uniform prior distributions are used for the two shape parameters, while the mixing parameter is also assumed to have a uniform prior. Jeffreys' "vague priors" are used for the two scale parameters. The estimators obtained are compared with their maximum-likelihood counterparts using a numerical example. Pandey and Upadhyay (1988) derive the Bayes estimator for the scale parameters for a twofold exponential mixture with mixing parameter known. Salem (1988) deals with the Bayesian estimation of the four parameters in a mixed Weibull distribution with equal shape parameters under multistage censored type I sampling. Using a Bayesian approach, Sinha and Sloan (1989) study the predictive distribution and predictive interval of a future observation from a mixture of two Weibull distributions. Chen et al. (1989) derive a method for the Bayes' estimator for a twofold Weibull mixture.

Attia (1993) considers a mixture of two Rayleigh distributions and obtains estimates of model parameters using the method of maximum likelihood and the Bayesian approach with censored sampling. Perlstein and Welch (1993) develop a Bayesian methodology for burn-in of manufactured items to differentiate items from a mixture involving two subpopulations. The life distribution of items is assumed to be a twofold exponential mixture. Singh and Bhattacharya (1993) carry out a Bayesian reliability analysis for a mixture of two exponential distributions. Attia (1993) obtains an estimate of model parameters using the Bayesian approach for a mixture of two Rayleigh distributions. Leonard et al. (1994) consider both Bayesian and likelihood estimation for equally weighted Weibull mixtures.

Kececioglu and Sun (1994) discuss parameter estimation for a twofold Weibull mixture. They present a Bayesian approach for both grouped and ungrouped burn-in test data. A numerical comparison is made by conducting the Kolmogorov–Smirnov goodness-of-fit test on the parameter estimates obtained using the Jensen graphical method and the Bayesian approach based on the conventional separation plotting method and the fractional rank plotting method, respectively. The Bayesian approach with the fractional rank plotting method yields better results. Ahmad et al. (1997) study approximate Bayes estimation by considering a five-dimensional vector of parameters for a mixture of two Weibull distributions under type 2 censoring.

Hybrid Methods

Hybrid methods are methods that combine different methods to obtain estimates of the model parameters.

Falls (1970) uses a compound method that combines the method of moments and the graphical method to estimate the five parameters of a twofold Weibull mixture. Yuan and Shih (1991) consider a twofold Weibull mixture with the two subpopulations well separated. They consider the maximum-likelihood method and the graphical method aided by the Bayesian approach. The methods are illustrated with an application to CMOS component failure data. Zhao and Zhang (1997) model the reliability of gears using a Weibull mixture model. The model parameters are estimated using regression and maximum-likelihood methods.

Kececioglu and Wang (1998) develop a compound approach to estimate the parameters of a Weibull mixture model. The population sample data is split into different subpopulation data sets using the posterior probability (based on concepts from fracture failure, mean order number, and the Bayes theorem). The parameters of each subpopulation are estimated using the corresponding divided data set. The approach is applicable for complete, censored, and grouped data samples. Its

Table 8.1 Applications in Reliability

Application	Reference
Radio transmitter receivers	Mendenhall and Hader (1958)
Electronic tubes	Kao (1959)
Air-conditioning system	Proschan (1963)
Optical waveguides aged in different environments	Liertz etc. (1977)
Motor vehicle degradation and safety inspection	Parker (1977)
Solid lubricant coatings	Radcliffe and Parry (1979)
Optimum repair limit policies	Kaio et al. (1982)
Electrical appliances	Lawless (1982)
Burn-in data	Jensen and Petersen (1982)
CMOS 4007 test	Moltoft (1983)
Shipboard gas generator failure data	Natesan and Jardine (1986)
Tensile strength of fibers/metal matrix composites	Ochiai et al. (1987)
Strength parameters of dental materials	de Rijk et al. (1988)
Stress corrosion cracking of Zircaloy-4 alloy	Hirao et al. (1988)
Ceramic reliability	Lamon (1988, 1990)
Instrument cluster	Kerscher III et al. (1989)
Microcrack initiation distribution of carbon steel	Goto (1989, 1991)
Mechanical components	Vannoy (1990)
Planar SiO_2 breakdown data	Sichart and Vollertsen (1991)
CMOS components	Yuan and Shih (1991)
Burn-in data	Perlstein and Welch (1993)
Load-carrying vehicles	Jiang and Murthy (1995a)
Environmental stress screening strategy to multicomponent electronic system	Pohl and Dietrich (1995)
Space-use traveling-wave tubes	Mita (1995)
Burn-in and stress screening of electronic products	Yan et al. (1995)
Failure time for oral irrigators	Jiang and Murthy (1997a)
Gears reliability	Zhao and Zhang (1997)
Accelerated aging tests on solid polymeric dielectric insulation	Grzbowski et al. (1998)
Optimal burn-in procedure applied to integrated circuit testing	Kim (1998)
Fatigue life testing	Ling and Pan (1998)
Fatigue cracks in medium carbon steel	Wu and Sun (1998)

superiority is particularly significant when the sample size is relatively small and the subpopulations are well mixed.

8.2.4 Modeling Data Set

The WPP plot serves as a good starting point to determine if a Weibull mixture model is appropriate to model a given data set. The plotting depends on the type of data available and is discussed in Section 4.5.2. If a smooth fit to the plotted data has a shape similar to that shown in Figure 8.1 or 8.2, one can tentatively assume that the twofold Weibull mixture model can be an appropriate model.

The model parameters can be estimated from the WPP plot as indicated in Section 8.2.3. These are crude estimates and suffer from the usual limitations discussed in Chapter 4. When the subpopulations are well separated and/or the size of the data set is not large, then fitting the left or right asymptotes might be difficult and hence seriously affect the parameter estimates. For further discussion of this, see Jiang and Murthy (1995a).

Goodness-of-Fit Tests

Kececioglu and Sun (1994) and Ling and Pan (1998) deal with Kolmogorov–Smirnov goodness-of-fit test and Schifani and Candela (1999) discuss the Cramer–Von Mises test.

8.2.5 Applications

Two- and threefold mixture models have been extensively used in reliability theory and in many other areas. Table 8.1 gives a list of applications in reliability and Table 8.2 gives a list of applications in other disciplines.

Table 8.2 Applications in Other Fields

Application	Reference
Atmospheric data	Falls (1970)
Molecular weight distribution of cellulose	Broido et al. (1977a, b)
Wind speed distribution	Mau and Chow (1980)
Time of death for irradiated mice	Elandt-Johnson and Johnson (1980)
Stationary waiting time of queue system	Harris (1985)
Occupational injury analysis	Chung (1986)
Crops	Woodward and Gunst (1987)
Traffic conflicts (accidents)	Chin et al. (1991)
Time of death for irradiated mice	Jiang and Murthy (1995a)
Partial discharge in insulation system	Contin et al. (1996)
of rotating machines	Schifani and Candela (1999)
Daily rainfall amounts	Chapman (1997)
AC superconducting wires and quench current	Hayakawa et al. (1997a, b)
Loading spectra	Nagode and Fajdiga (1999)

8.3. MODEL III(a)-2: INVERSE WEIBULL MIXTURE MODEL

Jiang et al. (1999a) is the first study dealing with the inverse Weibull mixture model. They deal with a multicomponent system where the individual components have inverse Weibull failure distributions with a common shape parameter and show that the system failure distribution is given by an inverse Weibull mixture allowing negative weights. Jiang et al. (2001a) deal with a twofold inverse Weibull mixture model and we discuss this in this section.

8.3.1 Model Structure

The model is given by

$$G(t) = pF_1(t) + (1-p)F_2(t) \qquad t \geq 0 \tag{8.29}$$

where $F_1(t)$ and $F_2(t)$ are two inverse Weibull distributions with scale and shape parameters (α_1, β_1) and (α_2, β_2), respectively, and p is the mixing parameter.

8.3.2 Model Analysis

Density Function

The different possible shapes (see Fig. 3.1) for the density function are as follows:

- Type 2: Unimodal
- Type 4: Bimodal

Hazard Function

The different possible shapes (see Fig. 3.2) for the hazard function are as follows:

- Type 5: Unimodal
- Type 9: Bimodal

IWPP Plot

The IWPP plot is discussed in Section 6.6.2. Let L_1 and L_2 denote the IWPP plots for the two subpopulations. They are two different straight lines given by (8.24) and (8.25), respectively. The IWPP plot for the mixture model is a complex curve. The shape of it depends on the two shape parameters. We need to consider the following two cases separately.

Case (i) The shape parameters are different and we can assume, without loss of generality, that $\beta_1 < \beta_2$. The lines L_1 and L_2 intersect and the IWPP plot has the following properties:

- Let I denote the intersection of L_1 and L_2. The IWPP plot passes through I and is always within the cone defined by L_1 and L_2.

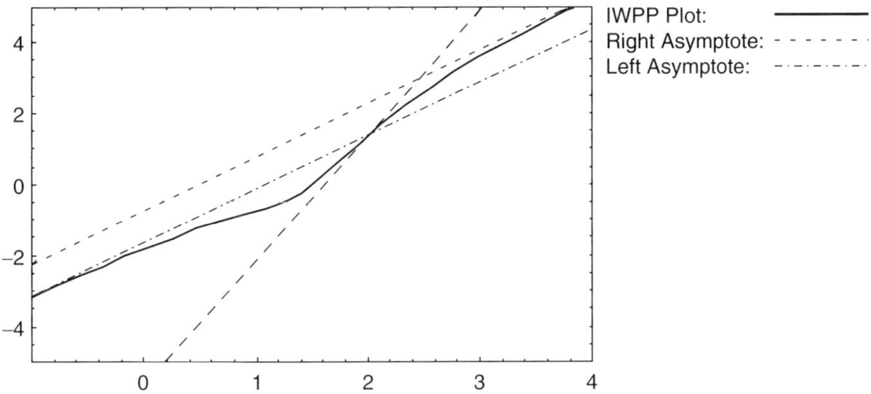

Figure 8.3 IWPP plot for inverse Weibull mixture model ($n = 2, \alpha_1 = 3, \beta_1 = 1.5, \alpha_2 = 5, \beta_2 = 3.5, p = 0.4$).

Let (x_I, y_I) denote the coordinates of I. These are given by (8.26). The slope of the IWPP at I is given by (8.27).

- The left asymptote is given by L_1, and the right asymptote is a straight line L_R given by

$$y = \beta_1[x - \ln(\alpha_1)] - \ln(p) \tag{8.30}$$

This is parallel to L_1 but is displaced vertically by $\ln(1/p)$.

One needs to consider the following two subcases: (a) well-mixed case ($\alpha_1 \approx \alpha_2$) and (b) well-separated case ($\alpha_1 \ll \alpha_2$ or $\alpha_1 \gg \alpha_2$). A typical IWPP plot for the well-mixed case is as shown in Figure 8.3 along with the asymptotes. This is a mirror image of the WPP plot in Figure 8.1 for the Weibull mixture. For the well-separated case, the IWPP plot is slightly different from that in Figure 8.3. This is due to the fact that the part to the right (left) of the intersection probably disappears or the corresponding feature becomes indistinct (e.g., the slope to the left (right) of the intersection point becomes flatter).

Case (ii) The two shape parameters are the same so that $\beta_1 = \beta_2 (= \beta)$ and without loss of generality we assume that $\alpha_1 < \alpha_2$. In this case L_1 and L_2 are parallel to each other.

The left asymptote is given by L_1 and the right asymptote is a straight line L_R given by

$$y = \beta[x - \ln(\alpha_1)] - \ln[p + (1 - p)(\alpha_2/\alpha_1)^{\beta}] \tag{8.31}$$

This is parallel to L_1 but is displaced vertically by the second term in the right-hand side of (8.31). A typical IWPP plot, along with the asymptotes, is shown in Figure 8.4.

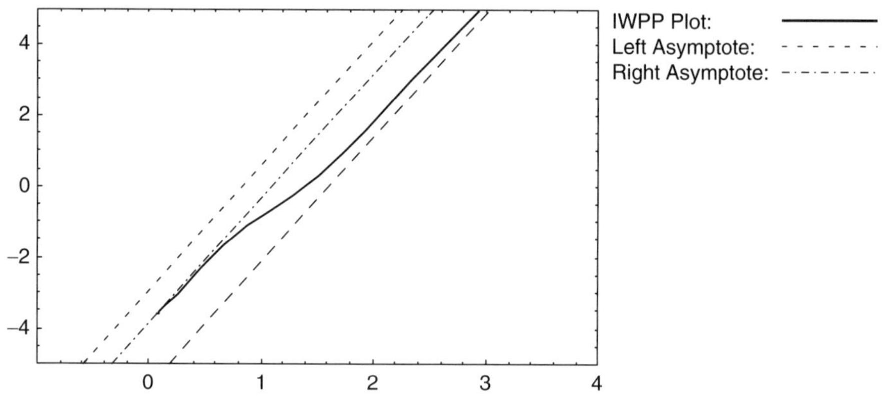

Figure 8.4 IWPP plot for inverse Weibull mixture model ($n = 2, \alpha_1 = 3, \alpha_2 = 5, \beta_1 = \beta_2 = 3.5$, $p = 0.4$).

8.3.3 Parameter Estimation

Graphical Method

This involves two parts (parts A and B) as discussed in Section 4.5.1. Part A (involving five steps) deals with plotting the data and is dependent on the type of data available. The steps involved are the same as in Section 4.5.1. If a smooth fit to the plotted data has a shape similar to that in Figure 8.3, then one can proceed to part B for estimation of model parameters. We outline the steps involved in part B and need to consider the following three cases separately:

Case 1: Well-Mixed Case ($\beta_1 \neq \beta_2$ and $\alpha_1 \approx \alpha_2$) Fit the left and right asymptotes, and from these one can obtain estimates of β_1, α_1, and p. Locate point I, the intersection of L_1 with the fitted IWPP plot, and obtain its coordinates. Using this in (8.26) yields estimates of β_2 and α_2.

Case 2: Well-Separated Case ($\beta_1 \neq \beta_2$ and $\alpha_1 \gg \alpha_2$ or $\alpha_1 \ll \alpha_2$) When $\alpha_1 \ll \alpha_2$, fit the left asymptote, and this yields estimates of β_1 and α_1. Locate the intersection of L_1 with the fitted IWPP plot and obtain the coordinates of the point. Using this in (8.26) yields estimates of β_2 and α_2. Locate the left (or first) inflection of the plot and obtain y coordinate of the point, say y_1, then an estimate of p is given by $\exp[-\exp(y_1)]$.

When $\alpha_1 \gg \alpha_2$, locate the right (or third) inflection of the plot and obtain y coordinate of the point, say y_2, then an estimate of $(1 - p)$ is given by $\exp[-\exp(-y_2)]$. Fit the right asymptote, and this yields estimates of β_1 and α_1 using the estimate of p. Locate the intersection of $L_1[= L_R - \ln(\hat{p})]$ with the fitted IWPP plot and obtain the coordinates of the point. Using this in (8.26) yields estimates of β_2 and α_2.

Case 3: Common Shape Parameter ($\beta_1 = \beta_2$) If a smooth fit to the plot of a given data set has a shape indicated in Figure 8.4, then the data set can be modeled by a mixture model with the same shape parameter. The model parameters can be estimated from the plot using the following procedure:

From the left and right asymptotes, estimates of β_1, α_1, and $p + (1 - p)k$ [where $k = (\alpha_2/\alpha_1)^{\beta}$] can be obtained. Denote $p + (1 - p)k$ as c_1. Locate the intersection of the IWPP plot and x axis, and this yields an estimate of x_0, the x coordinate of this point. Define $c_2 = \exp[-(\alpha_1/e^{x_0})^{\beta}]$. Then

$$y(x_0) = -\ln\{-\ln[pc_2 + (1 - p)c_2^k]\} = 0$$

Then p and k can be obtained by solving the following equations:

$$\begin{cases} p + (1 - p)k = c_1 \\ pc_2 + (1 - p)c_2^k = e^{-1} \end{cases} \tag{8.32}$$

Estimate of α_2 is then obtained from the estimate of k.

Discussion If $\alpha_1 \ll \alpha_2$, it may be hard to fit the right asymptote L_R and, hence the parameter p cannot be estimated by the above method. In this case, one can locate the inflection of the plot and obtain y coordinate of the point, say y_3, then estimate of p is given by $\exp[-\exp(-y_3)]$.

8.4 MODEL III(a)-3: HYBRID WEIBULL MIXTURE MODELS

There are a number of hybrid mixture models involving Weibull and non-Weibull distributions for the subpopulations. Majeske and Herrin (1995) consider a twofold mixture model involving a Weibull distribution and a uniform distribution for predicting automobile warranty claims. Chang (1998) deals with an analysis of the changing pattern of mortality and models it by a threefold mixture model involving Weibull, inverse Weibull, and Gompertz distributions to model three different age ranges. Landes (1993) deals with a mixture of normal and Weibull distributions. Goda and Hamada (1995) consider time-dependent Weibull distribution. Yan et al. (1995) and English et al. (1995) deal with truncated and mixed Weibull distributions. Hirose (1997) deals with a mixture Weibull power law model.

8.5 NOTES

Yehia (1993) and McNolty et al. (1980) deal with some characterization results. The former deals with a twofold Weibull mixture model with a common shape parameter and the characterization is given in terms of the residual moment and the hazard function. The latter deals mainly with the exponential mixture models and briefly discusses the mixed Weibull case.

EXERCISES

Data Set 8.1 Complete Data: All 50 Items Put into Use at $t = 0$ and Failure Times Are in Weeks

0.013	0.065	0.111	0.111	0.163	0.309	0.426	0.535	0.684	0.747
0.997	1.284	1.304	1.647	1.829	2.336	2.838	3.269	3.977	3.981
4.520	4.789	4.849	5.202	5.291	5.349	5.911	6.018	6.427	6.456
6.572	7.023	7.087	7.291	7.787	8.596	9.388	10.261	10.713	11.658
13.006	13.388	13.842	17.152	17.283	19.418	23.471	24.777	32.795	48.105

Data Set 8.2 Complete Data: Failure Times of 50 Components[a]

0.036	0.058	0.061	0.074	0.078	0.086	0.102	0.103	0.114	0.116
0.148	0.183	0.192	0.254	0.262	0.379	0.381	0.538	0.570	0.574
0.590	0.618	0.645	0.961	1.228	1.600	2.006	2.054	2.804	3.058
3.076	3.147	3.625	3.704	3.931	4.073	4.393	4.534	4.893	6.274
6.816	7.896	7.904	8.022	9.337	10.94	11.02	13.88	14.73	15.08

[a]Unit: 1000 h.

Data Set 8.3 The data is the censored data from Data Set 8.1 with the data collection stopped after 15 weeks.

Data Set 8.4 The data is the censored data from Data Set 8.2 with the data collection stopped after 1000 hours.

8.1. Carry out WPP and IWPP plots of Data Set 8.1. Can the data be modeled by a Weibull mixture or an inverse Weibull mixture model?

8.2. Suppose that Data Set 8.1 can be modeled by a Weibull mixture model. Estimate the model parameters based on (i) the WPP plot, (ii) the method of moments, and (iii) the method of maximum likelihood. Compare the estimates.

8.3. Consider the hypothesis that Data Set 8.1 can be adequately modeled by a Weibull mixture model with the following parameter values: $\alpha_1 = 4.0$, $\beta_1 = 0.8$, $\alpha_2 = 8.0$, $\beta_2 = 1.6$, and $p = 0.6$. Plot the P–P and Q–Q plots and discuss whether the hypothesis should be accepted or not?

8.4. How would you test the hypothesis of Exercise 8.3 based on A–D and, K–S tests for goodness of fit?

8.5. Repeat Exercise 8.1 with Data Set 8.2.

8.6. Suppose that Data Set 8.2 can be modeled by an inverse Weibull mixture model. Estimate the model parameters based on (i) the IWPP plot, (ii) the method of moments, and (iii) the method of maximum likelihood. Compare the estimates.

8.7. Repeat Exercises 8.1 to 8.4 with Data Set 8.3.

8.8. Repeat Exercises 8.5 to 8.6 with Data Set 8.4.

8.9. How would you generate a set of simulated data from a twofold Weibull mixture model? Can the method be used for a general n-fold Weibull mixture model?

8.10. Consider a twofold mixture model involving a two-parameter Weibull distribution and an inverse Weibull distribution. Study the WPP and IWPP plots for the model.*

8.11. For the model in Exercise 8.10 derive expressions for the various moments.

8.12. For the model in Exercise 8.10 derive the maximum-likelihood estimator for (i) complete data and (ii) censored data.

8.13. Consider a twofold mixture model involving a two-parameter Weibull distribution and an exponentiated Weibull distribution. Study the WPP plots for the model.*

* Research problem.

CHAPTER 9

Type III(b) Weibull Models

9.1 INTRODUCTION

A Type III(b) model involves n distributions and is derived as follows. Let T_i denote an independent random variable with a distribution function $F_i(t)$, $1 \le i \le n$, and let $Z = \min\{T_1, T_2, \ldots, T_n\}$. Then the distribution function for Z is given by

$$G(t) = 1 - \prod_{i=1}^{n} [1 - F_i(t)] \tag{9.1}$$

This model is commonly referred to as the *competing risk model* and is discussed in this chapter. It has also been called the *compound model*, *series system model*, and *multirisk model*. The risk model has a long history and Nelson (1982, pp. 162–163) traces its origin to 200 years ago.

In the reliability context, the model characterizes an n-component series system with T_i and Z representing the lifetime of component i, and of the system, respectively. The system fails at the first instance a component fails.

Note that (9.1) can be rewritten as

$$\bar{G}(t) = \prod_{i=1}^{n} \bar{F}_i(t) \tag{9.2}$$

where $\bar{G}(t) = 1 - G(t)$ and $\bar{F}_i(t) = 1 - F_i(t)$ for $1 \le i \le n$.

The competing risk model given by (9.1) is characterized by n (the number of subpopulations) and the form of the distribution function for each of the subpopulations. Some special cases of the model are as follows:

Model III(b)-1 Here the n subpopulations are either the standard two-parameter Weibull distributions given by (8.2) or the three-parameter Weibull distributions given by (8.3).

Weibull Models, by D.N.P. Murthy, Min Xie, and Renyan Jiang.
ISBN 0-471-36092-9

Model III(b)-2 Here all the n subpopulations are the inverse Weibull distributions given by (8.4).

Model III(b)-3 Here only some of subpopulations are either the two- or three-parameter Weibull distributions and the remaining are non-Weibull distributions. A more generalized competing risk model is given by

Model III(b)-4

$$G(x) = 1 - \prod_{i=1}^{n} [1 - p_i F_i(x)] \tag{9.3}$$

with $0 < p_i \leq 1$, $1 \leq i \leq n$. Note that if all the p_i's are one, then (9.3) gets reduced to (9.1). We can have many different models depending on the forms for the subpopulations.

The outline of the chapter is as follows. Section 9.2 deals with Model III(b)-1. The results for the general case are limited. In contrast, the special case $n = 2$ has been studied more thoroughly. Section 9.3 deals with Model III(b)-2 and examines the special case of $n = 2$. Section 9.4 deals with Model III(b)-3 with $n = 2$ and one subpopulation being a Weibull distribution and the other an inverse Weibull distribution. Section 9.5 looks at the generalized competing risk model.

9.2 MODEL III(b)-1: WEIBULL COMPETING RISK MODEL

9.2.1 Model Structure

Here $F_i(t)$, $i = 1, 2, \ldots, n$, are either two- or three-parameter Weibull distributions. When two of the subpopulations are two-parameter Weibull distributions with $\beta_i = \beta_j = \beta$, then $F_i(t)$, $i \neq j$, can be merged into one subpopulation with shape parameter β and scale parameter $(\alpha_i^{-\beta} + \alpha_j^{-\beta})^{-1/\beta}$. As a result, the model can be reduced to a $(n - 1)$-fold Weibull competing risk model. Hence, we assume, without loss of generality, that $\beta_i < \beta_j$ for $i < j$.

When two of the subpopulations are three-parameter Weibull distributions with $\beta_i = \beta_j = \beta$ and $\tau_i = \tau_j = \tau$, then they can be merged into one subpopulation with a common location parameter τ, shape parameter β, and scale parameter $(\alpha_i^{-\beta} + \alpha_j^{-\beta})^{-1/\beta}$. As a result, n can be reduced to $n - 1$. Hence, we can assume, without loss of generality, that $\tau_i < \tau_j$ for $i < j$, and $\beta_i < \beta_j$ for $i < j$ if $\tau_i = \tau_j$. Note that we allow $\beta_i = \beta_j$ as long as $\tau_i \neq \tau_j$.

The density function is given by

$$g(t) = \sum_{i=1}^{n} \left\{ \prod_{\substack{j=1 \\ j \neq i}}^{n} [1 - F_j(t)] \right\} f_i(t) \tag{9.4}$$

and can be rewritten as

$$g(t) = \bar{G}(t)\left\{\sum_{i=1}^{n}\left[\frac{f_i(t)}{\bar{F}_i(t)}\right]\right\} \tag{9.5}$$

From (9.5) it follows that the hazard function is given by

$$h(t) = \sum_{i=1}^{n} h_i(t) \tag{9.6}$$

where $h_i(t)$ is the hazard function associated with the ith subpopulation.

Shooman (1968) and Lawless (1982, pp. 254–255) discuss a model with hazard function given by

$$h(t) = \sum_{i=0}^{n-1} k_i t^i \tag{9.7}$$

When $k_i > 0$, the model is an n-fold Weibull competing risk model with the shape parameters being the set of positive integers 1 through n. Note that when one or more $k_i < 0$, the model is no longer a Weibull competing risk model.

Special Case $(n = 2)$
In this case, the model is given by

$$G(t) = F_1(t) + F_2(t) - F_1(t)F_2(t) \qquad t \geq 0 \tag{9.8}$$

This model has received considerable attention in the literature and has been studied extensively.

9.2.2 Model Analysis

Density Function
The density function is given by (9.4). When all the subpopulations are the two-parameter Weibull distributions, Jiang and Murthy (1999b) show that there are $2n$ different shapes for the density function. These are as follows:

- Type $2k - 1$ (decreasing followed by $k - 1$ modal), $0 < k \leq (n - 1)$
- Type $2k$, (k modal), $0 < k \leq n$

Special Case $(n = 2)$
In this case, the different possible shapes are as follows:

Type 1: Decreasing
Type 2: Unimodal

Type 3: Decreasing followed by unimodal
Type 4: Bimodal

Jiang and Murthy (1997c) give a complete parametric characterization of the density function in the three-dimensional parameter space.

Hazard Function
The hazard function is given by (9.6). Jiang and Murthy (1999b) show that the hazard rate has the following asymptotic properties:

$$h(t) \approx h_1(t) \quad \text{as } t \to 0 \qquad h(t) \approx h_n(t) \quad \text{as } t \to \infty \tag{9.9}$$

This implies that for small t, the hazard rate is the same as the hazard function for the subpopulation with the smallest shape parameter and for large t, it is the same as the hazard function for subpopulation with largest shape parameter.

Jiang and Murthy (1999b) also show that the hazard rate can have only three possible shapes:

Type 1: Decreasing when $\beta_1 < \beta_2 < \cdots < \beta_n < 1$
Type 3: Increasing when $1 < \beta_1 < \beta_2 < \cdots < \beta_n$
Type 4: Bathtub-shaped when $\beta_1 < \cdots < \beta_{j-1} < 1 < \beta_j \cdots < \beta_n$

Special Case ($n = 2$)
Jiang and Murthy (1997c) give a complete parametric characterization of the hazard function in the three-dimensional parameter space.

WPP Plot
Under the Weibull transformation given by (1.7), (9.1) with $F_i(t)$ given by (8.2) gets transformed into

$$y = \ln\left(\sum_{i=1}^{n} z_i e^x\right) \tag{9.10}$$

where

$$z_i(e^x) = (t/\alpha_i)^{\beta_i} = (e^x/\alpha_i)^{\beta_i} = \exp[\beta_i(x - \ln(\alpha_i))] \tag{9.11}$$

Jiang and Murthy (1999b) show that the WPP plot is convex. The left asymptote, L_1, is a straight line given by

$$y = \beta_1(x - \ln(\alpha_1)) \tag{9.12}$$

and the right asymptote, L_n, is another straight line given by

$$y = \beta_n(x - \ln(\alpha_n)) \tag{9.13}$$

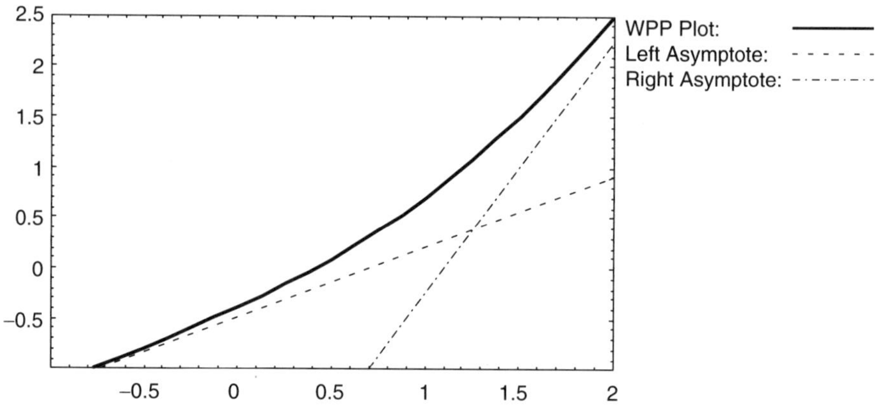

Figure 9.1 WPP plot for the Weibull competing risk model ($\alpha_1 = 2.0, \beta_1 = 0.7, \alpha_2 = 3.0, \beta_2 = 2.5$).

Note that these are the WPP plots for the subpopulations with the smallest and the largest shape parameters.

Special Case ($n = 2$)
The left asymptote of the WPP plot is given by (9.12). The right asymptote is a straight line L_2 given by (9.13) with $n = 2$. Let (x_I, y_I) denote the coordinates of the intersection of L_1 and L_2. Then we have

$$y(x)|_{x=x_I} = y_I + \ln(2) \qquad \left.\frac{dy(x)}{dx}\right|_{x=x_I} = \frac{\beta_1 + \beta_2}{2} \tag{9.14}$$

Figure 9.1 shows the WPP plot along with the two asymptotes.

9.2.3 Parameter Estimation

Graphical Method
For the general case, Jiang and Murthy (1999b) propose an iterative approach to estimating the parameters of the model, which is applicable when the subpopulations are well separated and the data set is large. It involves separating the subpopulations and estimating the parameters of one subpopulation during each stage of the iteration.

The basic idea is that the parameters of subpopulation with the smallest (largest) shape parameter are determined by fitting the left (or right) asymptote to the WPP plot of the data set. Then the data from this subpopulation is eliminated from the population, and the remaining data is viewed as coming from a new model with one less number of subpopulations.

We indicate the procedure for the first stage of the iteration process for the case of complete data.

Step 1: Let m_n denote the number of data points in the set and $R_N(t) = 1 - F_N(t)$ with $F_N(t)$ being the empirical distribution function obtained using the data. Carry out the WPP plot for the data as indicated in part A of Section 4.5.2.

Step 2: If the WPP plot is convex, then fit a straight line to the left (right) asymptote and estimate the parameters of the corresponding subpopulation.

Step 3: The data not deleted from the set [and modeled by a $(n - 1)$-fold model] is based on the following transformation:

$$R_{N-1}(t) = R_N(t)/\hat{R}_1(t) \tag{9.15}$$

or

$$R_{N-1}(t) = R_N(t)/\hat{R}_n(t) \tag{9.16}$$

depending on whether the parameters estimated are for subpopulation 1 or n with $R_i(t) = 1 - F_i(t)$, $1 \leq i \leq n$. Then $\hat{R}_1(t)$ or $\hat{R}_n(t)$ are computed using the estimated parameter values.

Let m_{n-1} denote the number of data not deleted from the set, and these data must satisfy the following conditions:

- $R_{N-1}(t_i) \in \left(\frac{1}{m_{n-1}+1}, \frac{m_{n-1}}{m_{n-1}+1}\right), i = 1, 2, \ldots, m_{n-1}$, and
- $R_{N-1}(t_i)$ is roughly monotonic

The procedure is repeated with the number of subpopulations in the model reduced by one at each stage of the iteration and is stopped when the final WPP plot is approximately a straight line. This allows one to obtain (a) a rough estimate $\hat{n}$ (of n) and (b) rough estimates of the model parameters. Note that the estimate $\hat{n}$ is informative, though it may not be accurate. As such, one might look at models with $n = (\hat{n} - 1, \hat{n}, \hat{n} + 1)$ to determine their suitability for modeling the given data set.

Special Case (Common Shape Parameter)
Nelson (1982, Chapter 5) considers the case where shape parameters are the same for all subpopulations. He discusses parameter estimation based on a Weibull hazard plot. Since the shape parameters are the same, this model can be reduced to a single Weibull distribution.

Special Case $(n - 2)$
Jiang and Murthy (1995b) discuss the estimation for this case. It involves the following steps:

Steps 1–5: As in Section 4.5.2 to obtain the WPP plot of the data.

Step 6: Fit the left asymptote to the WPP plot. The slope and intercept of this line yields estimates of β_1 and α_1.

Step 7: Fit the right asymptote to the WPP plot. The slope and intercept of this line yields the estimates of β_2 and α_2.

Step 8: Obtain the coordinates (x_I, y_I) of the intersection of the two asymptotes. Check to see if the first equation in (9.14) is roughly satisfied. If not, modify the fits to one or both of the asymptotes and repeat steps 6–8.

Method of Maximum Likelihood

Several authors have discussed the use of the maximum-likelihood estimation approach. Herman and Patell (1971) study the maximum-likelihood estimation for a model with a common shape parameter β ($= 1$). McCool (1976) discusses the maximum-likelihood estimation for a model with a common shape parameter β, and subpopulations are well separated. Friedman and Gertsbakh (1980) study the existence and properties of the maximum-likelihood estimators (MLEs) for a twofold competing risk model with exponential and Weibull subpopulations.

The competing risk model can be viewed as a model of an unreliable serial system. An interesting problem is to estimate component reliability and the parameter of the component lifetime distribution when only system-level failure is observed. This is called the "masked-data" problem. Usher et al. (1991) and Usher (1996) derive the likelihood function for the masked-data case and present an iterative procedure for obtaining maximum-likelihood estimates and confidence intervals for the component life distribution parameters. This topic has received considerable attention, and several Bayesian estimation procedures have been proposed and will be discussed later in the section.

Doganaksoy (1991) deals with confidence interval estimation where the data is singly time-censored and partially masked. He considers a three-component series system with exponentially distributed component failure times. The maximum-likelihood method is used to estimate the model parameters. The approximate confidence intervals are developed, illustrated, and assessed. Ishioka and Nonaka (1991) present a stable technique for obtaining the maximum-likelihood estimates of the parameters for a twofold model with $\beta_1 \neq \beta_2$ and illustrate their method through simulation studies. Cacciari et al. (1993) deal with a model with $n = 2$ and $\beta_1 \neq \beta_2$ and estimate the parameters using the maximum-likelihood approach.

Bayesian Method

Sinha and Kale (1980) deal with the case where the subpopulations are exponentially distributed. They discuss bounds for the parameter estimates using classical and Bayesian approaches and the data separation.

Badarinathi and Tiwari (1992) study the competing risk model with exponential distributions. They present a hierarchical Bayesian approach for parameter estimation and evaluate their approach through simulation studies. Basu et al. (1999) carry out a Bayesian analysis for a general n-component Weibull competing risk model. The n Weibulls are allowed to have different scale and shape parameters, and three different prior models are proposed. Kuo and Tae (2000) consider the Bayesian analysis of masked system failure data with Weibull component lifetimes.

General Curve Fitting Method

Here one matches the empirical distribution, density function, or hazard function with the corresponding curves from the competing risk Weibull family to determine the appropriateness of the models to model a given data set. The parameters are estimated by minimizing the sum of squared deviation or maximum absolute deviation.

Ling and Pan (1998) consider a twofold model involving three-parameter Weibull distributions. The model parameters are obtained by minimizing the maximum absolute difference between the observed probability of failure and the expected probability of failure. The Minimax algorithm in Matlab Toolbox *Optimization* is used to obtain the estimates.

Hybrid Method

Kanie and Nonaka (1985) consider a model with $n = 2$ and $\beta_1 \neq \beta_2$. They present an analytical technique for estimating the two shape parameters based on cumulative hazard functions. They fit two straight lines with slopes β_1 and β_2 to the WPP plot. The intercepts of the fitted straight lines yield estimates of the two scale parameters. Spivey and Gross (1991) consider a competing risk model with one risk following a Weibull distribution and the other a Rayleigh distribution.

Kundu and Basu (2000) extend an earlier study on exponential distribution to the case when the failure distributions are Weibull. Davison and Louzada-Neto (2000) study the problem of statistical inference for the poly-Weibull model. See also Berger and Sun (1993).

9.2.4 Modeling Data Set

If a smooth fit to the WPP plot of the data has a shape similar to that shown in Figure 9.1 (roughly convex), then one can tentatively assume that an n-fold Weibull competing risk model can be used to model the data set. The model parameters can be estimated from the WPP plot as indicated in Section 9.2.3.

9.2.5 Applications

Twofold competing risk models have been extensively used in the modeling of product reliability. Table 9.1 gives a list of the applications in reliability.

Table 9.1 Applications of Weibull Competing Risk Model

Applications	Reference
Radio transmitters	Herman and Patell (1971)
Bearing fatigue	McCool (1978)
Electron tubes	Kanie and Nonaka (1985)
Hydromechanical systems	Clark (1991)
Three-component series system	Doganaksoy (1991)
Insulation systems	Cacciari et al. (1993)
Naval main propulsion diesel engines	Jiang and Murthy (1995b)
System/component reliability prediction	Usher et al. (1991), Usher (1996)

9.3 MODEL III(b)-2: INVERSE WEIBULL COMPETING RISK MODEL

9.3.1 Model Structure

Only $n = 2$ has been studied by Jiang et al. (2001a) and we confine our discussion to this special case.

Special Case $(n = 2)$
The distribution function is given by (9.8) with $F_1(t)$ and $F_2(t)$ given by (8.4). Unlike the twofold Weibull competing risk model, we can allow $F_1(t) \equiv F_2(t)$. Therefore, we can assume without loss of generality that $\beta_1 \leq \beta_2$ and $\alpha_1 \leq \alpha_2$ when $\beta_1 = \beta_2$.

9.3.2 Model Analysis

Density Function
The possible shapes for the density function are as follows:

- Type 2: Unimodal
- Type 4: Bimodal

Hazard Function
The two possible shapes for hazard function are as follows:

- Type 5: Unimodal
- Type 9: Bimodal

IWPP Plot
Under the inverse Weibull transformation given by (6.57), the IWPP plot for the model is a smooth curve.

Jiang et al. (2001a) show that the IWPP plot is convex. The left asymptote, L_1, is a straight line given by (9.12) and the right asymptote, L_R, is given by the following straight line:

$$y = (\beta_1 + \beta_2)\left(x - \frac{\beta_1 \ln \alpha_1 + \beta_2 \ln \alpha_2}{\beta_1 + \beta_2}\right) \tag{9.17}$$

The two asymptotes intersect and the coordinates of this point are $(\ln(\alpha_2), \beta_1 \ln(\alpha_2/\alpha_1))$. A typical IWPP plot is shown in Figure 9.2. Note that the shape is similar to Figure 9.1.

9.3.3 Parameter Estimation

Graphical Approach
Once the IWPP plot of data set has been carried out, the model parameters are estimated as follows:

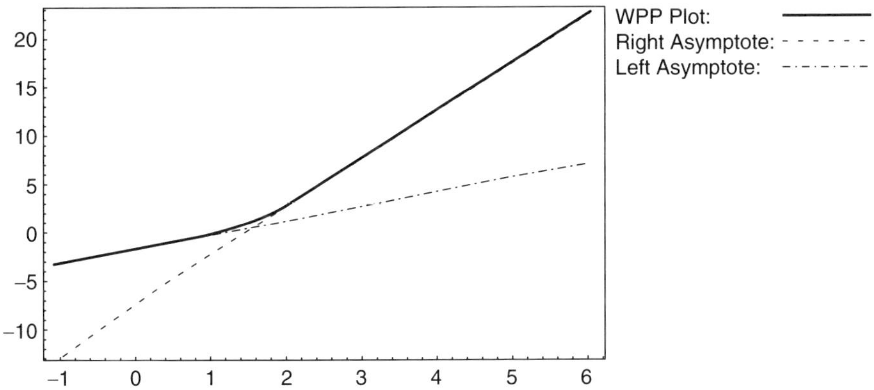

Figure 9.2 WPP plot for the inverse Weibull competing risk model ($\alpha_1 = 3.0, \beta_1 = 1.5, \alpha_2 = 5.0, \beta_2 = 3.5$).

Step 1: Fit the left asymptote and its slope, and intercept yields estimates of β_1 and α_1.

Step 2: Fit the right asymptote, and using its slope and intercept in (9.17) yields the estimates of β_2 and α_2.

Step 3: Determine the coordinates of the intersection point of the two asymptotes. Check to see if they are approximately close to $(\ln(\alpha_2), \beta_1 \ln(\alpha_2/\alpha_1))$, which are computed using the estimated values. If not, the asymptotes need to be adjusted until this is achieved.

9.4 MODEL III(b)-3: HYBRID WEIBULL COMPETING RISK MODEL

9.4.1 Model Structure

Gera (1995) proposes the following model:

$$G(t) = [\exp(-st^c)]\left[1 - \exp(-\alpha/t)^\beta\right] \qquad t \geq 0 \tag{9.18}$$

with $\alpha, \beta, s, c > 0$. This can be viewed as a competing risk model involving a two-parameter Weibull and a two-parameter inverse Weibull distribution.

The motivation for this model was to use it as a tool to evaluate the mean for the inverse Weibull distribution when the shape parameter $\beta \leq 1$. The parameter s is called the convergence factor and is chosen sufficiently small so that $\exp(-st_{(r)}) > 0.99$ where $t_{(r)}$ is the largest value in the ordered data set. The parameter c is assumed 1. As a result, the Weibull subpopulation can be viewed as a weighting function, and hence the estimated mean is the weighted mean.

9.5 MODEL III(b)-4: GENERALIZED COMPETING RISK MODEL

The generalized competing risk model given by (9.3) was first proposed in Hoel (1972) in the context of modeling cohort mortality by a probabilistic combination of competing risks.

Chan and Meeker (1999) consider a similar model, and it can be viewed as a special case of (9.3) with $n = 2$ in the context of the reliability of a component. The two subpopulations correspond to two different modes of failures (which are statistically independent of each other). As a result, their model is given by

$$G(t) = 1 - [1 - pF_1(t)][1 - F_2(t)] \tag{9.19}$$

They call the model a general limited failure population model as one mode of failure is common to all components and the other only to a fraction p of the population.

From (9.19) we have

$$\bar{G}(t) = p\bar{F}_1(t)\bar{F}_2(t) + q\bar{F}_2(t) \tag{9.20}$$

where $q = 1 - p \in (0, 1)$. This can be viewed as a mixture model. A possible physical interpretation for the model is as follows. The population can be divided into two subpopulations. The first (comprising of a fraction p of the total population) is subject to the influence of two competing risks and second (comprising of the remaining population) is only subject to the influence of one of the two risks. Consider the case where both $F_1(t)$ and $F_2(t)$ are two-parameter Weibull distributions. Three special cases of (9.20) are as follows:

1. When $p = 0$, the model reduces to a simple Weibull distribution.
2. When $p = 1$, the model reduces to a twofold Weibull competing risk model.
3. When the two shape parameters are identical, that is, $\beta_1 = \beta_2 = \beta$, we have

$$\bar{F}_1(t)\bar{F}_2(t) = \exp[-(t/\alpha_0)^{\beta}] \qquad \alpha_0 = \left(\alpha_1^{-\beta} + \alpha_2^{-\beta}\right)^{-1/\beta} \tag{9.21}$$

As a result, the model reduces to a twofold Weibull mixture model with a common shape parameter.

9.5.1 Model Analysis

The following results are for the model given by (9.20) and are from Jiang and Ji (2001).

Density Function

The shape of the density function depends on the four parameters: β_1, β_2, k $[= (\alpha_2/\alpha_1)^{\beta}]$, and p. For a set of given parameter combinations, the different possible shapes (for both $\beta_1 < \beta_2$ and $\beta_1 > \beta_2$) are as follows:

- Type 1: Decreasing
- Type 2: Unimodal

- Type 3: Decreasing followed by unimodal
- Type 4: Bimodal

Hazard Function
The different possible shapes are as follows:

- Type 1: Decreasing
- Type 3: Increasing
- Type 4: Bathtub shaped
- Type 5: Unimodal
- Type 6: Unimodal followed by increasing
- Type 7: Decreasing followed by unimodal
- Type 8: W shaped

WPP Plot
The left asymptote of the WPP plot is a straight line L_a given by

$$y = \beta_1[x - \ln(\alpha_1)] + \ln(p) \tag{9.22}$$

when $\beta_1 < \beta_2$ and by L_2 (which is the WPP plot of subpopulation 2) when $\beta_1 > \beta_2$. The right asymptote of the WPP plot is always L_2. Note that when $\beta_1 > \beta_2$, the two asymptotes are overlapping.

Generally speaking, the WPP plot has three different shapes depending on the parameter combination. They are as follows:

- Convex: As in Figure 9.1.
- S shaped: A typical plot is shown in Figure 9.3.
- Bell shaped: A typical plot is shown in Figure 9.4.

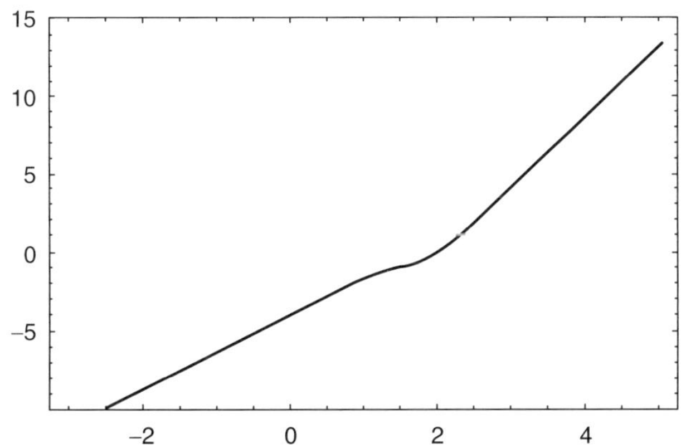

Figure 9.3 WPP plot for the generalized Weibull competing risk model ($\alpha_1 = 3.0, \beta_1 = 2.5, \alpha_2 = 8.0, \beta_2 = 4.5, p = 0.3$).

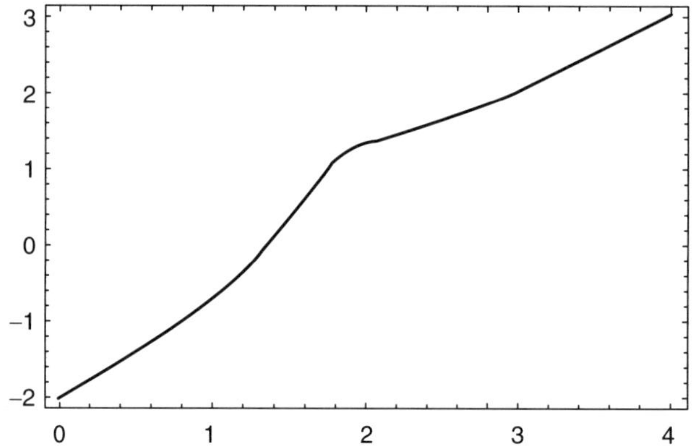

Figure 9.4 WPP plot for the generalized Weibull competing risk model ($\alpha_1 = 5.0, \beta_1 = 5.0, \alpha_2 = 5.0, \beta_2 = 1.5, p = 0.9$).

9.5.2 Parameter Estimation

Graphical Parameter Estimation

The WPP plot of the data indicates whether $\beta_1 < \beta_2$ or $\beta_1 > \beta_2$. As a result, the parameter estimation based on the WPP plot is as follows:

Case 1: $\beta_1 < \beta_2$

Step 1: Plot the WPP plot of the data. The left asymptote of the WPP plot yields the estimates of β_1 and $\alpha_1^{\beta_1}/p$, and the right asymptote yields the estimates of β_2 and α_2.

Step 2: Estimate the parameter p (or q) by one of the following three approaches:

Approach 1 If the failure causes are known for the failed units and the data set is complete data, then p can be estimated by n_1/n, where n_1 and n are the numbers of the data corresponding to the first failure mode and all failure modes, respectively.

Approach 2 If data is masked, then q can be estimated by $\min[\hat{R}(t_i), \hat{R}_2(t)]$.

Approach 3 If data is masked and the WPP plot of the data is S shaped, then the y coordinate of the inflection point, y_I, can be used to obtain an estimate of q since

$$\ln[-\ln(q)] \approx y_I \tag{9.23}$$

Step 3: From estimates of β_1, $\alpha_1^{\beta_1}/p$ (obtained in step 1) and q (obtained in step 2), one can obtain estimates of p and α_1.

Case 2: $\beta_1 > \beta_2$

Step 1: Plot the WPP plot of the data. The right asymptote yields the estimates of β_2 and α_2.

Step 2: Estimate the parameter p (or q) by one of the three approaches discussed above.

Step 3: Transform the data set by the following transformation:

$$R_1(t) \approx \frac{\hat{R}(t) - q\hat{R}_2(t)}{\hat{p}\hat{R}_2(t)} \tag{9.24}$$

Delete some of the transformed data so that the remaining data set satisfies

- $R_1(t_i) \in (0, 1)$, and
- $R_1(t_i)$ vs. t_i is roughly monotonically decreasing.

Plot the WPP plot using the remaining data and fit a straight line. The slope and intercept yield estimates of β_1 and α_1.

Method of Maximum Likelihood

Hoel (1972) discusses the estimation of the parameters for the model given by (9.3). Chan and Meeker (1999) discuss the maximum-likelihood method to estimate the parameters for the model given by (9.19) and use it to construct statistical confidence intervals.

EXERCISES

Data Set 9.1 Complete Data: All 50 Items Put into Use at $t = 0$ and Failure Times Are in Years

0.008	0.017	0.058	0.061	0.084	0.090	0.134	0.238	0.245	0.353
0.374	0.480	0.495	0.535	0.564	0.681	0.686	0.688	0.921	0.959
1.022	1.092	1.260	1.284	1.295	1.373	1.395	1.414	1.760	1.858
1.892	1.921	1.926	1.933	2.135	2.169	2.301	2.320	2.405	2.506
2.598	2.808	2.971	3.087	3.492	3.669	3.926	4.446	5.119	8.596

Data Set 9.2 Complete Data: Failure Times of 50 Components[a]

0.061	0.073	0.075	0.084	0.086	0.087	0.088	0.089	0.089	0.089
0.099	0.102	0.117	0.118	0.119	0.120	0.123	0.135	0.143	0.168
0.183	0.185	0.191	0.192	0.199	0.203	0.213	0.215	0.257	0.258
0.275	0.297	0.297	0.298	0.299	0.308	0.314	0.315	0.330	0.374
0.388	0.403	0.497	0.714	0.790	0.815	0.817	0.859	0.909	1.286

[a]Unit: 1000 h.

Data Set 9.3 The data is the censored data from Data Set 9.1 with the data collection stopped after 3 years.

Data Set 9.4 The data is the censored data from Data Set 9.2 with the data collection stopped after 400 h.

9.1. Carry out WPP and IWPP plots of Data Set 9.1. Can the data be modeled by one of Type III(a) or Type III(b) models involving either two Weibull or two inverse Weibull distributions?

9.2. Suppose that Data Set 9.1 can be modeled by a twofold Weibull competing risk model. Estimate the model parameters based on (i) the WPP plot, (ii) the method of moments, and (iii) the method of maximum likelihood. Compare the estimates.

9.3. Suppose Data Set 9.1 can be adequately modeled by the Weibull competing risk model with the following parameter values: $\alpha_1 = 2.0$, $\beta_1 = 0.6$, $\alpha_2 = 10.0$, and $\beta_2 = 1.8$. Plot the P–P and Q–Q plots and discuss whether the hypothesis should be accepted or not?

9.4. How would you test the hypothesis of Exercise 9.3 based on A–D and K–S tests for goodness of fit?

9.5. Repeat Exercise 9.1 with Data Set 9.2.

9.6. Suppose that Data Set 9.2 can be modeled by an inverse Weibull competing risk model. Estimate the model parameters based on (i) the IWPP plot, (ii) the method of moments, and (iii) the method of maximum likelihood. Compare the estimates.

9.7. Repeat Exercises 9.1 to 9.4 with Data Set 9.3.

9.8. Repeat Exercises 9.5 and 9.6 with Data Set 9.4.

9.9. How would you generate a set of simulated data from a twofold Weibull competing risk model? Can the method be used for a general n-fold Weibull competing risk model?

9.10. Consider a twofold competing risk model involving a two-parameter Weibull distribution and an inverse Weibull distribution. Study the WPP and IWPP plots for the model.*

9.11. For the model in Exercise 9.10 derive the maximum-likelihood estimator for (i) complete data and (ii) censored data.

9.12. Consider a twofold competing risk model involving a two-parameter Weibull distribution and an exponentiated Weibull distribution. Study the WPP plots for the model.*

* Research problem.

CHAPTER 10

Type III(c) Weibull Models

10.1 INTRODUCTION

A Type III(c) model involves n distributions and is derived as follows. Let T_i denote an independent random variable with a distribution function $F_i(t)$, $1 \leq i \leq n$ and let $Z = \max\{T_1, T_2, \ldots, T_n\}$. Then the distribution function for Z is given by

$$G(t) = \prod_{i=1}^{n} F_i(t) \tag{10.1}$$

This model is commonly referred to as the multiplicative model for obvious reasons. In contrast to the competing risk model, this model has received very little attention. The model has received some attention in the reliability literature where it arises in the context of modeling a functionally parallel system with independent components [see, for example, Shooman (1968, p. 206) and Elandt-Johnson and Johnson (1980)]. Basu and Klein (1982) call this the *complementary risk model.*

The multiplicative risk model given by (10.1) is characterized by n (the number of subpopulations) and the form of the distribution function for each of the subpopulations. Some special cases of the model are as follows:

Model III(c)-1

Here the n subpopulations are either the two-parameter Weibull distributions given by (8.2) or the three-parameter Weibull distributions given by (8.3).

Model III(c)-2

Here all the subpopulations are inverse Weibull distributions given by (8.4).

The outline of the chapter is as follows. Section 10.2 deals with Model III(c)-1. The results for the general case are limited. In contrast, the special case $n = 2$ has

Weibull Models, by D.N.P. Murthy, Min Xie, and Renyan Jiang.
ISBN 0-471-36092-9

been studied thoroughly. Section 10.3 deals with Model III(b)-2 and examines the special case $n = 2$.

10.2 MODEL III(c)-1: MULTIPLICATIVE WEIBULL MODEL

10.2.1 Model Structure

For the general n-fold Weibull multiplicative model, when all the subpopulations are two-parameter Weibull distributions, we can assume, without loss of generality, that $\beta_1 \leq \beta_2 \leq \cdots \leq \beta_n$, and $\alpha_i \geq \alpha_j$ for $i < j$ if $\beta_i = \beta_j$. Finally, when all the $F_i(t)$'s are identical, the multiplicative model gets reduced to an exponentiated Weibull model given by (7.24) and discussed in Section 7.5 with $\nu = n$.

In this section we consider the general n with all the subpopulations being two-parameter Weibull distributions. The results presented are from Jiang et al. (2001b). Later, we look at the special case $n = 2$.

10.2.2 Model Analysis

Distribution Function

For small t, $G(t)$ can be approximated by

$$G(t) \approx F_0(t) \tag{10.2}$$

where $F_0(t)$ is a two-parameter Weibull distribution with parameters β_0 and α_0 given by

$$\beta_0 = \sum_{i=1}^{n} \beta_i \qquad \alpha_0 = \prod_{i=1}^{n} \alpha_i^{\beta_i/\beta_0} \tag{10.3}$$

For large t, $G(t)$ can be approximated by

$$G(t) \approx (1 - k) + kF_1(t) \tag{10.4}$$

where k is the number of subpopulations with distribution identical to $F_1(t)$.

These results are used in establishing the asymptotic properties of the density and hazard functions and the WPP plot.

Density Function

The density function is given by

$$g(t) = G(t) \sum_{i=1}^{n} \frac{f_i(t)}{F_i(t)} \tag{10.5}$$

The density function can be one of the following $(n + 1)$ shapes:

- Type 1: Decreasing
- Type $(2k)$: k Modal $(1 \leq k \leq n)$

Special Case $(n = 2)$
Jiang and Murthy (1997d) carry out a detailed parametric study in the three-dimensional parameter space.

Hazard Function
The hazard function is given by

$$h(t) = \frac{G(t)}{1 - G(t)} \sum_{i=1}^{n} \frac{f_i(t)}{F_i(t)} \tag{10.6}$$

The asymptotes of the hazard function are given by

$$h(t) \approx \begin{cases} h_0(t) = \frac{\beta_0}{\alpha_0}\left(\frac{t}{\alpha_0}\right)^{\beta_0 - 1} & \text{as } t \to 0 \\ h_1(t) = \frac{\beta_1}{\alpha_1}\left(\frac{t}{\alpha_1}\right)^{\beta_1 - 1} & \text{as } t \to \infty \end{cases} \tag{10.7}$$

where $h_0(t)$ and $h_1(t)$ are the hazard functions associated with $F_0(t)$ and $F_1(t)$, respectively.

The different possible shapes for the hazard function can be grouped into four groups:

- Type 1: Monotonically decreasing
- Type 3: Monotonically increasing
- Type $(4k + 2)$: k modal followed by increasing $(1 \le k \le n - 1)$
- Type $(4k + 3)$: Decreasing followed by k modal $(1 \le k \le n - 1)$

It is worth noting that failure rate can never be decreasing for small t if it is increasing for large t, hence ruling out the possibility for the failure rate to have a bathtub shape.

Special Case $(n = 2)$
Jiang and Murthy (1997d) carry out a detailed parametric study in the three-dimensional parameter space. The boundaries separating the regions with different shapes are complex.

WPP Plot
Under the Weibull transformation given by (1.7) and $F_i(t)$ given by (8.2), (10.1) gets transformed into

$$y = \ln\{-\ln[1 - G(e^x)]\} \tag{10.8}$$

The WPP plot is analytically intractable. Jiang et al. (2001b) propose an approximation (see Section 10.2.4 for more details), which makes the WPP analytically tractable. Based on this approximation, they show that the WPP plot is concave and verify this through extensive numerical studies.

Based on this approximation, we have the following asymptotic results for the WPP plot. As $x \to -\infty$ (or $t \to 0$), the left asymptote is a straight-line y_L given by

$$y_L = \beta_0(x - \ln(\alpha_0)) \tag{10.9}$$

where β_0 and α_0 given by (10.3). As $x \to \infty$ (or $t \to \infty$), the right asymptote is given by another straight line L_1, which is the WPP plot for subpopulation 1, and is given by

$$y = \beta_1(x - \ln(\alpha_1)) \tag{10.10}$$

Special Case $(n = 2)$
The WPP plot is discussed in detail in Jiang and Murthy (1995b). In this case

$$\beta_0 = \beta_1 + \beta_2 \quad \text{and} \quad \alpha_0 = (\alpha_1)^{\beta_1/\beta_0}(\alpha_2)^{\beta_2/\beta_0} \tag{10.11}$$

As a result, the left asymptote has a slope $(\beta_1 + \beta_2)$ and the right asymptote has a slope β_1. Figure 10.1 shows a typical WPP plot along with the two asymptotes.

10.2.3 Parameter Estimation

Graphical Method
If n is specified (see Section 10.2.4 regarding how to estimate n based on the data), then Jiang et al. (2001b) propose a method for estimating the model parameters based on the WPP plot. It involves estimating the parameters of one subpopulation based on the fit to the right (or left) asymptote. The data set is modified to remove

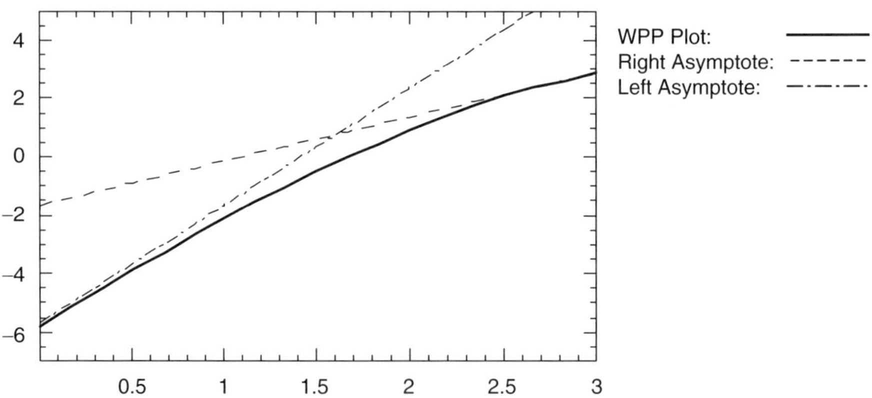

Figure 10.1 WPP plot for the Weibull multiplicative model ($\alpha_1 = 3.0, \beta_1 = 1.5, \alpha_2 = 5.0, \beta_2 = 2.5$).

the data from this subpopulation following a procedure similar to that used in Section 9.2.3. The WPP plot of the modified data is carried out and the process is repeated. The parameters of the last subpopulation are estimated by fitting a straight line to the WPP plot of the last modified data.

Note that this procedure only yields crude estimates. These are useful as they can be used as initial values for parameter estimation using more refined methods.

Special Case ($n = 2$)
Jiang and Murthy (1995b) discuss the estimation for this case. It involves the following steps:

Steps 1–5: As in Section 4.5.1 to obtain the WPP plot of the data.
Step 6: Fit the left asymptote to the WPP plot. The slope and intercept of this line yields estimates of β_0 and α_0.
Step 7: Fit the right asymptote to the WPP plot. The slope and intercept of this line yields the estimates of β_1 and α_1.
Step 8: Obtain estimates of β_2 and α_2 from (10.11).

Optimization Method
Ling and Pan (1998) consider a twofold multiplicative model involving three-parameter Weibull distributions. The model parameters are obtained by minimizing the maximum absolute difference between the observed probability of failure and the expected probability of failure. The Minimax algorithm in the Matlab Toolbox *Optimization* is used to obtain the estimates.

10.2.4 Modeling Data Set

Given a data set, one plots the data on the WPP plot. The plotting procedure depends on the type of data (complete, censored, grouped, etc.) as discussed in Chapter 5. If the WPP plot of the data is roughly concave, then an n-fold Weibull multiplicative model can be considered a potential model for the data set. Jiang et al. (2001b) suggest a two-part method, based on the WPP plot, as indicated below.

Part 1: Estimation of n
The basis for this is that for small t (and applicable for intermediate values) the Weibull multplicative model can be approximated by an exponentiated Weibull model with parameters α, β, and $\nu = n$. The WPP plot for the exponentiated Weibull model has the left asymptote given by

$$y_E = n\beta(x - \ln(\alpha)) \tag{10.12}$$

The WPP plot of the multplicative Weibull model has the left asymptote given by (10.9). If the selected value of n is appropriate, then the two straight lines should be close to each other.

Note that α_0 and β_0 in (10.9) can be estimated from the slope and intercept of the fit to the left asymptote of the WPP plot. For a given n, β and α in (10.12) can be obtained from the following linear regression equation using the data set:

$$\ln(-\ln\{1-[F(t)]^{1/n}\}) = \beta(x - \ln(\alpha)) \tag{10.13}$$

The following quantity measures the closeness between the lines y_L and y_E:

$$D(n) = \int_{x_1}^{x_2} [y_L(x) - y_E(x)]^2\, dx \tag{10.14}$$

The term $D(n)$ is evaluated for small values of x_1 and x_2. Jiang et al. (2001b) suggest $x_1 = \ln(t_{(1)}) - 0.5$ and $x_2 = \ln(t_{(1)}) + 0.5$, where $t_{(1)}$ is the smallest value in the order data set. Using (10.9) and (10.12) in (10.14) yields

$$D(n) = A^2/12 + (A\,\ln(t_{(1)}) + B)^2 \tag{10.15}$$

where

$$A = \beta_0 - n\beta \qquad B = \ln(\alpha^{n\beta}/\alpha_0^{\beta_0}) \tag{10.16}$$

The term $D(n)$ is small if the selected value of n is close to the true value of n. As such, we can use the following procedure to determine n:

Step 1: Fit a tangent line to the left end of the WPP plot of the data and obtain estimates of α_0 and β_0 using (10.9).

Step 2: For a specified value of n (= 2, 3, 4, ...) and using data points (t_i, $F(t_i)$) such that $F(t_i) \leq 0.8$ obtain the estimates ($\hat{\alpha}$, $\hat{\beta}$), which are functions of n, from the following regression:

$$y^{(i)} = \beta(x^{(i)} - \ln(\alpha)) \tag{10.17}$$

where $x^{(i)} = \ln(t_i), y^{(i)} = \ln(-\ln\{1-[\hat{F}(t_i)]^{1/n}\})$.

Step 3: Compute the $D(n)$ from (10.16) for $n = 2, 3, 4, \ldots$, using the estimates from steps 1 and 2. The value of n that yields the minimum for $D(n)$ is the estimate of n.

Part 2: Estimation of Subpopulation Parameters

This is done using the approach discussed in Section 10.2.3.

Ling and Pan (1998) propose another approach for model selection. They tentatively assume several models as candidate models and estimate the model parameters. Then they use a goodness-of-fit test to choose the most appropriate model from the set of models.

10.2.5 Applications

Jiang and Murthy (1995b) use a twofold Weibull multiplicative model to model several different data sets. These include failure times for transistors, shear strength of brass rivets, and pull strength of welds.

10.3 MODEL III(c)-2: INVERSE WEIBULL MULTIPLICATIVE MODEL

10.3.1 Model Structure

For the general n-fold inverse Weibull multiplicative model, if $\beta_i = \beta_j = \beta, i \neq j$, then the two subpopulations can be merged into one subpopulation with the common shape parameter β and with a new scale parameter $(\alpha_i^\beta + \alpha_j^\beta)^{1/\beta}$. Hence, we can assume, without loss of generality, that $\beta_i < \beta_j$ for $i < j$. Note that this differs from the n-fold Weibull multiplicative model where the parameters are not so constrained.

The general case is yet to be studied. The special case $n = 2$ has been studied by Jiang et al. (2001a), and we discuss this model in this section. The model is given by

$$F(t) = F_1(t)F_2(t) \qquad t \geq 0 \tag{10.18}$$

with the two subpopulations are the two-parameter inverse Weibull distribution given by (8.4).

10.3.2 Model Analysis

Distribution Function

For small t we have

$$F(t) \approx F_2(t) \tag{10.19}$$

and for large t we have

$$F(t) \approx F_1(t) \tag{10.20}$$

Density Function

The density function is given by

$$f(t) = f_1(t)F_2(t) + f_2(t)F_1(t) \tag{10.21}$$

Since the density functions for the two subpopulations are unimodal, one should expect the possible shapes of the density function to be (i) unimodal and (ii) bimodal. However, computer plotting of the density function for a range of parameter values show that the shape is always unimodal.

An explanation for this is as follows. For the density function to be bimodal, the two subpopulations must be well separated. Without loss of generality, assume that $t_{m1} \ll t_{m2}$, where t_{m1} and t_{m2} are the modes for the two subpopulations. Over the interval $t \leq t_{m1}$ and in the region close to $t = t_{m1}$, $F_2(t) \approx 0$ and $f_2(t) \approx 0$, and as a result $f(t) \approx 0$. Thus, the density function cannot have a peak in a region close to $t = t_{m1}$. This implies that $f(t)$ is unimodal.

The above results are in contrast to the twofold Weibull multiplicative model, which exhibits bimodal shape for the density function. This is because the Weibull model does not have a light left tail so that the condition $F_2(t) \approx 0$ and $f_2(t) \approx 0$ for $t \leq t_{m1}$ do hold.

Hazard Function
The failure rate function is given by

$$r(t) = r_1(t)p(t) + r_2(t)q(t) \tag{10.22}$$

where

$$\begin{aligned} p(t) &= \frac{F_2(t)[1 - F_1(t)]}{1 - F_1(t)F_2(t)} \\ q(t) &= \frac{F_1(t)[1 - F_2(t)]}{1 - F_1(t)F_2(t)} \end{aligned} \tag{10.23}$$

Since the failure rates of two subpopulations are unimodal, the possible shapes of the hazard functions are expected to be unimodal and bimodal. However, computer analysis indicate that the hazard function shape is always unimodal (type 5).

IWPP Plot
Under the inverse Weibull transform given by (6.57), the IWPP plot is given by

$$y = -\ln[z_1(x) + z_2(x)] \tag{10.24}$$

where

$$z_i(x) = (t/\alpha_i)^{-\beta_i} = (e^x/\alpha_i)^{-\beta_i} = \exp[-\beta_i(x - \ln(\alpha_i))] \tag{10.25}$$

It is easily shown that

$$y'(x) = \frac{\beta_1 z_1 + \beta_2 z_2}{z_1 + z_2} \qquad y'' = \frac{-(\beta_1 - \beta_2)^2 z_1 z_2}{(z_1 + z_2)^2} < 0 \tag{10.26}$$

so that the IWPP plot is concave.

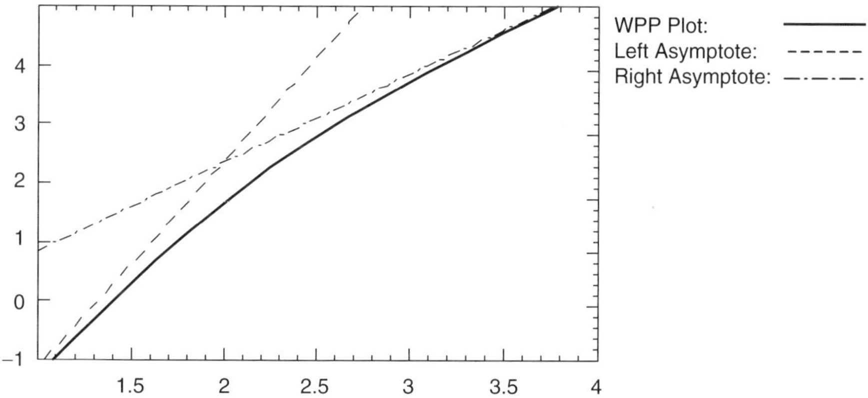

Figure 10.2 IWPP plot inverse Weibull for multiplicative model ($\alpha_1 = 3.0, \beta_1 = 1.5, \alpha_2 = 5.0, \beta_2 = 3.5$).

The two asymptotes for the IWPP plot are given by L_2 as $x \to -\infty$ (or $t \to 0$) and by L_1 as $x \to \infty$ (or $t \to \infty$) where L_2 and L_1 are the IWPP plots for the two subpopulations. Let (x_I, y_I) denote the coordinates of the intersection of L_1 and L_2. Then we have

$$y(x_I) = y_I - \ln(2) \qquad y'(x_I) = \frac{\beta_1 + \beta_2}{2} \tag{10.27}$$

Figure 10.2 shows a typical IWPP plot along with the two asymptotes.

An important observation is that the IWPP plot for the inverse Weibull multiplicative model is similar to the WPP plot for the twofold Weibull multiplicative model. Also, the left (right) asymptote for IWPP plot for the inverse Weibull multiplicative model is the same as the right (left) asymptote for the WPP plot for the Weibull competing risk model.

10.3.3 Parameter Estimation

Graphical Approach

The estimation based on the IWPP plot involves the following steps:

Steps 1–5: Use data to plot the IWPP plot.

Step 6: Fit the left asymptote to the IWPP plot. The slope and intercept of this line yields estimates of β_2 and α_2.

Step 7: Fit the right asymptote to the IWPP plot. The slope and intercept of this line yields the estimates of β_1 and α_1.

Note: One should ensure that the two asymptotes are selected so as to satisfy (10.27).

10.3.4 Modeling Data Set

If the IWPP plot of a given data set has a shape similar to that shown in Figure 10.2, then one can model the data by a twofold inverse Weibull multiplicative model. The parameters can be estimated following the procedure outlined in Section 10.3.3.

EXERCISES

Data Set 10.1 Complete Data: All 50 Items Put into Use at $t = 0$ and Failure Times Are in Weeks

1.578	1.582	1.858	2.595	2.710	2.899	2.940	3.087	3.669	3.848
4.740	4.848	5.170	5.783	5.866	5.872	6.152	6.406	6.839	7.042
7.187	7.262	7.466	7.479	7.481	8.292	8.443	8.475	8.587	9.053
9.172	9.229	9.352	10.046	11.182	11.270	11.490	11.623	11.848	12.695
14.369	14.812	15.662	16.296	16.410	17.181	17.675	19.742	29.022	29.047

Data Set 10.2 Failure Times of 50 Items[a]

0.061	0.073	0.075	0.084	0.086	0.087	0.088	0.089	0.089	0.089
0.099	0.102	0.117	0.118	0.119	0.120	0.123	0.135	0.143	0.168
0.183	0.185	0.191	0.192	0.199	0.203	0.213	0.215	0.257	0.258
0.275	0.297	0.297	0.298	0.299	0.308	0.314	0.315	0.330	0.374
0.388	0.403	0.497	0.714	0.790	0.815	0.817	0.859	0.909	1.286

[a] Unit: 1000 h.

Data Set 10.3 The data is the censored data from Data Set 9.2 with the data collection stopped after 15 weeks.

Data Set 10.4 The data is the censored data from Data Set 9.2 with the data collection stopped after 300 h.

10.1. Carry out WPP and IWPP plots of Data Set 10.1. Can the data be modeled by one of Type III(a), Type III(b), or Type III(c) models involving either two Weibull or two inverse Weibull distributions?

10.2. Suppose that Data Set 10.1 can be modeled by a twofold Weibull multiplicative model. Estimate the model parameters based on (i) the WPP plot, (ii) the method of moments, and (iii) the method of maximum likelihood. Compare the estimates.

10.3. Suppose that the data in Data Set 10.1 can be adequately modeled by the Weibull multiplicative model with the following parameter values:

$\alpha_1 = 2.0$, $\beta_1 = 0.6$, $\alpha_2 = 8.0$, and $\beta_2 = 1.6$. Plot the P–P and Q–Q plots and discuss whether the hypothesis should be accepted or not?

10.4. How would you test the hypothesis of Exercise 10.3 based on A–D and K–S tests for goodness of fit?

10.5. Repeat Exercise 10.1 with Data Set 10.2.

10.6. Suppose that Data Set 10.2 can be modeled by an inverse Weibull multiplicative model. Estimate the model parameters based on (i) the IWPP plot, (ii) the method of moments, and (iii) the method of maximum likelihood. Compare the estimates.

10.7. Repeat Exercises 10.1 to 10.4 with Data Set 10.3.

10.8. Repeat Exercises 10.5 and 10.6 with Data Set 10.4.

10.9. How would you generate a set of simulated data from a twofold Weibull multiplicative model? Can the method be used for a general n-fold Weibull multiplicative model?

10.10. Consider a twofold multiplicative model involving a two-parameter Weibull distribution and an inverse Weibull distribution. Study the WPP and IWPP plots for the model.*

10.11. For the model in Exercise 10.10 derive the maximum-likelihood estimator for (i) complete data and (ii) censored data.

10.12. Consider a twofold multiplicative model involving a two-parameter Weibull distribution and an exponentiated Weibull distribution. Study the WPP plots for the model.*

* Research problem.

CHAPTER 11

Type III(d) Weibull Models

11.1 INTRODUCTION

In a sectional model (also known as composite model, piecewise model, and step-function model), the failure distributions over different time intervals are given by different distribution functions. As a result, a general n-fold sectional model can be defined in terms of $\bar{F}(t)$ as

$$\bar{F}(t) = \begin{cases} (1-k_1) + k_1\bar{F}_1(t) & t \in (0, t_1) \\ k_2\bar{F}_2(t) & t \in (t_1, t_2) \\ \quad \cdots & \\ k_n\bar{F}_n(t) & t \in (t_{n-1}, \infty) \end{cases} \tag{11.1}$$

or in terms of $F(t)$ as

$$F(t) = \begin{cases} k_1F_1(t) & t \in (0, t_1) \\ \quad \cdots & \\ k_{n-1}F_{n-1}(t) & t \in (t_{n-2}, t_{n-1}) \\ (1-k_n) + k_nF_n(t) & t \in (t_{n-1}, \infty) \end{cases} \tag{11.2}$$

where $F_i(t)$, $1 \le i \le n$, are the n subpopulations. The time instants t_i, $1 \le i \le (n-1)$, are referred to as the break (partition) points, and they form an increasing sequence. The k_i's are all >0 and they define a family of models. Table 11.1 shows the six Weibull sectional models that have been studied in detail. For the distribution and density functions to be continuous at the break points, the parameters need to be constrained. We discuss this for the six models in later sections of this chapter.

Weibull Models, by D.N.P. Murthy, Min Xie, and Renyan Jiang.
ISBN 0-471-36092-9

Table 11.1 Six Weibull Sectional Models[a]

Model	n	k_1	k_2	k_3	SP-1	SP-2	SP-3	Form
1	2	1	1	—	2PW	3PW	—	(11.1)
2	2	1	k	—	2PW	2PW	—	(11.1)
3	2	k	1	—	2PW	2PW	—	(11.2)
4	2	k	k	—	2PW	2PW	—	(11.2)
5	3	1	1	1	2PW	3PW	3PW	(11.1)
6	3	1	k_2	k_3	2PW	2PW	2PW	(11.1)

[a] SP, subpopulation; 2PW, two-parameter Weibull; 3PW, three-parameter Weibull.

Some reasons for the use of sectional models are as follows:

1. They are mathematically more tractable than other Type III models.
2. They exhibit a diverse range of shapes for the density and hazard functions. As such, they provide the flexibility to model complex data sets.

The sectional model can be broadly grouped into the following two categories:

Type III(d)-1: Weibull Sectional Models

In this case the n subpopulations are either the two-parameter Weibull distributions given by (8.2) or the three-parameter Weibull distributions given by (8.3).

Kao (1959), Mann et al. (1974), and Elandt-Johnson and Johnson (1980) discuss the sectional model involving two Weibull distributions. In these models the density function and the failure rate function are discontinuous at each partition point.

Jiang and Murthy (1995b, 1997b) and Jiang et al. (1999) deal with six models (labeled Models 1–6) with $n = 2$ for Models 1–4 and $n = 3$ for Models 5 and 6. Aroian and Robinson (1966) and Colvert and Boardman (1976) discuss the sectional exponential model, which is a special case of the Weibull sectional model.

Type III(d)-2: Hybrid Sectional Models

Here some of the subpopulations are either the two- or three-parameter Weibull distributions [given by either (8.2) or (8.3)] and the remaining are non-Weibull distributions.

Geist et al. (1990) deal with a sectional model involving a normal and an exponential distribution. Ananda and Singh (1993) and Mukherjee and Roy (1993) propose sectional hazard rate models. Griffith (1982) and Kunitz (1989) propose sectional models to model the bathtub failure rate function.

The outline of the chapter is as follows. The analysis and parameter estimation for the six models are discussed in Sections 11.2 and 11.3, respectively. In Section 11.4, we discuss modeling of data sets based on WPP plots, and in Section 11.5 we give a list of applications of the Weibull sectional models.

11.2 ANALYSIS OF WEIBULL SECTIONAL MODELS

The results presented in this section for Models 1 and 2 are from Jiang and Murthy (1995b), for Models 3 and 4 are from Jiang et al. (1999b), and for Models 5 and 6 are from Jiang and Murthy (1997b).

The shape and scale parameter are β_1 and α_1 for $F_1(t)$ and β_2 and α_2 for $F_2(t)$ when they are two-parameter Weibill distributions. When one or both are three-parameter distributions, we have the additional location parameter, which we indicate when required. Define $x_1 = \ln(t_1)$ for the first four models.

11.2.1 Model 1

This model is given by (11.1) with $n = 2$, $k_1 = k_2 = 1$, $F_1(t)$ is a two-parameter Weibull distribution and $F_2\ (t)$ is a three-parameter Weibull distribution with location parameter τ. The continuity conditions at t_1 require

$$F_1(t_1^-) = F_2(t_1^+) \qquad f_1(t_1^-) = f_2(t_1^+) \tag{11.3}$$

From (11.3) we have

$$t_1 = \left[\frac{\alpha_1^{\beta_1}}{\alpha_2^{\beta_2}}\left(\frac{\beta_2}{\beta_1}\right)^{\beta_2}\right]^{1/(\beta_1-\beta_2)} \qquad \tau = \left(1 - \frac{\beta_2}{\beta_1}\right)t_1 \tag{11.4}$$

Although the model has six parameters, (11.4) implies that the number of independent parameters is four.

If $\beta_1 = \beta_2$, then either $\tau = 0$ (and this leads to $\alpha_1 = \alpha_2$) or $t_1 = \infty$. In either case, the model reduces to a single Weibull distribution. Hence, we need $\beta_1 \neq \beta_2$. Thus, there are two different cases: $\beta_1 < \beta_2$ (or $\tau < 0$) and $\beta_1 > \beta_2$ (or $\tau > 0$).

Density Function
The possible shapes (see Fig. 3.1) for the density function are as follows:

- Type 1: Decreasing
- Type 2: Unimodal
- Type 3: Decreasing followed by unimodal
- Type 4: Bimodal.

Murthy and Jiang (1997) present a complete parametric study of the density function.

Hazard Function
The possible shapes (see Fig. 3.2) for the hazard function are as follows:

- Type 1: Decreasing (which also includes nonincreasing)
- Type 3: Increasing (which also includes nondecreasing)

- Type 4: Bathtub shape
- Type 5: Unimodal

Murthy and Jiang (1997) present a complete parametric study of the hazard function.

WPP Plot

Under the Weibull transformation given by (1.7), the model gets transformed into

$$y = \begin{cases} \beta_1(x - \ln(\alpha_1)) & -\infty < x \leq x_1 \\ \beta_2[\ln(e^x - \tau) - \ln(\alpha_2)] & x_1 < x < \infty \end{cases} \tag{11.5}$$

where $x_1 = \ln t_1$. The WPP plot is the straight line L_1 (the WPP plot for subpopulation 1) to the left of x_1. To the right of x_1, it is concave for $\beta_1 > \beta_2$ and convex curve for $\beta_1 < \beta_2$. Typical WPP plots for the two cases are shown in Figures 11.1 and 11.2, respectively.

The right asymptote (as $x \to \infty$) is the straight line L_2 (the WPP plot for a two-parameter Weibull distribution with the parameters β_2 and α_2). The left asymptote is given by L_1. Let x_I denote the x coordinate of the intersection point of L_1 and L_2. Then it satisfies the following relationship:

$$x_I - x_1 = \frac{\beta_2 \ln(\beta_2/\beta_1)}{\beta_2 - \beta_1} \tag{11.6}$$

11.2.2 Model 2

This model is given by (11.1) with $n = 2$, $k_1 = 1$, $k_2 = k > 0$, and $F_1(t)$ and $F_2(t)$ are two-parameter Weibull distributions. The continuity conditions at t_1 require

$$\bar{F}_1(t_1^-) = k\bar{F}_2(t_1^+) \qquad f_1(t_1^-) = kf_2(t_1^+) \tag{11.7}$$

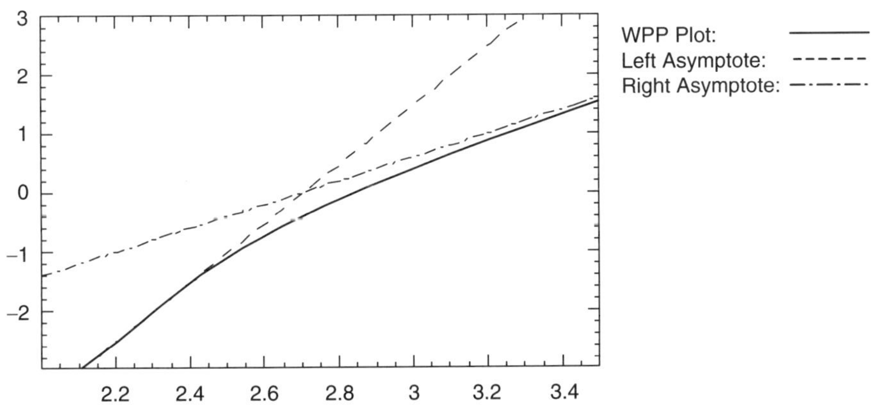

Figure 11.1 WPP plot for the Weibull sectional model 1 ($\beta_1 > \beta_2$).

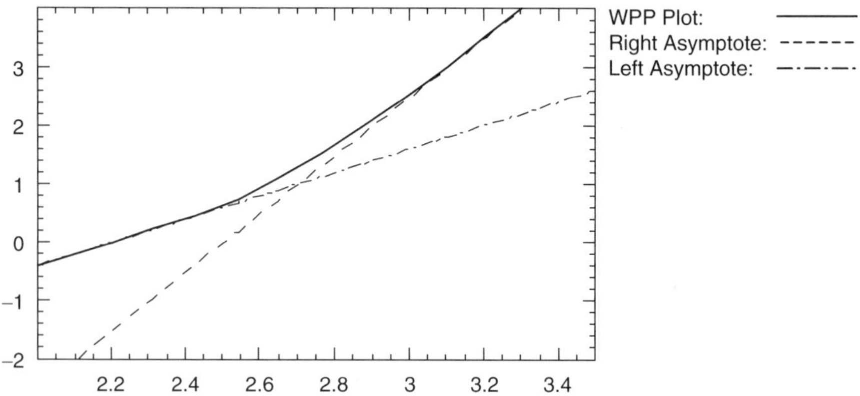

Figure 11.2 WPP plot for the Weibull sectional model 1 ($\beta_1 < \beta_2$).

From (11.7) we have

$$t_1 = \left(\frac{\beta_2 \alpha_1^{\beta_1}}{\beta_1 \alpha_2^{\beta_2}}\right)^{1/(\beta_1 - \beta_2)} \qquad k = \exp\left[\left(1 - \frac{\beta_2}{\beta_1}\right)\left(\frac{t_1}{\alpha_2}\right)^{\beta_2}\right] \tag{11.8}$$

This implies that the number of independent parameters is four.

If $\beta_1 = \beta_2$, then $k = 1$, and either $\alpha_1 = \alpha_2$ or $t_1 = 0$. In either case, the model reduces to a single Weibull distribution. Hence, we need $\beta_1 \neq \beta_2$.

Density Function

The different possible shapes are similar to that for Model 1. Murthy and Jiang (1997) present a complete parametric study of the density function.

Hazard Function

The different possible shapes are similar to that for Model 1. Murthy and Jiang (1997) present a complete parametric study of the density function.

WPP Plot

The WPP plot is convex, but the asymptotes are the same as in Model 1. The coordinates of the intersection point satisfy the relationship given by

$$x_I - x_1 = \frac{\ln(\beta_2/\beta_1)}{\beta_2 - \beta_1} \tag{11.9}$$

which is different from (11.6). This difference is useful in discriminating between the two models in the context of modeling a given data set.

11.2.3 Model 3

The model is given by (11.2) with $n = 2$, $k_1 = k > 0$, $k_2 = 1$, and $F_1(t)$ and $F_2(t)$ are two-parameter Weibull distributions. The continuity conditions at t_1 require

$$kF_1(t_1^-) = F_2(t_1^+) \qquad kf_1(t_1^-) = f_2(t_1^+) \tag{11.10}$$

From (11.10), one can obtain t_1 and k as functions of independent parameters α_1, β_1, α_2, and β_2, and this is given by the solution of the following equation:

$$\exp(cz^{\beta}) - c\beta z^{\beta-1}(e^z - 1) - 1 = 0 \tag{11.11}$$

where $c = (\alpha_1/\alpha_2)^{\beta_2}$, $\beta = \beta_2/\beta_1$, and $z \equiv z_1 = (t/\alpha_1)^{\beta_1}$. Let $\gamma_0 = (t_1/\alpha_1)^{\beta_1}$, then k can be expressed as a function of γ_0 and is given by

$$k = \frac{1 - \exp(-c\gamma_0^{\beta})}{1 - \exp(-\gamma_0)} \tag{11.12}$$

If $\beta_1 = \beta_2$, then either $t_1 = 0$ or $c = 1$. In either case, $F(t)$ reduces to a single Weibull distribution. Thus, we need $\beta_1 \neq \beta_2$. When $\beta_1 < \beta_2$, (11.11) has a nonzero solution and $k > 1$. When $\beta_1 > \beta_2$, (11.11) has a nonzero solution and $k < 1$.

Density Function
The different possible shapes are similar to that for Model 1.

Hazard Function
The shapes for the hazard function include those for Model 1 as well as the following two additional shapes:

- Type 6: Unimodal followed by increasing
- Type 7: Decreasing followed by unimodal

WPP Plot
The WPP plot is concave and the shape is similar to that for Model 1. However, the asymptotes are slightly different. To the left of $x_1 (= \ln(t_1))$, the WPP plot is a smooth curve with the left asymptote (as $x \to -\infty$) given by the straight line

$$y = \ln k + \beta_1(x - \ln(\alpha_1)) = \beta_1 x + \ln(k/\alpha_1^{\beta_1}) \tag{11.13}$$

This is parallel to L_1 (the WPP plot of subpopulation 1) but displaced vertically by $|\ln(k)|$. To the right of x_1, the WPP is a straight line given by the WPP plot of subpopulation 2.

11.2.4 Model 4

The model is given by (11.2) with $n = 2$, $k_1 = k_2 = k > 0$, and $F_1(t)$ and $F_2(t)$ are two-parameter Weibull distributions. The continuity conditions at t_1 require

$$f_1(t_1) = f_2(t_1) \qquad k = 1/[F_1(t_1) + R_2(t_1)] \tag{11.14}$$

Unlike the earlier three models, here we do not require $\beta_1 \neq \beta_2$. However, when $\beta_1 = \beta_2$, we need to have $\alpha_1 \neq \alpha_2$; otherwise the model reduces to a single Weibull model. Note also that (11.14) has a unique solution if $\beta_1 = \beta_2$ and $\alpha_1 \neq \alpha_2$. When $\beta_1 \neq \beta_2$, it results in two solutions for the independent model parameters.

Density Function

The different possible shapes are the same as that for Model 3.

Hazard Function

The different possible shapes are the same as that for Model 3.

WPP Plot

The WPP plot is slightly different from the earlier models. To the left of $x_1[= \ln(t_1)]$, it is a curve with the left asymptote (as $x \to -\infty$) given by (11.13). The curve is concave when $\beta_1 < \beta_2$ and convex when $\beta_1 > \beta_2$. To the right of x_1, the WPP plot is also a curve with the right asymptote (as $x \to \infty$) given by L_2. The curve is concave when $\beta_1 > \beta_2$ and convex when $\beta_1 < \beta_2$.

As a result, the WPP plot is S shaped with an inflection at $x = x_1$ (or $t = t_1$). Figure 11.3 shows the possible shapes depending on the shape parameters of the two subpopulations. For the first case, the WPP is concave for $x < x_1$ and convex for $x > x_1$. For the second case, it is the reverse.

11.2.5 Model 5

This model is given by (11.1) with $n = 3$ with $k_1 = k_2 = k_3 = 1$. The term $F_1(t)$ is a two-parameter Weibull distribution and $F_2(t)$ and $F_3(t)$ are three-parameter Weibull distributions with parameters (β_2, α_2, τ_2) and (β_3, α_3, τ_3), respectively. The continuity conditions are given by:

$$\begin{aligned} F_1(t_1^-) &= F_2(t_1^+) \qquad f_1(t_1^-) = f_2(t_1^+) \\ F_2(t_2^-) &= F_3(t_2^+) \qquad f_2(t_2^-) = f_3(t_2^+) \end{aligned} \tag{11.15}$$

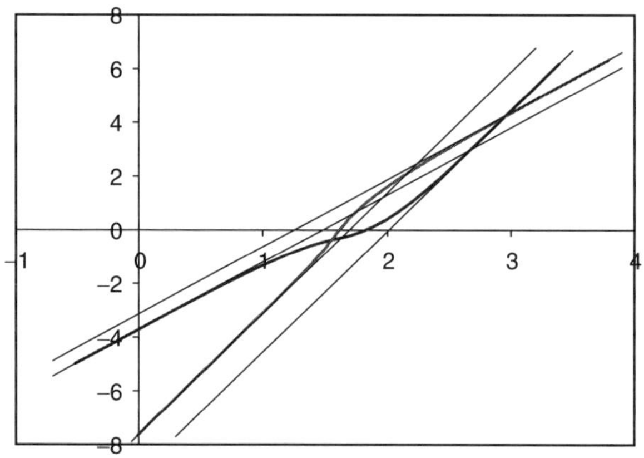

Figure 11.3 Different shapes for the WPP plot for Weibull sectional model 4.

From (11.15), we obtain t_i and τ_{i+1}, $i = 1, 2$, as follows:

$$t_i = \tau_i + \left[\frac{\alpha_i^{\beta_i}}{\alpha_{i+1}^{\beta_{i+1}}} \left(\frac{\beta_{i+1}}{\beta_i} \right)^{\beta_{i+1}} \right]^{1/(\beta_i - \beta_{i+1})}$$
$$\tau_{i+1} = \tau_i + \left(1 - \frac{\beta_{i+1}}{\beta_i} \right) (t_i - \tau_i) \tag{11.16}$$

with $\tau_1 = 0$. We need $\beta_1 \neq \beta_2$, $\beta_2 \neq \beta_3$, and $t_1 < t_2$ or else the number of subpopulations gets reduced from three to two.

Density Function

The possible different shapes for the density function are as follows:

- Type $2k$ ($k = 1 - 3$) so that we have unimodal, bimodal, and trimodal shapes
- Type $2k - 1$ ($k = 1 - 3$) so that we have decreasing, decreasing followed by unimodal, and decreasing followed by bimodal

Hazard Function

The possible different shapes for the hazard function are as follows:

- Type 1: Nonincreasing
- Type 3: Nondecreasing
- Type 5: Unimodal
- Type 4: Bathtub shape
- Types 6 and 7: Roller-coaster shapes

WPP Plot

The WPP plot can be any one of the many different shapes for Models 1–4 (shown in Figs. 11.1–11.3). To the left of $x_1 (= \ln(t_1))$, the WPP plot is a straight line L_1. To the right of $x_2 (= \ln(t_2))$, the WPP plot is a curve and has an asymptote given by L_3 (the WPP plot of the Weibull distribution with the parameters β_3 and α_3). Over the interval (x_1, x_2), the WPP plot can either be convex on the left and concave on the right or the reverse.

11.2.6 Model 6

This model differs from Model 5 in the sense that only $k_1 = 1$ and all three subpopulations are two-parameter Weibull distributions. The continuity conditions are given by

$$\begin{aligned} R_1(t_1^-) &= k_2 R_2(t_1^+) \qquad & r_1(t_1^-) &= r_2(t_1^+) \\ k_2 R_2(t_2^-) &= k_3 R_3(t_2^+) \qquad & r_2(t_2^-) &= r_3(t_2^+) \end{aligned} \tag{11.17}$$

From (11.17), we obtain t_i and k_{i+1}, $i = 1, 2$, as follows:

$$t_i = \left(\frac{\alpha_i^{\beta_i}}{\alpha_{i+1}^{\beta_{i+1}}} \times \frac{\beta_{i+1}}{\beta_i} \right)^{1/(\beta_i - \beta_{i+1})}$$
$$k_{i+1} = \frac{k_i R_i(t_i)}{R_{i+1}(t_i)} \tag{11.18}$$

where $k_1 = 1$. We need $\beta_1 \neq \beta_2$, $\beta_2 \neq \beta_3$, and $t_1 < t_2$ or else the number of subpopulations reduce from three to two.

Density Function
The different possible shapes are the same as that for Model 5.

Hazard Function
The different possible shapes are the same as that for Model 5.

WPP Plot
The different possible shapes for the WPP plot are the same as that for Model 5.

11.3 PARAMETER ESTIMATION

The two main methods that have been used for estimating the model parameters are the graphical method (based on the WPP plot) and the method of maximum likelihood. In this section we carry out a review of the two methods.

11.3.1 Methods Based on WPP

Kao (1959) uses a twofold two-parameter Weibull sectional model to approximate a twofold Weibull mixture model. The distribution function for the model is continuous, but the density and failure rate functions are discontinuous. The method involves WPP plotting of the data and fitting two straight lines—one for each subpopulation. The WPP plot for a twofold Weibull mixture is S shaped (with three inflections) whereas for the Weibull sectional model it is either concave or convex. Hence, Kao's approximation is not appropriate.

Elandt-Johnson and Johnson (1980) deal with a sectional model with the distribution function continuous at all the partition points. They discuss the use of Weibull hazard plot to model the data set.

Models 1 and 2
Jiang and Murthy (1995b) propose a method to estimate the parameters of Models 1 and 2. The basic idea is to fit straight lines to the left and right asymptotes. The slope and intercepts of these lines yield the estimates for the four independent parameters.

The procedure for Model 1 is as follows:

Steps 1–5: Use the data to plot the WPP plot. See, Section 4.5.1.

Step 6: Fit a straight line to the left part of the WPP plot. From its slope and intercept obtain estimates of β_1 and α_1.

Step 7: Fit the right asymptote. From its slope and intercept obtain estimates of β_2 and α_2.

Step 8: Obtain estimates of t_1 and τ from (11.4).

Step 9: Test whether (11.6) is satisfied. If not, adjust the location of the asymptote in step 7 and repeat steps 7 and 8.

The procedure for Model 2 is as follows:

Steps 1–7: Same as for Model 1.

Step 8: Obtain estimates of t_1 and k from (11.8).

Step 9: Test whether (11.9) is satisfied. If not, adjust the location of the asymptote in step 7 and repeat steps 7 and 8.

Models 3 and 4

Jiang et al. (1999b) propose a method to estimate the parameters of Models 3 and 4. The procedure for Model 3 is as follows:

Steps 1–5: Same as for Models 1 and 2.

Step 6: Draw a tangent line to the left end of the WPP plot. The slope and intercept yield estimates of β_1 and $\ln(k/\alpha_1^{\beta_1})$.

Step 7: Fit a straight line at the right end of the plot, and using its slope and intercept yields the estimates of β_2 and α_2.

Step 8: Locate the partition point x_1 where the fitted WPP curve starts deviating from the right asymptote. This is done through visual inspection. An estimate of t_1 is obtained from x_1.

Step 9: Define $\Gamma_0 = (t_1/\alpha_2)^{\beta_2}$. An estimate of Γ_0 can be obtained from the estimates of t_1, β_2, and α_2. From (11.10), we have

$$\frac{\exp(\gamma_0) - 1}{\gamma_0} = \frac{\exp(\Gamma_0) - 1}{\beta\Gamma_0} \tag{11.19}$$

and this yields an estimate of γ_0.

Step 10: Since $\Gamma_0 = c\gamma_0^{\beta}$, this yields an estimate of c. Using this, one can obtain an estimate of α_1 from the relation $c = (\alpha_1/\alpha_2)^{\beta_2}$ and an estimate of k using (11.12).

Step 11: Compute k from the estimate of the intercept of the tangent line obtained in step 6. Compare it with the one obtained in step 5 to determine if they are close to each other or not. If not, go to step 8 to adjust x_1, reestimate t_1 and Γ_0, and repeat steps 9 through 11.

The procedure for Model 4 is as follows:

Steps 1–5: Same as for Models 1 and 2.

Step 6: Draw a tangent line to the left end of the WPP plot. The slope and intercept yield estimates of β_1 and $\ln(k/\alpha_1^{\beta_1})$. Compute

$$\hat{\eta} = \ln(k/\alpha_1^{\beta_1}) \tag{11.20}$$

Step 7: Draw a tangent line to the right end of the plot. Its slope and intercept yield the estimates of β_2 and α_2.

Step 8: Locate the partition point x_1, which is the inflection point of the fitted WPP curve. This is done through visual inspection; t_1 can be obtained from the estimate of x_1.

Step 9: Estimates of k and α_1 are obtained from (11.14) and (11.20).

Models 5 and 6

Jiang and Murthy (1997b) propose graphical estimation methods for estimating the parameters of Models 5 and 6. The procedure for Model 5 is follows:

Steps 1–5: Same as for Models 1 and 2.

Step 6: Fit a straight line at the left end of the WPP plot of the data. From its slope and intercept obtain estimates of β_1 and α_1.

Step 7: Fit the asymptote to the right end of the WPP plot. From its slope and intercept obtain estimates of β_3 and α_3.

Step 8: Draw a tangent to the WPP plot in the region close to the middle. This is an approximation to the WPP plot of subpopulation 2. As a result, its slope and intercept yield estimates for β_2 and α_2.

Step 9: Estimate t_1, t_2, τ_2, and τ_3 using (11.16).

Step 10: Test whether the following relations are satisfied:

$$y_1 = \beta_2(x_1 - \ln(\alpha_2)) + \beta_2 \ln(\beta_2/\beta_1) \tag{11.21}$$

and

$$y_2 = \beta_2(x_2 - \ln(\alpha_2)) + \beta_2 \ln(1 - \tau_2/t_2) \tag{11.22}$$

If not, adjust the location of the line in step 4 and repeat steps 4 and 5 until these equations are satisfied.

The procedure for Model 6 is as follows:

Steps 1–8: The same as for Model 5.

Step 9: Estimate t_1, t_2, k_2, and k_3 using (11.18).

Step 10: Test whether the following relations are satisfied:

$$y_1 = \beta_2(x_1 - \ln(\alpha_2)) + \ln(\beta_2/\beta_1) \tag{11.23}$$

and

$$y_2 = \beta_2(x_2 - \ln(\alpha_2)) + \ln[1 - (1 - \beta_2/\beta_1)(t_1/t_2)^{\beta_2}] \tag{11.24}$$

If not, adjust the location of the line in step 4 and repeat steps 4 and 5 until these equations are satisfied.

11.3.2 Maximum-Likelihood Method

Aroian and Robison (1966) deal with an exponential distribution with changing scale parameter and estimate the parameters using the method of maximum likelihood. Colvert and Boardman (1976) also deal with a similar model.

11.4 MODELING DATA SET

If a smooth fit to the WPP plot of the data has a shape similar to one of those shown in Figures 11.1–11.3, then one can assume that one of the Weibull sectional models can be used to model the data set. The model parameters can be estimated from the WPP plot as indicated in Section 11.3.

11.5 APPLICATIONS

The Weibull sectional models have been used in reliability modeling of products, and Table 11.2 gives a list of the applications.

Table 11.2 Applications of Weibull Sectional Models

Applications	Reference
Electron tubes	Kao (1959)
Life tests	Aroian and Robison (1966)
Oral irrigators	Colvert and Boardman (1976)
Life data for white males	Elandt-Johnson and Johnson (1980)
Pull strength of welds	Jiang and Murthy (1995b)
Reanalysis of oral irrigators data	Jiang and Murthy (1997b)
Shear strength of brass rivets, failure times in solid tantalum capacitors	Jiang, Zuo and Murthy (1999b)

EXERCISES

Data Set 11.1 Complete Data

0.032	0.035	0.104	0.169	0.196	0.260	0.326	0.445	0.449	0.496
0.543	0.544	0.577	0.648	0.666	0.742	0.757	0.808	0.857	0.858
0.882	1.005	1.025	1.472	1.916	2.313	2.457	2.530	2.543	2.617
2.835	2.940	3.002	3.158	3.430	3.459	3.502	3.691	3.861	3.952
4.396	4.744	5.346	5.479	5.716	5.825	5.847	6.084	6.127	7.241
7.560	8.901	9.000	10.482	11.133					

Data Set 11.2 Complete Data

0.032	0.035	0.104	0.169	0.196	0.260	0.326	0.445	0.449	0.496
0.543	0.544	0.577	0.648	0.666	0.742	0.757	0.808	0.857	0.858
0.882	1.138	1.163	1.256	1.283	1.484	1.897	1.944	2.201	2.365
2.531	2.994	3.118	3.424	4.097	4.100	4.744	5.346	5.479	5.716
5.825	5.847	6.084	6.127	7.241	7.560	8.901	9.000	10.482	11.133

11.1. Carry out WPP and IWPP plots of Data Set 11.1. Can the data be modeled by one of Type III models involving either two Weibull or two inverse Weibull distributions?

11.2. Suppose that Data Set 11.1 can be modeled by a Type III(d)-1 model. Determine which one from Models 1 to 6 is appropriate to model the data based on the WPP plot?

11.3. Suppose that the data can be adequately modeled by Model 1 of Section 11.2. Estimate the model parameters based on (i) the WPP plot, (ii) the method of moments, and (iii) the method of maximum likelihood. Compare the estimates.

11.4. Suppose that the data in Data Set 11.1 can be adequately modeled by Model 2 of Section 11.2. Estimate the model parameters based on (i) the WPP plot, (ii) the method of moments, and (iii) the method of maximum likelihood.

11.5. Suppose that the data in Data Set 11.1 can be adequately modeled by Model 3 of Section 11.2. Estimate the model parameters based on (i) the WPP plot, (ii) the method of moments, and (iii) the method of maximum likelihood.

11.6. Suppose that the data in Data Set 11.1 can be adequately modeled by Model 4 of Section 11.2. Estimate the model parameters based on (i) the WPP plot, (ii) the method of moments, and (iii) the method of maximum likelihood.

11.7. How would you test the hypothesis that Data Set 11.1 is from Model 1 with $t_1 = 1.0$?

11.8. Repeat Exercises 11.1 to 11.6 with Data Set 11.2.

11.9. Test the hypothesis that Data Set 11.2 is from Model 1 with $t_1 = 5.0$.

PART E

Types IV to VII Models

CHAPTER 12

Type IV Weibull Models

12.1 INTRODUCTION

The parameters of the standard Weibull model are usually assumed to be constant. For models belonging to Type IV, this is not the case. The parameters are either a function of the independent variable (t) or some other variables (such as stress level) or are random variables. Type IV models can be divided into five subgroups—Types IV(a) to (e).

Type IV(a) models are models with time-varying parameters. Here either the scale and/or the shape parameter of two-parameter Weibull distribution are functions of the independent variable (t), which in most cases represents time.

Types IV(b) to (d) are models where the parameters are functions of some other auxiliary (variables such as electrical, mechanical, thermal stresses), and these are often called *covariates*. Type IV(b) is the accelerated life model where the auxiliary variable appears as a coefficient of the independent variable (t) of the model. Type IV(c) is the proportional hazard model where the auxiliary variables appear as the coefficients of base hazard function. Type IV(d) model is a general model that combines both IV(b) and IV(c).

In Type IV(e) models the model parameters are random variables. This implies that they need to be modeled probabilistically.

The outline of the chapter is as follows. Section 12.2 deals with Type IV(a) models, and we discuss five different models. Sections 12.3 to 12.5 discusses Types IV(b) to (d) models. Section 12.6 looks at Types IV(e) models and reviews the literature as there are many different such models that have been studied. Finally, in Section 12.7 we briefly discuss the Bayesian approach to parameter estimation as it is closely linked to the models in the previous section.

Weibull Models, by D.N.P. Murthy, Min Xie, and Renyan Jiang.
ISBN 0-471-36092-9

12.2 TYPE IV(a) MODELS

In these models the scale parameter (α) and/or the shape parameter (β) of the standard Weibull model are functions of the independent variable t.

12.2.1 Model IV(a)-1

Model Formulation

Srivasatava (1974) considers the case where $\alpha(t)$ alternates between two values. As a result, at any given time t it can be either α_1 or α_2. The hazard function $h(t)$ is given by

$$h(t) = \frac{\beta[\tau_1(t)]^{\beta-1}}{\alpha_1^{\beta}} + \frac{\beta[\tau_2(t)]^{\beta-1}}{\alpha_2^{\beta}} \tag{12.1}$$

where $\tau_1(t)$ and $\tau_2(t)$ are the cumulative times (since $t = 0$) for which $\alpha(t) = \alpha_1$ and $\alpha(t) = \alpha_2$.

Note that the model can be viewed as a competing risk model with two subpopulations with independent variables $\tau_1(t)$ and $\tau_2(t)$ rather than t.

Parameter Estimation

Srivastava (1974) discusses the estimation of model parameters based on the method of maximum likelihood with censored data.

12.2.2 Model IV(a)-2

Model Formulation

Zuo et al. (1999) consider a model where

$$\beta(t) = a\left(1 + \frac{1}{t}\right)^{b} e^{c/t} \tag{12.2}$$

and

$$\alpha(t) = at^{b}e^{c/t} \tag{12.3}$$

As a result, the cumulative hazard function is given by

$$H(t) = \left(\frac{D}{\alpha(t)}\right)^{\beta(t)} \tag{12.4}$$

where D is a constant. Note that $\alpha(t)$ and $\beta(t)$ need to be constrained so that $H(t)$ is a nondecreasing function that goes to infinity as $t \to \infty$.

From (12.4) one can derive the distribution and density function using (3.1).

Parameter Estimation

Zuo et al. (1999) examine parameter estimation based on regression.

12.2.3 Model IV(a)-3

Model Formulation

Zacks (1984) considers a model where the shape parameter $\beta = 1$ for $t < t_0$ and changes to $\beta > 1$ for $t > t_0$. The model is called the *exponential Weibull* distribution. Note that this can also be viewed as a sectional model discussed in Chapter 11.

Parameter Estimation

Zacks (1984) deals with Bayesian estimation of the change point t_0.

12.2.4 Model IV(a)-4

Model Formulation

The distribution function has the following general form:

$$F(t) = 1 - e^{-\Lambda(t)} \tag{12.5}$$

where $\Lambda(t)$ is a nondecreasing function with $\Lambda(0) = 0$ and $\Lambda(\infty) = \infty$. Srivastava (1989) calls this the *generalized Weibull* distribution and deals with the case where

$$\Lambda(t) = \lambda_i \phi(t) \tag{12.6}$$

for $1 \leq i \leq m$, to model the cumulative hazard functions for m different machines. Note (12.5) yields the two-parameter Weibull distribution when

$$\Lambda(t) = \left(\frac{t}{\alpha}\right)^{\beta} \tag{12.7}$$

12.3 TYPE IV(b) MODELS: ACCELERATED FAILURE TIME (AFT) MODELS

In AFT models the scale parameter α is a function of some supplementary variables S, which are often called *explanatory variables* or *covariates*. In reliability applications S represents the stress (electrical, mechanical, thermal, etc.) on an item and the life of the item (a random variable) is a function of S. This model has also been called parametric regression models [see Lawless (1982)]. Note that S can be either scalar or vector.

12.3.1 Model Formulations

Model IV(b)-1: Weibull AFT Models

The distribution function, for a given S, is given by

$$F(t|S) = 1 - \exp\left[-\left(\frac{t}{a(S)\alpha}\right)^{\beta}\right] \qquad t \geq 0 \tag{12.8}$$

where $a(S)$ is called the *acceleration factor*, which decreases with S increasing. Depending on the form of $a(S)$, we have a family of models. The following three (with S scalar) have been used extensively in reliability theory:

1. **Arrhenius–Weibull Model**: $a(s) = \exp(\gamma_0 + \gamma_1/S)$.
2. **Power Weibull Model**: $a(S) = \gamma_0/S^{\gamma_1}$.
3. **Eyring Model**: $a(s) = \gamma_0 S^{\gamma_2} \exp(\gamma_2 S)$.

In the reliability context the γ's need to be constrained so as to ensure that $a(S)$ decreases as S increases so as to model the accelerated failure resulting from higher stress.

Jensen (1995) discusses several accelerated life models for electronic components, in particular models involving the Arrhenius model for temperature and other stress factors.

Other AFT Weibull Models

These can be any of the Weibull models discussed in Chapters 5 to 11 with the scale parameter $a(S)$ a function of the supplementary variable and having the different forms as indicated above. Nelson (1990) discusses the competing risk AFT model. Townson (2002) deals with AFT models for the Weibull mixture and the Weibull competing risk models with two subpopulations.

12.3.2 Model Analysis

Model IV(b)-1: Weibull AFT Models

From (12.8), the hazard function is given by

$$h(t|S) = \frac{\beta}{a(S)\alpha}\left[\frac{t}{a(S)\alpha}\right]^{\beta-1} \tag{12.9}$$

Note that this increases as S increases.

Expressions for the moments are the same as for the two-parameter Weibull distribution (see Chapter 3) with the scale parameter α replaced by $\alpha a(S)$.

Under the Weibull transformation (1.7), (12.8) gets transformed into

$$y = \beta x - \beta \ln(\alpha) - \beta \ln[a(S)] \tag{12.10}$$

As a result, the WPP plot is a straight line for a given S. Different values for S result in straight lines that are parallel to each other.

For two different S_1 and S_2 we have the following results:

$$F(t|S_1) = F\left[\frac{a(S_1)}{a(S_2)}t|S_2\right] \tag{12.11}$$

The implication of this in the reliability context is as follows. If the life of a component has a life T_1 when $S = S_1$, then the same component at $S = S_2$ has a life T_2 given by

$$T_2 = \frac{a(S_2)}{a(S_1)} T_1 \tag{12.12}$$

If $S_2 > S_1$, then $T_2 < T_1$ implying that the time to failure decreases as S increases.

Expressions of moments are the same as that for the standartd Weibull model except that the scale parameter contains $a(S)$.

Other AFT Models

A study of the mean, variance, and percentiles as function of S and the WPP plots for the Weibull mixture and the Weibull competing risk models, with two subpopulations and different acceleration factors for each, can be found in Townson (2002).

12.3.3 Parameter Estimation

Model IV(b)-1: Weibull AFT Models

The parameter estimation for the Weibull AFT model has received a lot of attention in the literature.

Graphical Approach (WPP Plot)

At each stress level, a Weibull plot can be generated. If the plots are roughly straight lines, and parallel to each other, then the Weibull AFT model is appropriate to model the data under consideration. The slope (common to all) yields the shape parameter, and the intercept for different lines yields estimates of the scale parameters for different stress levels.

Method of Maximum Likelihood

This method has been studied extensively for the different cases (Arrhenius, power law, and Eyring models discussed earlier) and for both complete and censored data.

Nelson (1990) deals with both graphical as well as other statistical approaches for estimating the model parameters and illustrates with applications in many different areas. See also Lawless (1982) and Watkins (1994).

12.3.4 Model Selection and Validation

Model IV(b)-1: Weibull AFT Models

Graphical Approach (WPP Plot)

This has been discussed briefly in Section 12.3.3. Further discussion can be found in Lawless (1982).

Statistical Method

Some goodness-of-fit tests (based on the residuals) are discussed in Lawless (1982) and in Gertsbakh (1989) where readers can find other relevant references. See also Nelson (1990).

12.3.5 Varying-Stress Model

So far we have assumed that the supplementary variable S does not change with the independent variable t of the distribution function. In many cases, S can change with t, and one such is in accelerated life testing of unreliable components as indicated below.

The accelerated life testing reduces the time needed to complete the test to estimate the model parameters. The optimal stress level depends on the unknown model parameters. In this case one uses a multiple-step stress-accelerated testing where items are tested at an initial stress level S_1. After a predetermined time τ_1, the items that are still working are subjected to a higher stress level S_2. This is continued for a time τ_2, and items that have survived are subjected to still higher stress level S_3, and the process continues on.

Step-stress modeling requires taking into account the cumulative effect of different stress levels on the failure of an item. A common approach used is the following. If a component has survived for a period t_1 at a constant stress S_1 and is subjected to stress S_2 at time t_1, then it is treated as having survived for a period $\tilde{t}_1$ under a constant stress S_2. The relationship linking $\tilde{t}_1$ with t_1 is given by

$$F(\tilde{t}_1|S_2) = F(t_1|S_1) \tag{12.13}$$

Based on this interpretation, the distribution function for the Weibull ALT model under multiple-step stress-accelerated life testing [see Gouno and Balkrishnan (2001)] is given by

$$F(t) = \begin{cases} 1 - \exp\left[-\left(\frac{t}{\alpha_1}\right)^{\beta}\right] & 0 \le t \le \tau_1 \\ 1 - \exp\left[-\left(\frac{\tau_1}{\alpha_1} + \frac{t-\tau_1}{\alpha_2}\right)^{\beta}\right] & \tau_1 \le t \le \tau_2 \\ \cdots & \cdots \\ 1 - \exp\left[-\left(\frac{t-\tau_{i-1}}{\alpha_i} + \sum_{j=1}^{i-1} \frac{\tau_i - \tau_{i-1}}{\alpha_{i-1}}\right)^{\beta}\right] & \tau_{i-1} \le t \le \tau_i \end{cases} \tag{12.14}$$

where α_j is the scale parameter corresponding to stress S_j, $1 \le j \le (i-1)$. Note that this can be viewed as an n-fold sectional Weibull model similar to models 1 and 5 in Chapter 11.

For the special case of two-step accelerated life testing with simple power acceleration function $\alpha_i = CS_i^p$, $1 \le i \le 2$, the probability density function is given by

$$f(t) = \begin{cases} \frac{\beta}{\alpha_1}\left(\frac{t}{\alpha_1}\right)^{\beta-1} \exp\left[-\left(\frac{t}{\alpha_1}\right)^{\beta}\right] & 0 \le t \le \tau \\ \frac{\beta}{\alpha_2}\left(\frac{\tau}{\alpha_1} + \frac{t-\tau}{\alpha_2}\right)^{\beta-1} \exp\left[-\left(\frac{\tau}{\alpha_1} + \frac{t-\tau}{\alpha_2}\right)^{\beta}\right] & \tau \le t < \infty \end{cases} \tag{12.15}$$

where τ is the change point for the stress level.

The estimation of parameters for the step-stress model has received considerable attention. See, for example, Nelson (1990).

We consider the testing of n items with Type II censoring. The testing is stopped when the number of failures reaches some prespecified value r. Let the number of failures at each of two stress levels be n_1 and n_2, respectively. The log-likelihood [based on (12.15)] is given by

$$\begin{aligned}\ln L(C,p,\beta) &= r\ln\beta - n_1\beta(\ln C + p\ln S_1) - n_2(\ln C + p\ln S_2)\\ &\quad + (\beta-1)\left[\sum_{j=1}^{n_1}\ln(t_{1,j}) - n_2\ln C + \sum_{j=1}^{n_2}\ln\left(\frac{t_{2,j}-\tau}{S_2^p}\cdot\frac{\tau}{S_1^p}\right)\right]\\ &\quad - \frac{1}{C^\beta}\left[\sum_{j=1}^{n_1}\frac{t_{1,j}}{s_1^p} + \sum_{j=1}^{n_2}\left(\frac{t_{2,j}-\tau}{S_2^p}+\frac{\tau}{S_1^p}\right)^\beta\right.\\ &\quad \left. + \left(n-r\right)\sum_{j=1}^{n_2}\left(\frac{t_{2,n_2}-\tau}{S_2^p}+\frac{\tau}{S_1^p}\right)^\beta\right] \end{aligned} \tag{12.16}$$

where $t_{i,j}$ is the failure time of unit i operating under stress s_j. Gouno and Balakrishnan (2001) discuss the simplification of the likelihood equations so that the estimates can be obtained easily.

Nelson (1990) considers the estimation problem with Type I censoring and derives the confidence limits for the model parameters and for the reliability function.

12.4 TYPE IV(c) MODELS: PROPORTIONAL HAZARD (PH) MODELS

An alternate approach to modeling the effect of the supplementary variable on the distribution function is through its hazard function.

12.4.1 Model Formulations

In the PH model,

$$h(t|S) = \psi(S)h_0(t) \tag{12.17}$$

where $h_0(t)$ is called the baseline hazard function. The only restriction on the scalar function $\psi(S)$ is that it be a positive nondecreasing function. Depending on the form of $\psi(S)$, one has a family of different models associated with a particular base hazard function. This model is also called as the *multiplicative model* by Fahrmeir and Klinger (1998). A review of the literature on PH models up to 1994 can be found in Kumar and Klefsjo (1994).

The case where the baseline hazard function is the hazard function associated with a two-parameter Weibull distribution has received considerable attention. In contrast, the study of PH models for other forms of baseline hazard function has not received much attention. The only exception is the competing risk model discussed in Kalbefleisch and Prentice (1980).

Many different forms for ψ ( ) have been proposed. One such is the following:

$$\psi(S) = \exp\left(b_0 + \sum_{i=1}^{k} b_i S_i\right) \tag{12.18}$$

Model IV(c)-1: Weibull PH Model

The Weibull PH model is given by

$$h(t|S) = \psi(S)\frac{\beta}{\alpha}\left(\frac{t}{\alpha}\right)^{\beta-1} \tag{12.19}$$

and the distribution function is given by

$$F(t|S) = 1 - \exp\left[-\psi(S)\left(\frac{t}{\alpha}\right)^{\beta}\right] \tag{12.20}$$

Note that this is identical to (12.8) when $[a(S)]^{\beta} = 1/\psi(S)$. As a result, the Weibull PH model and the Weibull AFT model are identical. In fact, it can be shown (Cox and Oakes, 1984) that Weibull distribution is the only distribution for which the AFT and the proportional hazard models coincide. See also Newby (1994).

12.4.2 Model Analysis

The survivor functions at two stress levels S_1 and S_2 are related as follows:

$$\bar{F}(t|S_2) = [\bar{F}(t|S_1)]^{\psi(S_2)/\psi(S_1)} \tag{12.21}$$

The ratio of the hazard functions is

$$\frac{h(t|S_1)}{h(t|S_2)} = \frac{\psi(S_1)}{\psi(S_2)} \tag{12.22}$$

which is independent of t.

Since the Weibull PH model is identical to the Weibull ATF model, the analysis of Model IV(b)-1 is applicable here.

12.4.3 Parameter Estimation

Graphical Approach

Under the transformation given by (1.7), for a given S, (12.22) gets transformed into

$$y = \beta x - \beta \ln(\alpha) + \ln[\psi(S)] \tag{12.23}$$

This implies that the plots for different values of S are straight lines parallel to each other. If the data is roughly scattered along parallel straight lines, then the model parameters can be estimated from the plots in the usual way ensuring that the fitted lines are parallel to each other.

Method of Maximum Likelihood
The parameters of $\psi(S)$ (also called the *regression* parameters) need to be estimated in addition to the parameters of the baseline hazard function $h_0(t)$, which in the case of the Weibull PH model are α and β.

Cox (1975) suggests an approach to estimate the regression parameters without the knowledge of the baseline hazard function based on *partial* likelihood, which is not a likelihood in the normal sense [see Kalbafleisch and Prentice (1980) for further discussion].

The literature on the estimation of parameters with different types of censoring has received a lot of attention. They can be found in many books such as Kalbafleisch and Prentice (1980), Lawless (1982), and Cox and Oakes (1984). See also Cox and Oakes (1984) and Kalbfleisch and Prentice (1980). Love and Guo (1991) and Newby (1994) discuss the problem of model fitting. Gertsbakh (1989) provides a detailed discussion of estimation for proportional hazard models involving Weibull and exponential distributions.

12.5 MODEL IV(d)-1

12.5.1 Model Formulation

Distribution Function
The distribution function is given by

$$F(t|S) = 1 - \exp\left(\frac{g_1(a^T S)}{g_2(b^T S)} \ln\left\{1 - F_0\left[g_2(b^T S)t\right]\right\}\right) \tag{12.24}$$

where S is the vector of covariates, a and b are vectors of regression coefficients, $g_1(\cdot), g_2(\cdot)$ are general positive function with $g_1(0) = g_2(0) = 1$, and $F_0(t)$ is the baseline distribution function. Note that when $a = 0$ the model reduces to the AFT model, and when $b = 0$ it reduces to the PH model.

The hazard function is given by

$$h(t|S) = g_1(a^T S) \cdot h_0[g_2(b^T S)t] \tag{12.25}$$

where $h_0(t)$ is the hazard function associated with $F_0(t)$.

Special Case Shyur et al. (1999) deal with the special case $g_1(a^T S) = \exp(a^T S)$ and $g_2(b^T S) = \exp(b^T S)$.

12.5.2 Model Analysis

Moments

For the special case, the MTTF can be computed as

$$\text{MTTF} = \frac{1}{\exp(b^T S)} \cdot \int_0^\infty R(u)^{\exp((a-b)^T S)}\, du \tag{12.26}$$

with $u = \exp(b^T S)t$ and $R_0(t) = \exp\left[-\int_0^u h_0(s)\, ds\right]$.

12.5.3 Parameter Estimation

Maximum-Likelihood Estimation

Since the baseline hazard function is not independent of the covariates, the advantage of applying a partial-likelihood function does not exist in this model. A general-likelihood function is used in Shyur et al. (1999). For the special case, the log-likelihood function is given by

$$L = \sum_{i=1}^{N} \delta_i \left[\alpha^T S + \ln\left(h_0(u_i; \zeta)\right)\right] - \sum_{i=1}^{N} \exp\left[\frac{\exp(a^T S)}{\exp(b^T S)} \cdot \int_0^{u_i} h_0(t|S)\, dt\right] \tag{12.27}$$

where N is the number of items tested, t_i is the observation time, $\delta_i = 1$ if the data i is a failure time and 0 otherwise, $u_i = t_i \exp(b^T S)$, and ζ denotes the vector of parameters defining $h_0(\cdot)$ when a spline function is used for the estimation of the baseline hazard function.

When $F_0(t)$ is a two-parameter Weibull distribution, then one does not need to use a spline function to approximate $h_0(\cdot)$, the baseline hazard function.

12.6 TYPE IV(e) MODELS: RANDOM PARAMETERS

Here the one or both of the parameters ($\theta \equiv \alpha, \beta$) in the standard two-parameter Weibull model are random variables. If θ can assume only a finite set of values, then

$$F(t) = \sum_{i}^{K} F_0(t|\theta = u) p_i(u) \tag{12.28}$$

is the distribution function of a variable T, which, conditional on $\theta = u$, has a distribution $F_0(t|\theta = u)$. The expression $F_0(t|\theta = u)$ is a two-parameter Weibull distribution, and p_i is the probability $\theta = \theta_i$, $1 \leq i \leq K$. As a result, the model is a finite mixture model discussed in Chapter 8. Here we consider the case where the parameters are continuous valued random variables.

Harris and Singpurwalla (1968) consider the following cases:

1. The shape parameter β has a two-point prior and the scale parameter is constant.
2. The scale parameter α has a two-point prior and the shape parameter is constant.

They discuss the Bayesian approach to estimate the model parameters.

In this section we focus our attention on the case where θ is a continuous valued random variable. Let $F_\theta(\cdot)$ denote the distribution function for the parameters. Then the model is given by

$$F(t) = \int F_0(t|\theta = u)\, dF_\theta(u) \tag{12.29}$$

and the model is also referred to as a *continuous mixture* or *compound* model.

12.6.1 Model Formulations

Model IV(e): Compound Weibull Model

Here only the scale parameter is a random variable with a distribution function $F_\alpha(\cdot)$ and the shape parameter is a constant. It is called a *compound Weibull model* [see Dubey (1968)]. Different forms for $F_\alpha(\cdot)$ lead to different models.

Model IV(e)-1: Gamma Prior

Dubey (1968) considers the case where the probability density function for $\lambda = 1/\alpha^\beta$ is a gamma density function given by

$$f_\alpha(a) = \frac{\delta^\gamma a^{\gamma-1} e^{-\delta a}}{\Gamma(\gamma)} \qquad a \geq 0 \tag{12.30}$$

with $\gamma, \delta > 0$. Let

$$F_0(t|\lambda = a) = 1 - \exp(-at^\beta) \tag{12.31}$$

Then

$$f(t) = \int_0^\infty f_0(t|\alpha = a) f_\alpha(a)\, da = \frac{\beta\gamma\delta^\gamma t^{\beta-1}}{(t^\beta + \delta)^{\gamma+1}} \tag{12.32}$$

When $\delta = 1$, the compound Weibull distribution reduces to the Burr distribution, and hence the compound Weibull distribution can be considered as a generalized Burr distribution.

Model IV(e)-2: Uniform Prior

Harris and Singpurwalla (1968) consider the case where the probability density function for $\lambda = 1/\alpha^\beta$ is a uniform density function [over (b_1, b_2)] given by

$$F_a(a) = \begin{cases} 0 & a \le b_1 \\ (a - b_1)/(b_2 - b_1) & b_1 < a \le b_2 \\ 1 & a > b_2 \end{cases} \tag{12.33}$$

Using (12.33) and (12.31) in (12.29) yields

$$F(t) = 1 - \frac{\exp(-b_1 t^\beta) - \exp(-b_2 t^\beta)}{(b_2 - b_1)t^\beta} \tag{12.34}$$

12.6.2 Model Analysis

Model IV(e)-1: Gamma Prior

Moments

The kth moment is given by

$$E(T^k) = \gamma\delta^{k/\beta} B\left(\gamma - \frac{k}{\beta}, \frac{k}{\beta} + 1\right) \tag{12.35}$$

when $\beta\gamma > k$ $B(,)$ is the beta function [see Abramowitz and Stegun (1964)]. Specifically, the mean and variance are given by

$$E(T) = \frac{\delta^{1/\beta}\Gamma(\gamma - 1/\beta)\Gamma(1/\beta + 1)}{\Gamma(\gamma)} \tag{12.36}$$

and

$$\mathrm{Var}(T) = \frac{\delta^{2/\beta}\left[\Gamma\left(\gamma - \frac{2}{\beta}\right)\Gamma\left(\frac{2}{\beta} + 1\right) - \Gamma^2\left(\gamma - \frac{1}{\beta}\right)\Gamma^2\left(\frac{1}{\beta} + 1\right)/\Gamma(\gamma)\right]}{\Gamma(\gamma)} \tag{12.37}$$

Order Statistics

The following result is from Dubey (1968). Let $X_1, X_2, \ldots, X_n$ be n independent and identically distributed (i.i.d.) random variables from the compound distribution with parameters $(\beta, \lambda, \gamma, \delta)$. Then $Y_n = \min(X_1, X_2, \ldots, X_n)$; then Y_n obeys the compound Weibull law with parameters $(\beta, \lambda, n\gamma, \delta)$. Conversely, if Y_n has the compound Weibull distribution with the parameters $(\beta, \lambda, n\gamma, \delta)$, then each X_i obeys the compound Weibull law with the parameters $(\beta, \lambda, \gamma, \delta)$.

Model IV(e)-2: Uniform Prior

Moments

The first and second moments are given by

$$E[T] = \Gamma\left(\frac{1}{\beta}+1\right)\frac{\beta}{(b_2-b_1)(\beta-1)}\left(b_2^{1-1/\beta}-b_1^{1-1/\beta}\right) \tag{12.38}$$

$$E[T^2] = \Gamma\left(\frac{2}{\beta}+1\right)\frac{\beta}{(b_2-b_1)(\beta-1)}\left(b_2^{1-2/\beta}-b_1^{1-2/\beta}\right) \tag{12.39}$$

and the variance is given by

$$\begin{aligned}\text{Var}[T] &= \Gamma\left(\frac{2}{\beta}+1\right)\frac{\beta}{(b_2-b_1)(\beta-1)}\left(b_2^{1-2/\beta}-b_1^{1-2/\beta}\right)\\ &\quad - \Gamma^2\left(\frac{1}{\beta}+1\right)\frac{\beta^2}{(b_2-b_1)^2(\beta-1)^2}\left(b_2^{1-1/\beta}-b_1^{1-1/\beta}\right)^2\end{aligned} \tag{12.40}$$

12.6.3 Parameter Estimation

Model IV(e)-1: Gamma Prior

For the complete data given by $(t_1, t_2, \ldots, t_n)$, Harris and Singpurwalla (1968) suggest estimating the parameters (other than λ) by the method of moments and then using a Bayesian approach to estimating λ, which is given by

$$\hat{\lambda} = \frac{n+\hat{\gamma}+1}{\sum_{i=1}^{n} t_i^{\hat{\beta}} + 1/\hat{\delta}} \tag{12.41}$$

with estimates of other parameters obtained from the method of moments.

Model IV(e)-2: Uniform Prior

Harris and Singpurwalla (1968) suggest the following approach for the case of complete data given by $(t_1, t_2, \ldots, t_n)$. An estimate of β is obtained from the conditional skewness and is given by the solution of the following equation:

$$\frac{\Gamma(3/\beta+1) - 3\Gamma(3/\beta+1)\Gamma(1/\beta+1) + 2\Gamma^3(1/\beta+1)}{[\Gamma(2/\beta+1)-\Gamma^2(1/\beta+1)]^{3/2}} = \frac{\sum(t_i-\bar{t})^2}{s^3} \tag{12.42}$$

where $\bar{t}$ is the sample mean and s is the sample standard deviation. The estimates of b_1 and b_2 are obtained from (12.38) and (12.29) using sample estimates for the first two moments and a Bayesian approach is used to obtain the estimate of λ, which is given by

$$\hat{\lambda} = \frac{\Gamma\left(n+2, b_2 \sum_{i=1}^{n} t_i^{\beta}\right) - \Gamma\left(n+2, b_1 \sum_{i=1}^{n} t_i^{\beta}\right)}{\sum_{i=1}^{n} t_i^{\beta} \Gamma\left(n+1, b_2 \sum_{i=1}^{n} t_i^{\beta}\right) - \Gamma\left(n+1, b_1 \sum_{i=1}^{n} t_i^{\beta}\right)} \tag{12.43}$$

using the estimates of b_1, b_2, and β instead of the true values.

12.7 BAYESIAN APPROACH TO PARAMETER ESTIMATION

In the Bayesian approach, the unknown parameters of the model is viewed as random variables and characterized by an appropriate prior distribution function. Based on the data (censored and/or uncensored), one computes the posterior distribution, which is the conditional distribution for the parameters conditional on the observed data. As a result, the estimation uses random parameter models discussed in the previous section. Depending on the form of the prior distribution, one has a family of different estimators.

The Bayesian approach to estimating the parameters of the standard Weibull model has received a lot of attention. Some deal with one parameter where the shape parameter is known, and others deal with the case where both the shape and scale parameters are unknown. Some use discrete distributions, whereas others use continuous distributions to characterize the unknown parameters.

Bayesian point estimator for the parameters can be computed as the expected value of the posterior distribution in a traditional manner. Bayesian prediction interval can also be computed. For further details, see Mann et al. (1974), Martz and Waller (1982), Singpurwalla (1988), and Berger and Sun (1993).

EXERCISES

Data Set 12.1 Failure Times of Eight Components at Three Different Temperatures

Temp.	Lifetimes							
100	14.712	32.644	61.979	65.521	105.50	114.60	120.40	138.50
120	8.610	11.741	54.535	55.047	58.928	63.391	105.18	113.02
140	2.998	5.016	15.628	23.040	27.851	37.843	38.050	48.226

Data Set 12.2 Accelerated Life Testing of 40 Items with Change in Stress from 100 to 150 at $t = 15$

0.13	0.62	0.75	0.87	1.56	2.28	3.15	3.25	3.55	4.49
4.50	4.61	4.79	7.17	7.31	7.43	7.84	8.49	8.94	9.40
9.61	9.84	10.58	11.18	11.84	13.28	14.47	14.79	15.54	16.90
17.25	17.37	18.69	18.78	19.88	20.06	20.10	20.95	21.72	23.87

12.1. For each of the three different temperatures in Data Set 12.1, plot the WPP plot. Discuss if the Weibull AFT model is appropriate to model the effect of temperature on the lifetime of the item.

12.2. Suppose that the data in Data Set 12.1 can be modeled by the Weibull AFT model. Estimate the model parameters using the WPP plot and the method of maximum likelihood. Compare the two estimates.

12.3. Using the Weibull AFT model of Exercise 12.2, estimate the mean and median of the time to failure at $T = 160$.

12.4. Carry out a WPP plot of the data in Data Set 12.2. Interpret the plot and discuss if one of the models discussed in Chapter 11 can be used to model the data.

12.5. For the Arrhenius–Weibull model, derive expression for the hazard function as function of the stress s. Obtain the mean and variance of the time to failure as a function of s.

12.6. Repeat Exercise 12.5 for the power Weibull model.

12.7. Repeat Exercise 12.5 for the power Weibull model.

12.8. Prove that the PH and AFT models are equivalent for Weibull baseline hazard function.

12.9. Derive (12.35) to (12.37).

12.10. An item with two failure modes can be modeled by a Weibull mixture model [Model III(a)-1] with two subpopulations. The effect of stress on the two scale parameters is given by the power law relationship with different exponents. Discuss the shape of the WPP plot.*

12.11. Show that a Weibull mixture PH model is not equivalent to a Weibull mixture AFT model.*

12.12. Section 12.3 deals with the analysis of the Weibull distribution where the parameters are functions of stress. Carry out a similar analysis for the inverse Weibull distribution.*

* Research problem.

CHAPTER 13

Type V Weibull Models

13.1 INTRODUCTION

The models studied so far are appropriate for modeling a continuous random variable that can assume any value over the interval $[0, \infty)$. In this chapter we look at Weibull models for modeling discrete random variables that only assume nonnegative integer values. Such models are useful, for example, for modeling the number of cycles to failure when components are subjected to cyclical loading.

Discrete Weibull models can be obtained as the discrete counterparts of either the distribution function or the failure rate function of the standard Weibull model. These lead to different models, and in this chapter we discuss four discrete models that are the counterparts of the standard two-parameter Weibull distribution.

The outline of the chapter is as follows. In Section 13.2 we introduce some concepts and notations that will be used later in the chapter. The next four sections (Sections 13.3 to 13.6) deal with the four different discrete Weibull models.

13.2 CONCEPTS AND NOTATION

In this section we discuss some basic concepts needed for modeling a random variable X that can assume any nonnegative integer value:

Probability mass function: $p(x) = P(X = x)$ for $0 \leq x$ (integer) $< \infty$.

Probability distribution function: $F(x) = P(X \leq x)$ for $0 \leq x$ (integer) $< \infty$.

Survivor (reliability) function: $\bar{F}(x) = 1 - F(x) = P(X > x)$ for $0 \leq x$ (integer) $< \infty$.

Hazard (failure rate) function:

$$\lambda(x) = \Pr(X = x | X \geq x) = \frac{\Pr(X = x)}{\Pr(X \geq x)} = \frac{\bar{F}(x-1) - \bar{F}(x)}{\bar{F}(x-1)} \qquad x = 1, 2, \ldots \tag{13.1}$$

Weibull Models, by D.N.P. Murthy, Min Xie, and Renyan Jiang.
ISBN 0-471-36092-9

Define

$$\Lambda(x) = \sum_{i=0}^{x} \lambda(i) \tag{13.2}$$

Note that $\bar{F}(x) \neq \exp(-\Lambda(x))$. This is in contrast to the continuous case [see (3.1)].

Roy and Gupta (1992) propose an alternate discrete failure rate function $r(x)$ defined as follows:

$$r(x) = \ln\left[\frac{\bar{F}(x-1)}{\bar{F}(x)}\right] \qquad x = 1, 2, \ldots \tag{13.3}$$

and called it the second "rate of failure." Define

$$R(x) = \sum_{i=1}^{x} r(i) \tag{13.4}$$

Then it is easily seen that

$$\bar{F}(x) = \exp[-R(x)] \tag{13.5}$$

Note that this is similar to the relationship between the cumulative hazard function and the survivor function for the continuous case given in (3.1). Xie et al. (2002a) advocate that $r(x)$ be called the discrete failure rate because of this similarity with the continuous case. However, it is important to note that $r(x)$ is not a conditional probability whereas $\lambda(x)$ is a conditional probability. Hence, we shall use the term *pseudo-hazard function* for $r(x)$ so as to differentiate it from the hazard function $\lambda(x)$.

13.3 MODEL V-1

This model is due to Nakagawa and Osaki (1975).

13.3.1 Model Structure

The distribution function is given by

$$F(x) = \begin{cases} 1 - q^{x^\beta} & x = 0, 1, 2, 3 \ldots \\ 0 & x < 0 \end{cases} \tag{13.6}$$

where the parameters $0 < q < 1$ and $\beta > 0$. Note that it is similar to the two-parameter Weibull distribution given by (1.3).

The probability mass function is given by

$$p(x) = q^{(x-1)^\beta} - q^{x^\beta} \qquad x = 0, 1, 2, 3 \ldots \tag{13.7}$$

The hazard function is given by

$$\lambda(x) = 1 - q^{x^{\beta}-(x-1)^{\beta}} \tag{13.8}$$

From (13.3) we have

$$r(x) = \ln\left(\frac{q^{(x-1)^{\beta}}}{q^{x^{\beta}}}\right) = \ln\left(q[(x-1)^{\beta} - x^{\beta}]\right) \tag{13.9}$$

Consider a geometric random variable Y with $P(Y \geq x) = q^k$ (i.e., Y is distributed according to the geometric distribution). Then the random variable $X = Y^{1/\beta}$, $\beta > 0$, has the property $P(X \geq x) = q^{x^{\beta}}$, implying that X is distributed according the discrete Weibull distribution. When $\beta = 1$, the discrete Weibull distribution reduces to the geometric distribution. This is the counterpart of the power law relationship linking the exponential and the Weibull distribution in the continuous case.

13.3.2 Model Analysis

Moments

The kth moment is given by

$$E(X^k) = \sum_{x=1}^{\infty} x^k \left(q^{(x-1)^{\beta}} - q^{x^{\beta}}\right) \tag{13.10}$$

It is difficult to get closed-form analytical expressions for the moments. One needs to evaluate this numerically, given specific parameter values.

Khan et al. (1989) compare the means for the discrete and the continuous case. For the continuous case, let the distribution be given by (1.6). Let $q = \exp\{-\lambda'\}$ and μ_d and μ_c denote the means for the discrete and continuous case. Then Khan et al. (1989) show that $\mu_d - 1 < \mu_c < \mu_d$ using the following result:

$$\sum_{x=1}^{\infty} q^{x^{\beta}} < \int_0^{\infty} e^{-\lambda' x^{\beta}}\, dx < \sum_{x=0}^{\infty} q^{x^{\beta}} \tag{13.11}$$

Hazard Function

The hazard function can be increasing, decreasing, or constant depending on whether β is greater than, equal to, or less than 1. This is similar to the continuous distribution.

The pseudo-hazard function can be increasing or decreasing depending on the value of β. Note that when $\beta = 2$, it increases linearly, which is similar to the continuous case.

13.3.3 Parameter Estimation

Method of Moments

Moment estimators for the discrete Weibull distribution can be obtained as follows for the case of complete data given by $(x_1, x_2, \ldots, x_n)$. The first and second moments about the origin, μ_1 and μ_2, are given by

$$\mu_1 = \sum_{x=0}^{\infty} q^{x^\beta} \qquad \mu_2 = 2\sum_{x=0}^{\infty} xq^{x^\beta} - \mu_1^2 \tag{13.12}$$

The sample moments are given by

$$\hat{\mu}_1 = \frac{1}{n}\sum_{i=0}^{n} x_i \qquad \hat{\mu}_2 = \frac{1}{n}\sum_{i=0}^{n}(x_i - \hat{\mu})^2 \tag{13.13}$$

Khan et al. (1989) propose a method to estimate the parameters by minimizing

$$J(q, \beta) = (\hat{\mu}_1 - \mu_1)^2 + (\hat{\mu}_2 - \mu_2)^2 \tag{13.14}$$

with respect to (q, β).

Another simple estimation method proposed by Khan et al. (1989) is as follows. Let y be the number of 1's in the sample. The ratio y/n is an estimate of the probability $F(1) = 1 - q$. Hence

$$\hat{q} = 1 - y/n \tag{13.15}$$

is an estimate of q. Similarly, let the number of 2's in the sample be denoted by z. Then, from $F(2) = q - q^2$ we have the estimate

$$\hat{\beta} = \frac{\ln[\ln(q - z/n)/\ln(q)]}{\ln 2} \tag{13.16}$$

Method of Maximum Likelihood

For complete data set, $(x_1, x_2, \ldots, x_n)$, the log-likelihood function is

$$\ln(L) = \sum_{k=1}^{\infty} \ln\left(q^{(x_k-1)^\beta} - q^{x_k^\beta}\right) \tag{13.17}$$

The estimates can be obtained from the usual first-order conditions (setting the partial derivatives to zero).

Kulasekera (1994) investigates the case with right censored data. Let

$$\delta_i = \begin{cases} 1 & \text{if } x_i = x_i^0 \\ 0 & \text{otherwise} \end{cases} \tag{13.18}$$

where $x_i = \min(x_i^0, L_i), i = 1, 2, \ldots, n$, and x_i is the ith failure time, L_i is the ith censoring time, and x_i^0 is the failure time. The estimates are obtained as the solution of the following set of equations that need to be solved numerically:

$$\sum_{i=1}^{n}(1-\delta_i)x_i^{\beta} + \sum_{i=1}^{n}\delta_i(x_i-1)^{\beta} + \sum_{i=1}^{n}\delta_i \frac{(x_i-1)^{\beta} - x_i^{\beta}}{q^{(x_i-1)^{\beta}-x_i^{\beta}} - 1} = 0 \tag{13.19}$$

and

$$\sum_{i=1}^{n}(1-\delta_i)x_i^{\beta} \ln x_{i_i} + \sum_{X_i \neq 1}^{n}\delta_i(x_i-1)^{\beta} \ln(x_i-1) + \sum_{i=1}^{n}\delta_i \frac{(x_i-1)^{\beta} \ln(x_i-1) - x_i^{\beta} \ln(x_i)}{q^{(x_i-1)^{\beta}-x_i^{\beta}} - 1} = 0 \tag{13.20}$$

Kulasekera (1994) suggests an approximation that involves the simple estimator of Khan et al. (1989).

13.4 MODEL V-2

The model is due to Stein and Dattero (1984).

13.4.1 Model Structure

The distribution function is given by

$$F(x) = \sum_{k=1}^{x-1} ck^{\beta-1} \prod_{j=1}^{k-1}(1 - cj^{\beta-1}) \qquad x = 1, 2, \ldots, m \tag{13.21}$$

where m is a truncation value, given by

$$m = \begin{cases} \text{int}\{c^{-[1/(\beta-1)]}\} & \text{if } \beta > 1 \\ \infty & \text{if } \beta \leq 1 \end{cases} \tag{13.22}$$

and int$\{\cdot\}$ represents the integer part of the quantity inside the brackets.

The hazard function has a form similar to that of continuous Weibull distribution, and it is given by

$$\lambda(x) = \begin{cases} cx^{\beta-1} & x = 1, 2, \ldots, m \\ 0 & x \leq 0 \end{cases} \tag{13.23}$$

Note that the constraint on m ensures that $\lambda(x) \leq 1$.

The probability mass function is given by

$$p(x) = cx^{\beta-1}\prod_{j=1}^{x-1}(1 - cj^{\beta-1}) \qquad x = 1, 2, \ldots, m \tag{13.24}$$

When $\beta = 1$, the distribution reduces to the geometric distribution that can be viewed as the discrete counterpart of the exponential distribution.

The pseudo-failure rate function is given by

$$r(x) = \ln\{1/(1 - cx^{\beta-1})\} \qquad x = 1, 2, \ldots, m \tag{13.25}$$

Note that this is similar to the hazard function. Both of these can be increasing, decreasing, or constant, depending on the value of β. Also, they are a power function of $(\beta - 1)$, which is similar to the continuous case except that x is constrained by an upper limit m.

13.5 MODEL V-3

Padgett and Spurrier (1985) proposed this model.

13.5.1 Model Structure

The distribution function is given by

$$F(x) = 1 - \exp\left(-\sum_{i=1}^{x} ci^{\beta-1}\right) \qquad x = 0, 1, 2, \ldots \tag{13.26}$$

The probability mass function is given by

$$\begin{aligned} p(x) &= \exp\left(-c\sum_{i=1}^{x} i^{\beta-1}\right) - \exp\left(-c\sum_{i=1}^{x-1} i^{\beta-1}\right) \\ &= \exp\left(-c\sum_{i=1}^{x-1} i^{\beta-1}\right)\left[1 - \exp(-cx^{\beta-1})\right] \end{aligned} \tag{13.27}$$

The hazard function is given by

$$\lambda(x) = \exp\left(-c\sum_{i=1}^{x} i^{\beta-1}\right) \tag{13.28}$$

and the pseudo-hazard rate function is given by

$$r(x) = cx^{\beta-1} \qquad x = 1, 2, \ldots \tag{13.29}$$

Again, the hazard and the psuedo-hazard functions can be increasing, decreasing, or constant depending on β.

13.5.2 Model Analysis

Moments

The mean is given by

$$E(X) = \sum_{k=1}^{\infty} \exp\left(-c\sum_{j=1}^{k} j^{\beta-1}\right) \tag{13.30}$$

and this exists if $\beta > 0$, or $\beta = 0$ and $c > 1$.

13.5.3 Parameter Estimation

When the data is complete, the log-likelihood function is given by

$$\ln L = \sum_{i=1}^{n} \ln\left\{1 - \exp\left[-c(x_i + 1)^{\beta-1}\right]\right\} - c\sum_{i=1}^{n}\sum_{j=1}^{x_i} j^{\beta-1} \tag{13.31}$$

However, this has to be solved numerically. Padgett and Spurrier (1985) discuss this further. For example, when all $x_i = 0$, the log-likelihood function (13.31) is an increasing function of c, and it does not involve β. If at least one $x_i > 0$, then the estimates can be found numerically.

13.6 MODEL V-4

This model is briefly mentioned in Salvia (1996) and the hazard function is given by

$$\lambda_k = 1 - e^{-c(k+1)^{\beta}} \qquad k = 0, 1, \ldots \tag{13.32}$$

and $c > 0$ and $\beta > -1$ are the model parameter.

For small c, it can be seen that $\lambda_k \approx c(k+1)^{\beta}$, and this model is hence reminiscent of the Weibull model. The hazard function is increasing when $\beta > 0$ and decreasing when $\beta < 0$.

For $\beta > 0$, the mean residual life, ν_k, is bounded by

$$e^{\{-c(k+1)^{\beta}\}} \leq \nu_k \leq \frac{1}{e^c - 1} \tag{13.32}$$

and by

$$\frac{1}{e^c - 1} \leq \nu_k \leq \exp\{c\}\nu_{k-1} \qquad \nu_k \leq \exp\{c\}E[X] \tag{13.33}$$

for $\beta < 0$.

EXERCISES

Data Set 13.1 Complete Data: Number of Shocks before Failure

2	3	6	6	7	9	9	10	10	11
12	12	12	13	13	13	15	16	16	18

Data Set 13.2 Censored Data: 20 Items Subjected to Shocks and Testing Stopped after 14 Shocks[a]

1	3	3	4	4	4	4	5
5	6	6	7	10	11	12	14

[a] The data comprises of the number of shocks to failure for failed items.

13.1. Discuss if one or more of the models of Chapter 13 are suitable for modeling Data Set 13.1.

13.2. Suppose that Data Set 13.1 can be modeled by Model V-1. Estimate parameters using the method of moments and the method of maximum likelihood.

13.3. Suppose that Data Set 13.1 can be modeled by Model V-3. Estimate parameters using the method of moments and the method of maximum likelihood.

13.4. Repeat Exercise 13.1 with Data Set 13.2.

13.5. Suppose that Data Set 13.2 can be modeled by Model V-1. Estimate parameters using the method of maximum likelihood.

13.6. Develop a graphical procedure to estimate the parameters in the Nakagawa–Osaki discrete Weibull distribution.*

13.7. Develop a goodness-of-fit test that can be used for validating the discrete Weibull models discussed in Chapter 13.*

13.8. Show that the hazard function defined in (13.1) satisfies $0 \leq \lambda(x) \leq 1$.

13.9. Show that the pseudo-hazard function $r(x)$ defined in (13.3) has the same monotonicity property as the hazard function defined in (13.1). That is, if one is increasing (decreasing) the other is also increasing (decreasing).

13.10. For Model V-3 plot (13.28) and (13.29) and compare the results.

13.11. Discuss possible generalization of discrete Weibull for some three- and four-parameter model such as those discussed in previous chapters.*

* Research problem.

CHAPTER 14

Type VI Weibull Models (Multivariate Models)

14.1 INTRODUCTION

This group is comprised of models that are extensions of the univariate Weibull distribution to bivariate ($n = 2$) and multivariate ($n > 2$) cases so that the model is given by an n-dimensional distribution function $F(t_1, t_2, \ldots, t_n)$. This is useful for modeling failures in a multicomponent system where the failure times are statistically dependent. (When component failures are independent, they can be analyzed separately by looking at the marginal univariate distributions.) In this chapter, we discuss a variety of multivariate Weibull models that are related to the standard Weibull distribution.

There are many different approaches to develop multivariate extensions of the one-dimensional Weibull distribution. One simple approach is to transform the bivariate or multivariate exponential distribution through power transformation. Another approach is to specify the dependence between two or more univariate Weibull variables so that bivariate or multivariate distribution has Weibull marginals. Yet another approach involves the transformation of multivariate extreme value distributions.

Many different bivariate and multivariate Weibull models have been proposed and studied. We confine our attention to those models that have been studied in detail.

The outline of this chapter is as follows. We start with some preliminary discussion where we introduce the notation and discuss alternate approaches to classify the different multivariate distributions in Section 14.2. Section 14.3 deals with the bivariate Weibull models and Section 14.4 with the multivariate models. In contrast to the univariate models, the results for the bivariate and multivariate models are

Weibull Models, by D.N.P. Murthy, Min Xie, and Renyan Jiang.
ISBN 0-471-36092-9

rather limited. For the models discussed, we review the results relating to model analysis and parameter estimation.

14.2 SOME PRELIMINARIES AND MODEL CLASSIFICATION

14.2.1 Preliminaries

A general multivariate model involves n random variables $(T_1, T_2, \ldots, T_n)$ with a distribution function $F(t_1, t_2, \ldots, t_n)$ so that

$$F(t_1, t_2, \ldots, t_n) = P(T_1 \leq t_1, T_2 \leq t_2, \ldots, T_n \leq t_n) \tag{14.1}$$

The survivor function, $\bar{F}(t_1, t_2, \ldots, t_n)$ is given by

$$\bar{F}(t_1, t_2, \ldots, t_n) = P(T_1 > t_1, T_2 > t_2, \ldots, T_n > t_n) \tag{14.2}$$

Note that in contrast to the univariate case, we have

$$F(t_1, t_2, \ldots, t_n) + \bar{F}(t_1, t_2, \ldots, t_n) \leq 1 \tag{14.3}$$

If $F(t_1, t_2, \ldots, t_n)$ is differentiable, then the n-dimensional density function is given by

$$f(t_1, t_2, \ldots, t_n) = \frac{\partial^n F(t_1, t_2, \ldots, t_n)}{\partial t_1 \ \partial t_2 \cdots \partial t_n} \tag{14.4}$$

The n-marginal distributions are given by

$$F_i(t_i) = F(\infty, \ldots, \infty, t_i, \infty, \ldots, \infty) \qquad 1 \leq i \leq n \tag{14.5}$$

One can define several conditional density and distribution functions. Let I and J denote two disjoint sets such that $J \cap I = \emptyset$ and $J \cup I = \{1, 2, \ldots, n\}$. Then the conditional density function is given by

$$f(t_i, i \in I | t_j, j \in J) = \frac{f(t_1, t_2, \ldots, t_n)}{f(t_j, j \in J)} \tag{14.6}$$

For the general case one can define several different hazard functions, and the details can be found in Shaked and Shanthikumar (1994). We confine our discussion to the bivariate case where we have the two marginal distribution functions $[F_1(t_1)$ and $F_2(t_2)]$ and the two conditional distribution functions $[F_1(t_1|t_2)$ and $F_2(t_2|t_1)]$. These are all univariate distributions and result in four different hazard functions associated with the bivariate distribution function. These notions are appropriate in the reliability context where the two random variables (T_1 and T_2)

represent the lifetimes of the two components. Yang and Nachlas (2001) define the bivariate hazard function as

$$h(t_1, t_2) = \frac{f(t_1, t_2)}{\overline{F}(t_1, t_2)} \tag{14.7}$$

where the two random variables denote the age and usage of a component at failure. This notion can be extended to the more general case. For further discussion on the hazard functions, see Block and Savits (1981), Zahedi (1985), and Roy and Mukherjee (1988).

14.2.2 Classification of Multivariate Distributions

Several different classification schemes have been proposed by classifying the different multivariate distributions. Since multivariate distributions are useful for the modeling of dependent random variables, a classification that has received some attention is the one based on the dependence structure between the random variables. Hougaard (1980) and Lu and Bhattacharyya (1990) have studied the dependence structure of some bivariate Weibull models. In reliability and survival analysis, multivariate distributions have been classified based on their aging properties. Various classes are discussed in Roy (1994). Hutchinson and Lai (1990) discuss various continuous distributions based on dependence and aging properties.

Lee (1979) comments that although there are several multivariate Weibull models that have been suggested, most of them appear to have little in common with the univariate Weibull except that the marginal distributions are Weibull. He then presents a detailed classification for a class of multivariate distributions having Weibull minimums after arbitrary scaling. Random variables $T_1, T_2, \ldots, T_n$ are said to have such a distribution if for arbitrary constants $a_i > 0, i = 1, 2, \ldots, n$; $\min_i(a_i T_i)$ has a one-dimensional Weibull distribution given by

$$P\left[\min_i(a_i T_i) > t\right] = \exp[-k(\mathbf{a})t^{\beta}] \qquad t \geq 0 \tag{14.8}$$

for some $\beta > 0$ and $k(\mathbf{a})$ is a scalar function of the vector $\mathbf{a} = (a_1, a_2, \ldots, a_n)$.

The hierarchy of classes of multivariate Weibull distributions presented in Lee (1979) is as follows:

A. $T_1, T_2, \ldots, T_n$ are independent and each has a Weibull distribution of the form $F_i(t) = \exp(-\lambda_i t^{\beta})$, $t \geq 0, i = 1, 2, \ldots, n$.

B. $T_1, T_2, \ldots, T_n$ have a multivariate Weibull distribution generated from independent Weibull distributions by letting

$$T_i = \min(Z_i; i \in J), i = 1, 2, \ldots, n$$

where the sets J are elements of a class $\Im$ of nonempty subsets of $\{1, 2, \ldots, n\}$ having the property that for each $i, i \in J$ for some $J \in \Im$, and the random variables $Z_J, J \in \Im$, are independent having Weibull distributions of the form $\bar{F}_j(t) = \exp(-\lambda_J t^\beta)$.

C. $T_1, T_2, \ldots, T_n$ have a joint distribution satisfying (14.8).

D. $T_1, T_2, \ldots, T_n$ have a joint distribution with Weibull minimums, that is,

$$P\big(\min_{i \in S} T_i > t\big) = \exp(-\lambda_S t^\beta)$$

for some $\lambda_S > 0$ and all nonempty subsets S of $\{1, 2, \ldots, n\}$.

E. Each $T_i, i = 1, 2, \ldots, n$ has a Weibull distribution of the form $\bar{F}_i(t) = \exp(-\lambda_i t^{\beta_i})$ with $\beta_i > 0, i = 1, 2, \ldots, n$.

As noted in Lee (1979), classes A to E are distinct since each class is seen to contain distributions not belonging to the class preceding it. He illustrates this by giving several examples of distributions proposed by other researchers.

14.3 BIVARIATE MODELS

14.3.1 Model VI(a)-1

This model was proposed by Marshall and Olkin (1967) and derived as an extension of bivariate exponential distribution. The model is given by

$$\bar{F}(t_1, t_2) = \exp\big\{ - \big[\lambda_1 t_1^{\beta_1} + \lambda_2 t_2^{\beta_2} + \lambda_{12} \max\big(t_1^{\beta_1}, t_2^{\beta_2}\big)\big]\big\} \tag{14.9}$$

It reduces to a simple bivariate exponential distribution when $\beta_1 = \beta_2 = 1$. Furthermore, with the transformation $Y_1 = T_1^{\beta_1}$ and $Y_2 = T_2^{\beta_2}$, (Y_1, Y_2) follows the bivariate exponential distribution.

When $\beta_1 \neq \beta_2$, the distribution satisfies class E, but not class D of the Lee classification scheme given in Section 14.2.2. On the other hand, when $\beta_1 = \beta_2$ and $\lambda_{12} > 0$, this distribution belongs to class B but not class A.

Model Analysis

The marginal distributions of (14.9) are given by

$$F_1(t_1) = 1 - \exp\big[- (\lambda_1 + \lambda_{12}) t_1^{\beta_1}\big] \tag{14.10}$$

and

$$F_2(t_2) = 1 - \exp\big[- (\lambda_2 + \lambda_{12}) t_2^{\beta_2}\big] \tag{14.11}$$

The bivariate distribution contains an absolutely continuous part and a singular part. The singular part is seen from the fact that $P(T_1^{\beta_1} = T_2^{\beta_2}) > 0$. In fact, when $\beta_1 = \beta_2 = \beta$, then the event $T_1 = T_2$ has positive probability. In the reliability context, this corresponds to both components failing at the same time due to a common cause.

It can be shown that

$$P(T_1^{\beta_1} > T_2^{\beta_2}) = \frac{\lambda_2}{\lambda_1 + \lambda_2 + \lambda_{12}} \tag{14.12}$$

$$P(T_1^{\beta_1} < T_2^{\beta_2}) = \frac{\lambda_1}{\lambda_1 + \lambda_2 + \lambda_{12}} \tag{14.13}$$

and

$$P(T_1^{\beta_1} = T_2^{\beta_2}) = \frac{\lambda_{12}}{\lambda_1 + \lambda_2 + \lambda_{12}} \tag{14.14}$$

When $\beta_1 = \beta_2 = \beta$, the following results are due to Moeschberger (1974):

1. (T_1, T_2) is bivariate Weibull if and only if there exist independent Weibull random variables X_1, X_2, X_3 such that $T_1 = \min(X_1, X_3)$ and $T_2 = \min(X_2, X_3)$.
2. $\min(T_1, T_2)$ is Weibull if (T_1, T_2) is bivariate Weibull.
3. If (T_1, T_2) is bivariate Weibull, then the distribution function has an increasing hazard rate according to Harris (1970).

Parameter Estimation

Moeschberger (1974) deals with the maximum-likelihood estimation. When the marginals have the same shape parameter, the problem gets reduced to estimating the scale parameters of three independent Weibull distributions with common shape parameters. When the marginals have different shape parameters, Moeschberger (1974) proposes a conditional-likelihood approach to estimating the parameter by conditioning on $\beta_1 > \beta_2$ or $\beta_1 < \beta_2$.

14.3.2 Model VI(a)-2

This model can be viewed as a slight modification of Model VI(a)-1 and was proposed by Lee and Thompson (1974) and studied in more detail by Lu (1989). The model is given by the distribution function

$$\bar{F}(t_1, t_2) = \exp\left[-\lambda_1 t_1^{\beta_1} - \lambda_2 t_2^{\beta_2} - \lambda_0 \max(t_1, t_2)^{\beta_0} \right] \tag{14.15}$$

Note that it differs from the earlier model with the exponent in the third term being the new additional parameter.

Model Analysis

The two marginal distributions are given by

$$F_1(t_1) = 1 - \exp\left(-\lambda_1 t_1^{\beta_1} - \lambda_0 t_1^{\beta_0} \right) \tag{14.16}$$

and

$$F_2(t_2) = 1 - \exp\left(- \lambda_2 t_2^{\beta_2} - \lambda_0 t_2^{\beta_0} \right) \tag{14.17}$$

respectively. As can be seen, they are different from the standard Weibull distribution. Note that the two marginals are each competing risk Weibull models.

The model is also mentioned in David (1974) and in Lee and Thompson (1974). Lee and Thompson show that when $\lambda_1 \neq \lambda_2$, the distribution does not belong to class E in the classification of Section 14.2.2. However, when $\lambda_1 = \lambda_2$, it belongs to class B but not to class A.

Special Cases

Lu (1989) studies some interesting special cases. One of them involves marginal distribution functions with linear hazard function so that they are of the form

$$F_1(t_1) = 1 - \exp(-\lambda_1 t_1^2 - \lambda_0 t_1) \tag{14.18}$$

A bivariate extension of the univariate linear hazard rate distribution is obtained with $\beta_1 = \beta_2 = 2$ and $\beta_0 = 1$, that is,

$$\bar{F}(t_1, t_2) = \exp[-\lambda_1 t_1^2 - \lambda_2 t_2^2 - \lambda_0 \max(t_1, t_2)] \tag{14.19}$$

Note that this model can also be obtained from (14.15) with $\beta_1 = \beta_2 = 2$ and $\beta_0 = 1$.

Two special cases of (14.15) are as follows. With $\beta_1 = \beta_2 = 1$, we have

$$\bar{F}(t_1, t_2) = \exp[-\lambda_1 t_1 - \lambda_2 t_2 - \lambda_0 \max(t_1, t_2)^{\beta}] \tag{14.20}$$

and with $\beta_0 = 1$, we have

$$\bar{F}(t_1, t_2) = \exp[-\lambda_1 t_1^{\beta_1} - \lambda_2 t_2^{\beta_2} - \lambda_0 \max(t_1, t_2)] \tag{14.21}$$

Both can be seen as bivariate extension to the univariate model of the form

$$F_1(t_1) = 1 - \exp(-\lambda_1 t_1^{\beta} - \lambda_0 t_1) \tag{14.22}$$

14.3.3 Model VI(a)-3

A general bivariate model, due to Lu and Bhattacharyya (1990), is as follows:

$$\bar{F}(t_1, t_2) = \exp\left[-\left(\frac{t_1}{\alpha_1}\right)^{\beta_1} - \left(\frac{t_2}{\alpha_2}\right)^{\beta_2} - \delta\psi(t_1, t_2) \right] \tag{14.23}$$

Different forms for the function of $\psi(t_1, t_2)$, which may even depend on other model parameters, yield a family of models.

Special Cases

Several interesting special cases are described in Lu and Bhattacharyya (1990). When

$$\psi(t_1, t_2) = \max\left(t_1^{\beta_1}, t_2^{\beta_2}\right) \tag{14.24}$$

the model reduces to Model VI(a)-1. When

$$\psi(t_1, t_2) = \left[\left(\frac{t_1}{\alpha_1}\right)^{\beta_1/m} + \left(\frac{t_2}{\alpha_2}\right)^{\beta_2/m}\right]^m \tag{14.25}$$

then

$$\bar{F}(t_1, t_2) = \exp\left\{-\left(\frac{t_1}{\alpha_1}\right)^{\beta_1} - \left(\frac{t_2}{\alpha_2}\right)^{\beta_2} - \delta\left[\left(\frac{t_1}{\alpha_1}\right)^{\beta_1/m} + \left(\frac{t_2}{\alpha_2}\right)^{\beta_2/m}\right]^m\right\} \tag{14.26}$$

By introducing the transformation $Y_1 = (T_1/\alpha_1)^{\beta_1}$ and $Y_2 = (T_2/\alpha_2)^{\beta_2}$, the joint survivor function of (Y_1, Y_2) is given by

$$\bar{F}(y_1, y_2) = \exp\left[-y_1 - y_2 - \delta\left(y_1^{1/m} + y_2^{1/m}\right)^m\right] \tag{14.27}$$

and it can be shown that the allowable range for the parameters are $\delta \geq 0$ and $0 < m \leq 1$.

Another specific formulation studied in Lu and Bhattacharyya (1990) is the following:

$$\begin{aligned}\bar{F}(t_1, t_2) = \exp\Bigg(&-\left(\frac{t_1}{\alpha_1}\right)^{\beta_1} - \left(\frac{t_2}{\alpha_2}\right)^{\beta_2} - \delta\left\{1 - \exp\left[-\left(\frac{t_1}{\alpha_1}\right)^{\beta_1}\right]\right\} \\ &\times \left\{1 - \exp\left[-\left(\frac{t_2}{\alpha_2}\right)^{\beta_2}\right]\right\}\Bigg)\end{aligned} \tag{14.28}$$

A model proposed by Lee (1979) is given by

$$\bar{F}(t_1, t_2) = \exp\left\{-\left[\lambda_1 c_1^{\beta} t_1^{\beta} + \lambda_2 c_2^{\beta} t_2^{\beta} + \lambda_{12} \max\left(c_1^{\beta} t_1^{\beta}, c_2^{\beta} t_2^{\beta}\right)\right]\right\} \tag{14.29}$$

and it also belongs to this category with $\beta_1 = \beta_2 = \beta$. When $c_1 \neq c_2$ and $\lambda_{12} > 0$, the distribution belongs to class C but not class B in the classification of Section 14.2.2.

14.3.4 Model VI(a)-4

A different type model presented in Lu and Bhattacharyya (1990) is as follows:

$$\bar{F}(t_1, t_2) = \left[1 + \left(\left\{\exp\left[\left(\frac{t_1}{\alpha_1}\right)^{\beta_1}\right] - 1\right\}^{1/\gamma} + \left\{\exp\left[\left(\frac{t_2}{\alpha_2}\right)^{\beta_2}\right] - 1\right\}^{1/\gamma}\right)^{\gamma}\right]^{-1} \tag{14.30}$$

This model has an interesting random hazard interpretation due to dependence between component lifetimes resulting from the effect of some common uncertain environmental stresses.

Model Analysis
Lu and Bhattacharyya (1990) show that

$$\{[\bar{F}(t_1, t_2)]^{-1} - 1\}^{1/\gamma} = \{[\bar{F}_1(t_1)]^{-1} - 1\}^{1/\gamma} + \{[\bar{F}_2(t_2)]^{-1} - 1\}^{1/\gamma} \tag{14.31}$$

which leads to the following result:

$$\frac{\bar{F}_1(t_1)\bar{F}_2(t_2)}{\bar{F}(t_1, t_2)} \leq 1 - \bar{F}_1(t_1)\bar{F}_2(t_2) \leq 1 \tag{14.32}$$

This implies that for no value of γ, the two random lifetimes are independent.

14.3.5 Model VI(a)-5

This model was proposed by Lee (1979) and is as follows:

$$\bar{F}(t_1, t_2) = \exp[-(\lambda_1 t_1^{\beta_1} + \lambda_2 t_2^{\beta_2})^{\gamma}] \tag{14.33}$$

where $\beta_i > 0, 0 < \gamma \leq 1$, $\lambda_i > 0, t_i \geq 0, i = 1, 2$.

Model Analysis
When $\beta_1 = \beta_2 = \beta$ and $\beta\gamma = 1$, the above distribution reduces to the third of the three distributions studied by Gumbel (1958). It has exponential marginals and exponential minimum after arbitrary scaling.

There is an alternative derivation of this model based on an accelerated life model. Consider the random variables defined as $T_j = X_j Z^{-1/\beta}$ where Z and $X_j, j = 1, 2$, are independent random variables with Z following $P(z)$ and X_j is Weibull (λ, β). Then (T_2, T_2) follow the bivariate Weibull distribution given by (14.33).

Lu and Bhattacharyya (1990) have studied this model in detail where they provide a physical basis for the model based on a general random hazards approach. They show that the distribution function is absolutely continuous and has Weibull marginals and minimum.

An interesting feature of this model is the following. Let (Y_1, Y_2) be random variables from (14.33). Define

$$(Z_1, Z_2) = \left(\lambda_1 Y_1^{\beta_1}, \lambda_2 Y_2^{\beta_2}\right) \tag{14.34}$$

Then the joint distribution of (Z_1, Z_2) is given by

$$\bar{G}(z_1, z_2) = 1 - \exp[-(z_1 + z_2)^{\gamma}] \tag{14.35}$$

and the joint density function by

$$g(z_1, z_2) = [\gamma(1-\gamma)(z_1 + z_2)^{\gamma-2} + \gamma^2(z_1 + z_2)^{2\gamma-2}]\exp[-(z_1 + z_2)^{\gamma}] \tag{14.36}$$

Furthermore, the covariance is given by

$$\mathrm{Cov}(z_1, z_2) = (1/\gamma)\Gamma(2/\gamma) - (1/\gamma^2)\Gamma^2(2/\gamma) \tag{14.37}$$

Define

$$(U_1, U_2) = \left[\frac{Z_1}{Z_1 + Z_2}, (Z_1 + Z_2)^{\gamma}\right] \tag{14.38}$$

The joint density function of (U_1, U_2) is given by

$$h(u_1, u_2) = [(1-\gamma) + \gamma u_2]e^{-u_2} \qquad 0 < u_1 < 1 \qquad 0 < u_2 < \infty \tag{14.39}$$

This implies (U_1, U_2) are independent random variables. The distribution of U_1 is uniform on the interval (0,1), and the distribution of U_2 is a mixture of gamma distribution with the density function given by

$$h(u_2) = [(1-\gamma) + \gamma u_2]e^{-u_2} \qquad u_2 > 0 \tag{14.40}$$

Parameter Estimation

Hougaard (1986) discusses the estimation problem for the symmetric bivariate case, that is, $\lambda_1 = \lambda_2 = \lambda$. Define $W_{i\bullet} = T_{i1}^{\beta} + T_{i2}^{\beta}$. The log-likelihood function is given by

$$\begin{aligned} L(\lambda, \beta, \gamma) = {} & \sum_{i,j} D_{ij} \ln(\gamma\lambda^{\gamma}\beta T_{ij}^{\beta-1} W_{i\bullet}^{\gamma-1}) \\ & + \sum_i D_{i1}D_{i2} \ln\{1 + (\gamma^{-1} - 1)\lambda^{-\gamma} W_{i\bullet}^{-\gamma}\} - \sum_i \lambda^{\gamma} W_{i\bullet}^{\gamma} \end{aligned} \tag{14.41}$$

where D_{ij} is binary valued variable assuming 0 or 1 depending on whether the data is censored or not. The maximum-likelihood estimates are obtained by differentiating with respect to the parameters and equating to zero. One needs to use an iterative approach to obtain the estimates.

14.4 MULTIVARIATE MODELS

Many of the bivariate models discussed in the last section can be extended to yield more general multivariate Weibull models. We describe a few of these.

14.4.1 Model VI(b)-1

The model, as stated in Lee (1979), is given by

$$\bar{F}(\mathbf{t}) = \exp\left[-\sum_{J} \lambda_J \max_{i \in J} (t_i^{\alpha})\right] \qquad \mathbf{t} > 0 \tag{14.42}$$

with $\alpha > 0$ and $\lambda_J > 0$ for $J \in \Im$ where the sets J are elements of the class $\Im$ of nonempty subsets of $\{1, 2, \ldots, n\}$ having the property that for each $i, i \in J$ for some $J \in \Im$.

This is an extension of the bivariate model of Marshall and Olkin (1967). The model has the property that the marginal distributions are all Weibull.

14.4.2 Model VI(b)-2

The model was proposed by Roy and Mukherjee (1988) and is given by

$$\bar{F}(\mathbf{t}) = \exp\left\{-\left[\sum_{i=1}^{p} (\lambda_i^{\nu} t_i^{\alpha\nu})\right]^{1/\nu}\right\} \qquad \mathbf{t} > 0 \tag{14.43}$$

This model also has the property that all its marginal distributions are Weibull. A similar model can also be found in Hougaard (1986) and Crowder (1989). This distribution has been referred to as the *multivariate extension of Weibull distribution* (MEWD).

Model Analysis

The so-called ith crude hazard rate, defined for $t_i = t$, is given by

$$h_i(t) = r_i(t, t, \ldots, t) = \alpha\lambda^{1-\nu}\lambda_i^{\nu} t^{\alpha-1} \tag{14.44}$$

The following results are from Roy and Mukherjee (1998).

1. If $\mathbf{T}$ follows MEWD $(\alpha, \nu, \lambda_1, \ldots, \lambda_p)$, then $Z = \min_{1 \le i < p} T_i$ follows Weibull with parameters (α, λ) where $\lambda = \left(\sum_{i=1}^{p} \lambda_i^{\nu}\right)^{1/\nu}$.
2. If $\mathbf{T}$ follows MEWD $(\alpha, \nu, \lambda_1, \ldots, \lambda_p)$ with $\alpha > 1$, then $[h_1(T_1), \ldots, h_p(T_p)]$ follows MEWD $(\alpha', \nu, \beta_1, \ldots, \beta_p)$ with $\alpha' > 1$, where $1/\alpha' + 1/\alpha = 1$ and $\beta_i = \lambda^{\alpha'(\nu-1)}\lambda_i^{1-\alpha'\nu}\alpha^{-\alpha}$.

3. If $\mathbf{T}$ follows MEWD, then, for all $\mathbf{t} \geq 0$ and $c \geq 0$, we have that $\bar{F}(ct_1, \ldots, ct_p)\bar{F}(1, \ldots, 1) = \bar{F}(c, \ldots, c)\bar{F}(t_1, \ldots, t_p)$.

Conversely, if the above condition is valid, then $\mathbf{T}$ follows the Weibull minimum property, with each marginal distribution being Weibull.

14.4.3 Model VI(b)-3

This model was proposed by Crowder (1989) and is given by

$$\bar{F}(\mathbf{t}) = \exp[\kappa^{1/\nu} - (\kappa + s)^{1/\nu}] \tag{14.45}$$

$\kappa > 0$ and $s = \sum_{i=1}^{p}(\xi_i t_i^{\phi_i})$. This is an extended version of the model given by

$$\bar{F}(\mathbf{t}) = \exp(-s^{1/\nu}) \tag{14.46}$$

which is a similar to (14.43) with Weibull marginals.

Model Analysis

Crowder (1989) gives a detailed analysis of the density function and the hazard functions and some interesting results for the case $\phi_j = \phi$ for all j. The expressions for the density function are cumbersome. The vector hazard rate defined as $h(\mathbf{t}) = -\nabla[\ln\,(F(\mathbf{t}))]$ has as the ith component

$$h_i(t_i|\mathbf{T} > \mathbf{t}) = \frac{(\kappa + s)^{1/\nu}}{\nu}\phi_i \xi_i y_i^{\phi_i - 1} \tag{14.47}$$

It can be seen that if $\nu \leq 1$ and $\phi_i > 1$, then $h_i(t_i|\mathbf{T} > \mathbf{t})$ is increasing in t_i. For the marginal hazard rate, $h_i(t_i) = -\partial[\ln\,(\bar{F}_i(t_i))]/\partial t_i$, we have that

$$\frac{h_i(t_i|\mathbf{T} > \mathbf{t})}{h_i(t_i)} = \left(\frac{\kappa + s}{\kappa + \xi_i t_i^{\phi_i}}\right)^{1/\nu - 1} \tag{14.48}$$

and it is greater than (less than) unity for $\nu < 1$ ($\nu > 1$).

The multivariate distribution given by (14.45) has a number of interesting properties relating to the minima. Let $U = \min\{Y_j : j = 1, \ldots, p\}$. Then U has the survival function

$$\bar{F}_u(u) = \exp[\kappa^{1/\nu} - (\kappa + s_u)^{1/\nu}] \tag{14.49}$$

with $s_u = \sum_{i=1}^{p}(\xi_i u^{\phi_i})$, which strongly resembles the distribution given in (14.45).

Parameter Estimation

Crowder (1989) briefly discusses the method of likelihood for estimating the model parameters. This involves expressions for the density function, and as mentioned earlier these are cumbersome to derive.

14.4.4 Model VI(b)-4

This model involves multivariate mixtures of Weibull distribution and was proposed by Patra and Dey (1999) in the context of a r-component system with each component failure modeled by a mixture of Weibull distributions. The model is given by

$$\bar{F}(\mathbf{t}) = \prod_{i=1}^{r} \sum_{j=1}^{k} \beta_{ij} \exp\left[-\left(\alpha_{ij} t_i^{\alpha_{ij}} + \frac{\theta_0 t_0}{r}\right)\right] \qquad \mathbf{t} > 0 \qquad (14.50)$$

with $t_0 = \max(t_1, \ldots, t_r)$ and $\alpha_{ij}, \theta_{ij} > 0$, for all i, j.

Model Analysis

Patra and Dey (1999) carry out the analysis of this model. They point out that the presence t_0 makes it difficult to calculate the multivariate density of $(T_1, T_2, \ldots, T_r)$ as it involves taking mixed derivatives over all possible partitions of the sample space. They discuss the special case with $r = k = 2$ (bivariate model involving two components and the failure of each modeled by a twofold Weibull mixture) in detail and also the more general competing risk model involving Weibull and exponential mixtures.

14.5 OTHER MODELS

Our discussion in this chapter has been very brief. As mentioned earlier, most multivariate models have little in common with the univariate Weibull model except that the marginal distributions are Weibull. Jaisingh et al. (1993) deal with a multivariate model generated by a Weibull and an inverse Gaussian mixture. Bjarnason and Hougaard (2000) deal with gamma frailty models involving Weibull hazards. Finally, the close link between the Weibull and extreme value distributions for the univariate also holds for the multivariate case. For more discussion on some earlier studies, see Johnson and Kotz (1970b).

EXERCISES

Data Set 14.1 Failure Times of Components for Two-Component System

T_1	77.2	74.3	9.6	251.6	134.9	115.7	195.7	42.2	27.8
T_2	156.6	108.0	12.4	108.0	84.1	51.2	289.8	59.1	35.5

14.1. Derive (14.10) and (14.11).

14.2. Show that Model VI(a)-2 belongs to class B but not to class A when $\lambda_1 = \lambda_2$.

14.3. Consider a system with two components with lifetimes T_1 and T_2, respectively. The joint distribution of (T_1, T_2) is given by the Marshall–Olkin bivariate Weibull distribution with parameters $\lambda_1 = \lambda_2 = 10$, $\lambda_{12} = 5$, and $\beta_1 = \beta_2 = 1.1$. Compute the probability that both components fail before $t = 5$.

14.4. A system consists of two modules. The joint distribution can be assumed to follow the bivariate Weibull distribution:

$$\bar{F}(t_1, t_2) = \exp[-2t_1 - 3t_2^{1.2} - \max(t_1, t_2)^{1.2}]$$

Compute the probability that module 1 will fail before module 2. What is the probability of system failure if the modules are connected in parallel? What is the probability of system failure if the modules are connected in series?

14.5. A system consists of two components connected in parallel so that the system fails when both components fail. The failure times of components 1 and 2 for nine system failures are given in Data Set 14.1. How would you test whether the component failure times are statistically independent or not?

14.6. Assume that the component failure times in Exercise 14.5 are statistically independent and given by two Weibull distributions. Based on the WPP plots of the data, estimate the model parameters.

14.7. Derive for the maximum-likelihood estimator for Models VI(a)-1 to VI(a)-4.

14.8. Suppose that Data Set 14.1 can be modeled by Model VI(a)-4. How would you estimate the parameters using the method of maximum likelihood?

14.9. For Model VI(a)–1, derive interval estimators for the model parameters.

14.10. For a univariate distribution function, the hazard function characterizes the notion of aging. How can one extend this concept to a multivariate distribution function?

14.11. A system is comprised of two components. The failure times are dependent and influenced by a set of covariates. Discuss alternate approaches to modeling the effect of covariates on the failure times.

14.12. The Weibull transformation for the one-dimensional case is given by (1.7). An extension of this concept to the two-dimensional case is as follows:

$$y(x_1, x_2) = \ln[-\ln(\bar{F}(t_1, t_2)] \qquad x_1 = \ln(t_1) \quad \text{and} \quad x_2 = \ln(t_2)$$

Carry out a study of $y(x_1, x_2)$ versus x_1 and x_2 for Model VI(a)-1.*

14.13. Let (T, X) denote the age and usage of a nonrepairable component at failure. It can be modeled by a bivariate distribution function $F(t, x), t \geq 0, x \geq 0$. The failed components are replaced immediately by a new one, and the time for replacement is negligible. In this case, failures occur according to a two-dimensional renewal process. The item is sold with a warranty that requires the manufacturer to replace all failures occurring with the warranty period characterized by a rectangle $[0, W) \times [0, U))$ so that the warranty expires when the total time exceeds W or the total usage exceeds U. Show that the expected number of replacements under warranty is given by the following renewal equation:

$$M(W, U) = F(W, U) + \int_0^U \int_0^W M(W - t, U - x)f(t, x)\, dt\, dx$$

[see Hunter (1984) for more on two-dimensional renewal processes.]

* Research problem.

CHAPTER 15

Type VII Weibull Models

15.1 INTRODUCTION

A stochastic point process is a continuous time stochastic process characterized by events that occur randomly along the time continuum (or a one-dimensional spatial coordinate). A counting process is a special case of point process where the focus is on the number of events over a time interval, and this is a random variable and depends on the interval. In this chapter we discuss three groups of counting process models that are related to the standard Weibull model. They are as follows:

Type VII(a) models: Weibull intensity models
Type VII(b) models: Weibull renewal process models
Type VII(c) model: Power law–Weibull renewal process

The outline of the chapter is as follows. Section 15.2 deals with model formulations. We start with a general discussion of counting process and discuss different approaches to modeling a counting process. Sections 15.3 to 15.5 deal with the Weibull intensity models. In Sections 15.6 to 15.8 we discuss the Weibull renewal process models, and Section 15.9 deals with the power law–Weibull renewal process.

15.2 MODEL FORMULATIONS

15.2.1 Counting Process

A point process $\{N(t), t \geq 0\}$ is a counting process if it represents the number of events that have occurred until time t. It must satisfy:

1. $N(t) \geq 0$.
2. $N(t)$ is integer valued.

Weibull Models, by D.N.P. Murthy, Min Xie, and Renyan Jiang.
ISBN 0-471-36092-9

3. If $s < t$, then $N(s) \leq N(t)$.
4. For $s < t$, $\{N(t) - N(s)\}$ is the number of events in the interval $(s, t]$.

We shall confine ourselves to $t \geq 0$. The behavior of $N(t)$, for $t \geq 0$, depends on whether or not $t = 0$ corresponds to the occurrence of an event. The analysis of the case with $t = 0$ corresponding to the occurrence of an event is simpler than the alternate case. Also, we assume that $N(0) = 0$.

15.2.2 Alternate Approaches to Modeling Counting Process

Let T_i denote the time at which the ith ($i \geq 1$) event occurs. This is a random variable, and $T_i = t_i$ implies that $N(t) \geq i$ for $t > t_i$. Let $X_i = T_i - T_{i-1}$ with $T_0 = 0$. This is a continuous random variable and represents the time between events $(i - 1)$ and i.

The three different approaches to modeling are as follows:

Approach 1 This involves specifying the probability distribution for the discrete random variable $[N(t) - N(s)]$ as a function of t and s. In the simplest case this is given by

$$p_n(t, s) = P[N(t) - N(s) = n] \tag{15.1}$$

for $0 \leq s < t < \infty$ and $n = 0, 1, \ldots$.

Approach 2 This involves specifying the probability distribution functions for continuous random variable X_i for $i \geq 1$. In the simplest case this is given by

$$F_i(x) = P(X_i \leq x) \tag{15.2}$$

for $i \geq 1$.

Approach 3 This involves specifying the probability of an event occurring in a small interval $[t, t + \delta t)$ as a function of t. In the simplest case this is given by

$$P[N(t + \delta t) - N(t) = 1] = \lambda(t)\delta t + o(\delta t) \tag{15.3}$$

with

$$P[N(t + \delta t) - N(t) \geq 2] = o(\delta t) \tag{15.4}$$

This implies that

$$P[N(t + \delta t) - N(t) = 0] = 1 - \lambda(t)\delta t + o(\delta t) \tag{15.5}$$

Comment

1. The three model formulations are equivalent in the sense that one can be derived (in principle) in terms of the others. However, depending on the context, a particular formulation might be more suited as it results in either a simpler description of the model or easier for analysis.
2. If $p_n(t, s)$ (in approach 1), $F_i(x)$ (in approach 2), and $\lambda(t)$ (in approach 3) are functions of the past history of the counting process, then these are the conditional probabilities. At time t, let H_t denote the history of the process up to t. In general, the behavior of $N(t)$ depends on H_t. As a result, we have $p_n(t, s \mid H_s)$, $F_i(x \mid H_{t_i-1})$, and $\lambda(t \mid H_t)$.

15.2.3 Model VII(a): Weibull Intensity Models

Model VII(a)-1: Power Law Process

The occurrence of events is given by an intensity function

$$\lambda(t) = \frac{\beta t^{\beta-1}}{\alpha^{\beta}} \qquad t \geq 0 \tag{15.6}$$

with $\alpha, \beta > 0$. Note that the intensity function is a function of only t and not of the history of events up to t.

This model has been called by many different names: *Power law process*—Bassin (1973), Rigdon and Basu (1989), and Klefsjo and Kumar (1992); *Rasch–Weibull process*—Moller (1976); *Weibull intensity function*—Crow (1974); *Weibull–Poisson process*—Bain and Engelhardt (1986); and *Weibull process*—Bain (1978). Ascher and Feingold (1984) address the confusion and misconception that has resulted from this plethora of terminology.

Note that $\lambda(t)$ given by (15.6) is identical to the hazard function for the standard Weibull distribution [given by (3.12)] and hence the name *Weibull intensity function*.

Model VII(a)-2: Modulated Power Law Process

This can be viewed as a thinning of the power law process where only every kth event of the power law process is recorded as an event. As a result, the joint density function for the time instants $(t_1 < t_2 < \cdots < t_n)$ of the first n events for the modulated power law process is given by

$$f(t_1, t_2, \ldots, y_n) = \left\{ \prod_{i=1}^{n} \lambda(t_i)[\Lambda(t_i) - \Lambda(t_{i-1})]^{k-1} \right\} \frac{\exp[-\Lambda(t_n)]}{[\Gamma(k)]^n} \tag{15.7}$$
$$0 < t_1 < t_2 < \cdots < t_n < \infty$$

where $\lambda(t)$ is given by (15.6) and

$$\Lambda(t) = \int_0^t \lambda(x)\, dx = \left(\frac{t}{\alpha}\right)^{\beta} \tag{15.8}$$

is the integrated intensity up to t for the power law process.

This model is due to Lakey and Rigdon (1993) and is a special case of the modulated gamma process proposed by Berman (1981).

Model VII(a)-3: Proportional Intensity Model

This is an extension of an earlier model along the lines of Model IV(c). As such, the model is given by

$$\lambda(t; S) = \lambda_0(t)\psi(S) \qquad t \geq 0 \tag{15.9}$$

where $\lambda_0(t)$ is of the form given by (15.6) and $\psi(S)$ is a function of explanatory variables S. The only restriction on $\psi(S)$ is that $\psi(S) > 0$. Various forms of $\psi(S)$ have been studied; see Cox and Oaks (1984) and Prentice et al. (1981).

15.2.4 Model VII(b): Weibull Renewal Process Models

Model VII(b)-1: Ordinary Renewal Process

In this model the occurrence of events is as follows. The interevent times $X_i, i = 1, 2 \ldots,$ are independent and identically distributed according to the two-parameter Weibull distribution given by (1.3) or the three-parameter Weibull distribution given by (1.1).

Note that for $t_i \leq t < t_{i+1}$ the intensity function $\lambda(t)$ is given by $\lambda(t) = h(t - t_i)$, where $h(x)$ is the hazard function associated with the distribution function $F(x)$.

Model VII(b)-2: Modified Renewal Process

Here the distribution function for the time to first event, $F_1(x)$, is different from that for the subsequent interevent times, which are identical and independent random variables with distribution function $F(x)$. Here either $F_1(x)$ and/or $F(x)$ are two- or three-parameter Weibull distributions. Note that when $F_1(x) = F(x)$, the model reduces to the ordinary renewal process model.

Model VII(b)-3: Alternating Renewal Process

Here the odd-numbered interevent times are a sequence of independent and identically distributed random variables with a distribution function $F_1(x)$. The even-numbered interevent times are another sequence of independent and identically distributed random variables with a distribution function $F_2(x)$. As before, either $F_1(x)$ and/or $F_2(x)$ are the two- or three-parameter Weibull distributions.

15.2.5 Model VII(c): Power Law–Weibull Renewal Process

For $t_i \leq t < t_{i+1}$, define $u(t) = t - t_i$. The occurrence of events is given by intensity function

$$\lambda(t) = \frac{\beta + \delta - 1}{\alpha^{\beta+\delta-1}} t^{\beta-1} [u(t)]^{\delta-1} \qquad t \geq 0 \tag{15.10}$$

with $\alpha > 0$ and $\beta + \delta > 1$. Note that when $\delta = 1$, the model reduces to the simple Weibull intensity model [Model VII(a)-1], and, when $\beta = 1$, it reduces to an ordinary Weibull renewal process model [Model VII(b)-1].

This model was proposed and studied by Calabria and Pulcini (2000). It is a special case of a more general class of processes introduced by Lawless and Thiagarajah (1996).

15.3 MODEL VII(a)-1: POWER LAW PROCESS

15.3.1 Model Analysis

Using (15.6) in (15.8) we have

$$\Lambda(t) = \left(\frac{t}{\alpha}\right)^{\beta} \tag{15.11}$$

We present various results without proofs, and they can be found in either Ross (1980) or Rigdon and Basu (2000).

1. The probability of i events in the interval $(s, t]$ is given by

$$P[N(t+s) - N(t) = i] = \frac{e^{-[\Lambda(t+s)-\Lambda(t)]}[\Lambda(t+s) - \Lambda(t)]^i}{i!} \tag{15.12}$$

   for $i \geq 0$ with $\Lambda(t)$ given by (15.8).
2. The probability density function (pdf) for T_n, the time of nth event to occur, is given by

$$g_n(t) = \frac{1}{\Gamma(n)}\frac{\beta}{\alpha}\left(\frac{t}{\alpha}\right)^{n\beta-1} \exp\left[-\left(\frac{t}{\alpha}\right)^{\beta}\right] \tag{15.13}$$

   which is the pdf of the generalized gamma distribution. When $n = 1$, this becomes the pdf for the standard two-parameter Weibull distribution.
3. The event times $T_1, T_2, \ldots, T_{n-1}$ conditioned on $T_n = t_n$ are distributed as $(n-1)$ order statistics from the distribution with CDF:

$$G(y) = \begin{cases} 0 & y \leq 0 \\ \Lambda(y)/\Lambda(t_n) & 0 < y \leq t_n \\ 1 & y > t_n \end{cases} \tag{15.14}$$

4. The expected number of events in $[0, t)$ is given by

$$M(t) = E[N(t)] = \Lambda(t) = \left(\frac{t}{\alpha}\right)^{\beta} \tag{15.15}$$

15.3.2 Parameter Estimation

The estimation of parameter depends on the type of data available and the method used. We consider the following three different cases:

Case 1: Event Truncation: The data is said to be event truncated when the data collection stops after a predetermined number (n) of events.

Case 2: Time Truncation: The data is said to be time truncated when the data collection stops at a predetermined time t.

Case 3: Grouped Data: Let $a_j j \geq 0$, denote an increasing sequence with $a_0 = 0$. The jth ($j \geq 1$) interval is given by $[a_{j-1}, a_j)$, and let N_j denote the number of events in interval j. The data available for estimation is the number of events $N_j = n_j$ in interval j, $1 \leq j \leq J$.

The results are presented without the details of the derivations. Interested readers can find them in Rigdon and Basu (2000).

Comment We confine our attention to the case where the data is generated from a single power law process. The results for the case of multiple processes (which are independent and identical) can be found in Rigdon and Basu (2000).

Graphical Approach

From (15.15) we have

$$E\left[\frac{N(t)}{t}\right] = \frac{t^{\beta-1}}{\alpha^{\beta}} \tag{15.16}$$

On taking the logarithm of both sides, we have the following linear equation:

$$y = (\beta - 1)x - \beta \ln(\alpha) \tag{15.17}$$

where

$$y = \ln\left(E\left[\frac{N(t)}{t}\right]\right) \quad \text{and} \quad x = \ln(t) \tag{15.18}$$

Duane (1964) suggests plotting $y_i = \ln[N(t_i)/t_i]$ versus $x_i = \ln(t_i)$ (commonly referred to as the *Duane plot*) and fitting a straight line to the plotted data. The slope of the line yields an estimate for $(\beta - 1)$ and the intercept as an estimate for $\beta \ln(\alpha)$. Molitor and Rigdon (1993) study the efficiency of this estimator and find that this least-squares estimator is almost as efficient as the maximum-likelihood estimator of β.

Rigdon and Basu (2000) discuss this approach more critically and show that $E\{\ln[N(T_i)/T_i]\}$ versus $E[\ln(T_i)]$ is not linear.

Recently, Akersten et al. (2001) study some graphical techniques based on the total time on test (TTT) plot for data from repairable systems. Instead of using the time between failures, the actual failure times since the beginning are used in generating the TTT plot (see Section 3.2.6). Different test statistics can then be used for the testing of the Duane model versus the homogeneous Poisson process.

Method of Maximum Likelihood (Point Estimates)

Case 1: Event Truncation
The likelihood function is given by

$$L(t_1, t_2, \ldots, t_n; \theta) = \left[\prod_{i=1}^{n} \lambda(t_i)\right] \exp\left[-\int_0^{t_n} \lambda(x)\, dx\right] \tag{15.19}$$

Point estimates are given by

$$\hat{\beta} = \frac{n}{\sum_{i=1}^{n-1} \ln(t_n/t_i)} \quad \text{and} \quad \hat{\alpha} = \frac{t_n}{n^{1/\hat{\beta}}} \tag{15.20}$$

The estimate $\hat{\beta}$ given by (15.20) is biased. The estimator

$$\bar{\beta} = \frac{n-2}{\sum_{i=1}^{N} \ln(T_n/T_i)} \tag{15.21}$$

is an unbiased estimator for β.

Case 2: Time Truncation
Let $N(t)$ denote the number of events observed in $[0, t)$. The likelihood function, conditional on $N(t) = n$, is given by

$$L[t_1, t_2, \ldots, t_n; \theta \mid N(t) = n] = n! \prod_{i=1}^{n} \frac{\lambda(t_i)}{\Lambda(t)} \tag{15.22}$$

Point estimates are given by

$$\hat{\beta} = \frac{N}{\sum_{i=1}^{N} \ln(t/t_i)} \quad \text{and} \quad \hat{\alpha} = \frac{t}{N^{1/\hat{\beta}}} \tag{15.23}$$

The estimate $\hat{\beta}$ (conditional on $N = n \geq 1$) given by (15.23) is biased. The estimator

$$\bar{\beta} = \frac{N-1}{\sum_{i=1}^{N} \ln(T/T_i)} \tag{15.24}$$

is the conditionally unbiased estimator for β.

Case 3: Grouped Data
The likelihood function is given by

$$L(n_1, \ldots, n_J; \theta) = \prod_{j=1}^{J} L_j(n_j) \tag{15.25}$$

where

$$L_j(n_j) = \frac{e^{-[\Lambda(a_j) - \Lambda(a_{j-1})]}[\Lambda(a_j) - \Lambda(a_{j-1})]^{n_j}}{n_j!} \tag{15.26}$$

Point estimates are obtained by minimizing (15.26) using a computational scheme.

Method of Maximum Likelihood: Interval Estimates

Case 1: Event Truncation
In this case, $2n\beta/\hat{\beta}$ has a chi-square distribution with $2(n-1)$ degrees of freedom. Using this, the $100 \times (1-\gamma)\%$ two-sided confidence interval is given by

$$\frac{\chi^2_{1-\gamma/2}[2(n-1)]\hat{\beta}}{2n} < \beta < \frac{\chi^2_{\gamma/2}[2(n-1)]\hat{\beta}}{2n} \tag{15.27}$$

A confidence interval for α can be obtained from the test statistic $W = (\hat{\alpha}/\alpha)^{\hat{\beta}}$. Finkelstein (1976) gives the percentage points for this statistics based on simulation studies. A $100 \times (1-\gamma)\%$ two-sided confidence interval for α is given by

$$\frac{\hat{\alpha}}{a^{1/\hat{\beta}}} < \alpha < \frac{\hat{\alpha}}{b^{1/\hat{\beta}}} \tag{15.28}$$

where a and b are obtained from the tables given in Finkelstein (1976).

Case 2: Time Truncation
Conditional on $N(t) = n$, $2n\beta/\hat{\beta}$ has a chi-square distribution with $2n$ degrees of freedom. From this, the $100 \times (1-\gamma)\%$ two-sided confidence interval is given by

$$\frac{\chi^2_{1-\gamma/2}(2n)\hat{\beta}}{2n} < \beta < \frac{\chi^2_{\gamma/2}(2n)\hat{\beta}}{2n} \tag{15.29}$$

There appears to be no method for obtaining exact confidence intervals for α. Bain and Engelhardt (1980) suggest an approximate method that uses the tables from Finkelstein (1976).

Bayesian Method
Let $p(\alpha, \beta)$ denote the prior distribution of the unknown parameters and $\underline{t} \equiv [t_1, t_2, \ldots, t_n]$ denote the observed data. Let $f(\underline{t}|\alpha, \beta)$ denote the likelihood

function. Then, the posterior distribution is

$$p(\alpha, \beta | \underline{t}) = cp(\alpha, \beta) f(\underline{t} | \alpha, \beta) \tag{15.30}$$

where c, the normalizing constant, is given by

$$c = \left[\int_0^\infty \int_0^\infty p(\alpha, \beta) f(\underline{t} | \alpha, \beta) \, d\alpha \, d\beta \right]^{-1} \tag{15.31}$$

Noninformative Prior
Guida et al. (1989) suggest the following improper prior density function (which does not integrate to 1):

$$p(\alpha, \beta) = \frac{1}{\alpha\beta} \qquad \alpha, \beta > 0 \tag{15.32}$$

For event-truncated data,

$$p(\alpha, \beta \mid \underline{t}) = c\beta^{n-1} P_n^{\beta-1} \alpha^{-n\beta-1} \exp\left[-\left(\frac{t_n}{\alpha} \right)^\beta \right] \tag{15.33}$$

where

$$P_n = \prod_{i=1}^n t_i \tag{15.34}$$

The marginal posterior density function for β is given by

$$p(\beta \mid \underline{t}) = c \int_0^\infty \beta^{n-1} P_n^{\beta-1} \alpha^{-n\beta-1} \exp\left[-\left(\frac{t_n}{\alpha} \right)^\beta \right] d\alpha = c_1 \beta^{n-2} \left(\frac{P_n}{t_n^n} \right)^\beta \tag{15.35}$$

The posterior mean of β yields the Baye's estimate and is given by

$$\hat{\beta}_B = E[\beta \mid \underline{t}] = \frac{n-1}{\sum_{i=1}^{n-1} \ln(t_n/t_i)} \tag{15.36}$$

The posterior mean of α is a more complicated function and can be found in Rigdon and Basu (2000).

Informative Prior Distribution
Guida et al. (1989) suggest choosing a time T and then specifying a gamma prior distribution $h(\eta)$ for $\eta = (T/\alpha)^\beta$ rather than for α. Let $g(\beta)$ be the prior distribution

for β. This yields the prior distribution for (η, β) given by

$$p(\eta, \beta) = g(\beta)h(\eta) \tag{15.37}$$

Expressions for the Bayes estimate can be found in Rigdon and Basu (2000).

For further details of the Bayesian approach, see Guida et al. (1989), Calabria et al. (1990), Bar-Lev et al. (1992), and Rigdon and Basu (2000).

15.3.3 Model Selection and Validation

Graphical Approach

Klefsjo and Kumar (1992) propose a test based on a TTT plot to determine whether a given event-truncated data $t_1 < t_2 < \cdots < t_{n-1} < t_n$ can be adequately modeled by the Weibull intensity model or not. This is based on the following result.

If the event times $T_1 < T_2 < \cdots < T_{n-1} < T_n$ are from a Weibull intensity model with parameters α and β, then the log-ratio transformed variables

$$W_j = -\ln\left(\frac{T_n}{T_{n-j}}\right) \qquad j = 1, 2, \ldots, n-1 \tag{15.38}$$

are distributed as $(n-1)$ order statistics from the exponential distribution with mean $1/\beta$ [see Moller (1976)].

The test involves the TTT plot (see Section 3.2.6) of the log-ratio transformed variables. If the plotted data lies close to the diagonal line in the unit square, then the Weibull intensity model may be an acceptable model for the data.

Goodness-of-Fit Test

For goodness-of-fit tests discussed in this section, the null hypothesis is that the Weibull intensity model with α and β is the correct model. The alternative hypothesis is that the Weibull model is not the correct model. The goodness-of-fit tests involve the transformation of the event times. Two types of transformation (for event-truncated data) are as follows:

1. Log-ratio transformation was discussed earlier and is given by (15.38).
2. Ratio-power transformation is defined by

$$R_i = \left(\frac{T_i}{T_n}\right)^{\beta} \qquad i = 1, 2, \ldots, n-1 \tag{15.39}$$

Note that

$$R_i = \frac{\Lambda(T_i)}{\Lambda(T_n)} \qquad i = 1, 2, \ldots, n-1 \tag{15.40}$$

are distributed as $(n - 1)$ order statistics from a uniform distribution over the interval [0, 1].

The goodness-of-fit test depends on the type of data available and whether or not the parameters (of the model in the null hypothesis) are completely specified or not.

Case 1: Event-Truncated Data

Tests Based on Ratio-Power Transformation When the parameters are completely specified, then β is known and any goodness-of-fit test for the uniform distribution could be applied to the R_i's given by (15.39). For a discussion of such tests, see for example, D'Agostino and Stephens (1986) and Park and Seoh (1994).

When β is unknown and needs to be estimated from the given (or observed) data, (15.21) provides an unbiased estimate $\bar{\beta}$. In this case, one computes

$$\hat{R}_i = \left(\frac{t_i}{t_n}\right)^{\bar{\beta}} \tag{15.41}$$

and the test statistics for the Cramér–von Mises test is given by

$$C_R^2 = \frac{1}{12(n-1)} + \sum_{i=1}^{n-1}\left[\hat{R}_i - \frac{2i-1}{2(n-1)}\right]^2 \tag{15.42}$$

Critical values of C_R^2 for different values of $(n - 1)$ are given in Rigdon and Basu (2000) and are used to determine whether the null hypothesis is to be accepted or rejected. Large values of C_R^2 indicate evidence of departure from the Weibull intensity model.

Tests Based on Log-Ratio Transformation When the parameters are specified (so that β is known), then from (15.38) we have

$$U_i = \beta \ln\left(\frac{T_n}{T_{n-i}}\right) \qquad i = 1, 2, \ldots, (n-1) \tag{15.43}$$

distributed as order statistics from the exponential distribution with mean = 1. Any goodness-of-fit test for the exponential distribution could be applied to the u_i's obtained from (15.56) using t_i in place of T_i.

When the parameter β is not specified, then any goodness-of-fit test for the exponential distribution with unknown mean can be applied using the transformed data given by (15.38). Rigdon and Basu (2000) list several different tests and provide references where further details can be found. Rigdon (1989) investigates the power of these tests.

Case 2: Time-Truncated Data

The tests are similar to the event-truncated case.

Tests Based on Ratio-Power Transformation The ratio transformation is given by

$$\hat{R}_i = \left(\frac{t_i}{t}\right)^{\bar{\beta}} \tag{15.44}$$

with $\bar{\beta}$ given by (15.24). The test statistics for the Cramér–von Mises test is given by

$$C_R^2 = \frac{1}{12n} + \sum_{i=1}^{n-1}\left(\hat{R}_i - \frac{2i-1}{2n}\right)^2 \tag{15.45}$$

Critical values of C_R^2 for different values of $(n - 1)$ are given in Rigdon and Basu (2000) and are used to determine whether the null hypothesis is to be accepted or rejected.

Tests Based on Log-Ratio Transformation In this case

$$W_i = \ln\left(\frac{t}{T_{n-i+1}}\right) \qquad i = 1, 2, \ldots, (n-1) \tag{15.46}$$

are distributed as n-order statistics from an exponential distribution with mean $1/\beta$. Any goodness-of-fit test for the exponential distribution could be applied to the w_i's obtained from (15.46) using t_{n-i+1}. For more details and the goodness-of-fit tests for the case where the parameters are not specified, see Rigdon and Basu (2000).

15.4 MODEL VII(a)-2: MODULATED POWER LAW PROCESS

15.4.1 Model Analysis

The distribution function of the nth event depends only on the time of the $(n - 1)$st event and as a result, following Black and Rigdon (1996), the intensity function is given by

$$\lambda(t \mid t_{n-1}) = \frac{\frac{\beta}{\alpha^{\beta}\Gamma(k)} t^{\beta-1} a(t)^{\beta-1} \exp[-a(t)]}{\int_t^{\infty} \frac{\beta}{\alpha^{\beta}\Gamma(k)} x^{\beta-1} \exp[-a(x)]\, dx} \tag{15.47}$$

where

$$a(t) = \left(\frac{t}{\alpha}\right)^{\beta} - \left(\frac{t_{n-1}}{\alpha}\right)^{\beta} \tag{15.48}$$

15.4.2 Parameter Estimation

Method of Maximum Likelihood

The likelihood function is given by

$$L(\alpha, \beta, k \mid t_1, t_2, \ldots, t_n) = \left\{ \prod_{i=1}^{n} \lambda(t_i)[\Lambda(t_i) - \Lambda(t_{i-1})]^{k-1} \right\} \frac{\exp[-\Lambda(t_n)]}{[\Gamma(k)]^n} \tag{15.49}$$

Using (15.6) and (15.11) in (15.49) yields the following expression for the log-likelihood function:

$$\ln[L(\alpha, \beta, k \mid t_1, t_2, \ldots, t_n)] = -\left(\frac{t_n}{\alpha}\right)^{\beta} + n \ln(\beta) - n \ln[\Gamma(k)] - n\beta k\, ln(\alpha) + (\beta - 1)\sum_{i=1}^{n} \ln(t_i) + (k-1)\sum_{i=1}^{n} \ln(t_i^{\beta} - t_{i-1}^{\beta}) \tag{15.50}$$

and the estimates are obtained by using a computational approach. Black and Rigdon (1996) and Rigdon and Basu (2000) give more details as well as the confidence limits for the parameters.

15.4.3 Hypothesis Tests

The testing of the following hypothesis

1. H_0: $k = 1$ versus H_1: $k \neq 1$,
2. H_0: $\beta = 1$ versus H_1: $\beta \neq 1$, and
3. H_0: $\beta = 1$ and $k = 1$ versus H_1: $\beta \neq 1$ or $k \neq 1$,

based on the likelihood ratio test is discussed in Black and Rigdon (1996).

15.5 MODEL VII(a)-3: PROPORTIONAL INTENSITY MODEL

The baseline intensity function is the same as the power law process model (15.6), and the scale parameter a is a function of explanatory variables, S, which could be a vector. A common model is

$$\psi(S) = \exp(\zeta^T S)$$

and the overall intensity function is

$$\lambda(t; S) = \lambda_0(t)\exp[(\zeta^T S)]$$

where $\lambda_0(t) = \beta\lambda_0 t^{\beta-1}$ is the baseline intensity function.

The distribution for the time to first failure is identical to the proportional Weibull hazard model discussed in Section 12.4.

Landers and Soroudi (1991) study this and other proportional intensity models with reliability applications. The plots of $\lambda(t; S)$ vs. $\ln(t)$ are linear and parallel, with the spacing determined by the regression coefficients. Hence graphical model validation and parameter estimation can be carried out in a traditional matter.

The maximum-likelihood estimators can also be obtained using standard algorithms. Lawless (1987) discusses this in detail.

15.6 MODEL VII(b)-1: ORDINARY WEIBULL RENEWAL PROCESS

15.6.1 Model Analysis

The number of events until time t, $N(t)$, is given by

(i) $N(0) = 0$, and

(ii) $N(t) = \sup(n : S_n \leq t)$, where

$$S_0 = 0, \;\; S_n = \sum_{i=1}^{n} X_i \qquad n \geq 1 \tag{15.51}$$

Note that S_n is the time instant for the nth event (or renewal) and is the sum of n independent and identically distributed random variables. Since the X_i's are distributed with distribution function $F(x)$, the distribution of S_n is given by the n-fold convolution of $F(x)$ with itself, that is,

$$P(S_n \leq x) = F^{(n)}(x) = F(x)^* F(x)^* \cdots {}^* F(x) \tag{15.52}$$

where * is the convolution operator.

Distribution of N(t)

$N(t) \geq n$ if and only if $S_n \leq t$. This implies that

$$\begin{aligned} P[N(t) = n] &= P[N(t) \geq n] - P[N(t) \geq (n+1)] \\ &= P(S_n \leq t) - P(S_{n+1} \leq t) \end{aligned} \tag{15.53}$$

for $n = 0, 1, \ldots$, with $S_0 = 0$. Since

$$P(S_n \leq t) = F^{(n)}(t) \tag{15.54}$$

where $F^0(t) \equiv 1$, we have

$$P(N(t) = n) = F^{(n)}(t) - F^{(n+1)}(t) \tag{15.55}$$

Moments of N(t)

From (15.55), expressions for the moments of $N(t)$ can be obtained. Of particular interest is the first moment $M(t)$, the expected number of renewals in $[0, t)$. This is given by the following integral equation:

$$M(t) = F(t) + \int_0^t M(t - x)f(x)\, dx \tag{15.56}$$

and is referred to as the renewal function associated with the distribution function $F(x)$. A direct and simpler derivation of (15.56), based on conditional expectation, can be found in Ross (1970).

In general, it is not possible to obtain $M(t)$ analytically for most $F(x)$ (including the Weibull distribution). Many different approaches to obtaining the renewal function numerically for the Weibull distribution have been proposed. For further details, see Smith and Leadbetter (1963), White (1964a), Lomnicki (1966), Baxter et al. (1982), Spearman (1989), Xie (1989), Constantine and Robinson (1997), and Xie et al. (2001).

For large t we have the following asymptotic result:

$$M(t) = \frac{t}{\mu} + \frac{\sigma^2 - \mu^2}{2\mu^2} + o(1) \tag{15.57}$$

where $\mu = E(X_i)$ and $\sigma^2 = E[(X_i - \mu)^2]$, and

$$\lim_{t \to \infty} P[N(t) < a(t)] = \Phi(y) \tag{15.58}$$

where $a(t) = t/\mu + y\sigma\sqrt{t/\mu^3}$ for all y, and $\Phi(\cdot)$ is the cumulative distribution function for the standard normal distribution (with mean = 0 and variance = 1). For a proof, see Rigdon and Basu (2000).

The renewal density function, $m(t)$, is given by

$$m(t) = \frac{dM(t)}{dt} \tag{15.59}$$

and satisfies the equation

$$m(t) = f(t) + \int_0^t m(t - x)f(x)\, dx \tag{15.60}$$

where $f(x)$ is the density function associated with $F(x)$.

The variance of $N(t)$ is given by

$$\text{Var}[N(t)] = \sum_{n=1}^{\infty} (2n - 1)F^{(n)}(t) - [M(t)]^2 \tag{15.61}$$

Excess (or Residual) Life at Time t

Let $B(t)$ denote the time from t until the next event, that is,

$$B(t) = S_{N(t)+1} - t \tag{15.62}$$

where $B(t)$ is the excess or residual life at t. The distribution function for $B(t)$ is given by

$$P[B(t) \leq x] = F(t+x) - \int_0^t [1 - F(t+x-y)]\, dM(y) \qquad x \geq 0 \tag{15.63}$$

where $M(x)$ is the renewal function associated with the distribution $F(x)$. In general, it is difficult to solve (15.63) analytically. In the limit as $t \to \infty$, this becomes

$$\lim_{t\to\infty} P\{B(t) \leq x\} = \frac{\int_0^x [1 - F(y)]\, dy}{\mu} \tag{15.64}$$

where $\mu = E(X_i)$. The details of the derivations of (15.63) and (15.64) can be found in Ross (1970).

Age at Time t

Let $A(t)$ denote the time from t since the last event, that is,

$$A(t) = t - S_{N(t)} \tag{15.65}$$

where $A(t)$ is the age of the item in use at time t and hence is called the age at t. The distribution function for $A(t)$ is given by

$$P[A(t) \leq x] = \begin{cases} F(t) - \int_0^{t-x} [1 - F(t-y)]\, dM(y) & 0 \leq x \leq t \\ 1 & x > t \end{cases} \tag{15.66}$$

where $M(t)$ is the renewal function associated with the interevent distribution function $F(x)$. The derivation of (15.66) can be found in Ross (1970).

Renewal-Type Equation

A renewal-type equation is an equation of the form

$$g(t) = h(t) + \int_0^t g(t-x)\, dF(x) \tag{15.67}$$

where $h(\cdot)$ and $F(\cdot)$ are known functions, and $g(\cdot)$ is the unknown function to be obtained as a solution to the integral equation. Then $g(t)$ given by

$$g(t) = h(t) + \int_0^t h(t-x)\, dM(x) \tag{15.68}$$

where $M(x)$ is the renewal function associated with $F(x)$.

Renewal Reward Theorem

Suppose that a reward of Z_i is earned at the time of the ith renewal. Then the total reward earned by time t is given by

$$Z(t) = \sum_{i=1}^{N(t)} Z_i \tag{15.69}$$

where $N(t)$ is the number of events in $[0, t)$; $Z(t)$ is called the cumulative process.

If $E|Z_i|$ and $E(X_i)$ are finite, then [see Ross (1970)]

(1) with probability 1, $\lim_{t\to\infty} \frac{Z(t)}{t} \to \frac{E(Z_i)}{E(X_i)}$

and

(2) $$\lim_{t\to\infty} E\left[\frac{Z(t)}{t}\right] \to \frac{E(Z_i)}{E(X_i)}$$

15.6.2 Parameter Estimation

As with the power law process, we need to consider different cases.

Case 1: Event Truncation

The data for estimation is the set of event times $\{t_1, t_2, \ldots, t_n\}$. Then $x_i = t_i - t_{i-1}, 1 \le i \le n$, with $t_0 = 0$ can be viewed as n-independent observed values from the distribution function $F(x; \theta)$, and the parameters can be obtained from the methods discussed in Chapter 4.

Case 2: Time Truncation

Let the data for estimation be the set of n event times $\{t_1, t_2, \ldots, t_n\}$. Then in this case $x_i = t_i - t_{i-1}, 1 \le i \le n$, with $t_0 = 0$ can be viewed as n-independent observed values and $(t - t_n)$ as the censored data from distribution function $F(x; \theta)$. Again, the parameters can be obtained from the methods discussed in Chapter 4.

15.7 MODEL VII(b)-2: DELAYED RENEWAL PROCESS

15.7.1 Model Analysis

Expected Number of Renewals in [0, t)

Let $M_d(t)$ denote the expected number of renewals over $[0, t)$. It is given by

$$M_d(t) = F_1(t) + \int_0^t (t - x) f_1(x)\, dx \tag{15.70}$$

where $M(t)$ is the renewal function associated with $F(t)$.

Excess (or Residual) Life at Time t
Let $B_d(t)$ denote the time from t until the next renewal. The distribution function for $B_d(t)$ is given by

$$P\{B_d(t) \le x\} = F_1(t+x) - \int_0^t [1 - F_1(t+x-y)]\, dM_d(y) \qquad x \ge 0 \qquad (15.71)$$

For a proof of this, see Cinlar (1975).

15.8 MODEL VII(b)-3: ALTERNATING RENEWAL PROCESS

15.8.1 Model Analysis

In an ordinary renewal process, the interevent times are independent and identically distributed. In an alternating renewal process, the interevent times are all independent but not identically distributed. More specifically, the odd-numbered interevent times $X_1, X_3, X_5, \ldots$ have a common distribution function $F_1(x)$, and the even-numbered ones $X_2, X_4, X_6, \ldots$ have a common distribution function $F_2(x)$.

State of Item at Given Time
At any given time the item can be either in its working state or in a failed state and undergoing repair. A variable of interest is the probability $P(t)$ that the item is in its working state at time t. $P(t)$ is given by

$$P(t) = 1 - F_1(t) + \int_0^t P(t-x)\, dH(x) \qquad (15.72)$$

where $H(x)$ is the convolution of $F_1(x)$ and $F_2(x)$, and given by

$$H(x) = F_1(x)^* F_2(x) \qquad (15.73)$$

15.9 MODEL VII(c): POWER LAW–WEIBULL RENEWAL PROCESS

Calabria and Pulcini (2000) deal with the statistical inference and testing for this model. They derive expressions for the maximum-likelihood estimator and discuss the likelihood ratio tests to test several different hypotheses. Several applications of the model to model real data are also presented.

EXERCISES

Data Set 15.1 Time Between Failures for Repairable Item

1.43	0.11	0.71	0.77	2.63	1.49	3.46	2.46	0.59	0.74
1.23	0.94	4.36	0.40	1.74	4.73	2.23	0.45	0.70	1.06
1.46	0.30	1.82	2.37	0.63	1.23	1.24	1.97	1.86	1.17

Data Set 15.2 Time to Failure in Hours for Nonrepairable Item

156.6	108.0	289.8	198.0	84.1	51.2	12.4	59.1	35.5	6.3

15.1. Show that the three approaches discussed in Section 15.2.2 are equivalent.

15.2. Derive (15.12) to (15.15).

15.3. Suppose that Data Set 15.1 can be modeled by Model VII(a)-1 (Weibull intensity model–power law process). Estimate the model parameters using the method of maximum likelihood.

15.4. Discuss whether Data Set 15.1 supports the null hypothesis $\beta = 1$ or not.

15.5. Derive (15.47).

15.6. How would you simulate a time history of the modulated power law process?

15.7. Derive (15.56).

15.8. Derive (15.63) and (15.65).

15.9. Data Set 15.2 can be modeled as an ordinary renewal process. Estimate the model parameters assuming that the time between failures can be modeled by a two-parameter Weibull distribution.

15.10. For the model in Exercise 15.4, calculate the probability that the number of replacements needed over a 500-h interval.

15.11. Write a program to solve the one-dimensional renewal equation given by (15.56).

15.12. A component is subjected to a reliability improvement process. It is tested to failure and the cause of the failure analyzed. Based on this design, changes are made and the process repeated. Suppose that the failure time of a component is given by an exponential distribution with the scale parameter (mean time to failure) changes with n, the number of times the item has been tested. The data generated by the test are the failure times $(x_i, 1 \leq i \leq n)$ with subscript i corresponding to the failure after $(i - 1)$ modifications.

The data given below is the order data $x_{(i)}$ as opposed to x_i:

4.8 5.1 10.2 16.0 27.2 31.4 32.1 38.5 47.2 47.6

Is it possible to draw any inference from this to state that the reliability of the component has improved during the development process? Can one estimate the reliability of the component at the end of testing based on the above data? If not, what kind of information should have been recorded?

PART F

Weibull Modeling of Data

CHAPTER 16

Weibull Modeling of Data

16.1 INTRODUCTION

The building of a model to solve a problem must take into account all the relevant information available to the modeler. This can vary from very limited to considerable, and we look at this in the context of reliability theory in the next two chapters. In this chapter we focus our attention on model building based solely on data comprising failure and censored times. The resulting model is called an *empirical model* (or *data-dependent model* or *black-box model*) as the data available forms the basis for the model building. This type of model is also appropriate when there is insufficient understanding so that no other approach can be used to model the data.

The data that we will be looking at are univariate, continuous valued, and statistically independent. The Weibull models that are appropriate for modeling the data are Types I to III discussed in Chapters 3 to 11. We confine our attention to the models listed in Table 16.1.

Empirical modeling involves the following steps:

Step 1: Model selection
Step 2: Estimation of model parameters
Step 3: Model validation

We suggest a two-stage approach as indicated below:

Stage 1: Preliminary model selection and parameter estimation based on the WPP (or IWPP) plot of the data.
Stage 2: Final model selection based on a more rigorous statistical approach.

The outline of the chapter is as follows. We start with a brief discussion relating to the data aspects. This is done in Section 16.2. Sections 16.3 and 16.4 deal with

Weibull Models, by D.N.P. Murthy, Min Xie, and Renyan Jiang.
ISBN 0-471-36092-9

Table 16.1 Weibull Models for Modeling

Model	Name	Section
Model 1	Two-parameter Weibull	3.3
Model I(a)-2	Three-parameter Weibull	3.4
Model I(b)-3	Inverse Weibull	6.6
Model II(b)-1	Extended Weibull	7.4
Model II(b)-2	Exponentiated Weibull	7.5
Model II(b)-3	Modified Weibull	7.6
Model II(b)-9	Four-parameter Weibull	7.10
Model II(b)-13	Weibull extension	7.13
Model III(a)-1	Twofold Weibull mixture	8.2
Model III(a)-2	Twofold inverse Weibull mixture	8.3
Model III(b)-1	Twofold Weibull competing risk	9.2
Model III(b)-2	Twofold inverse Weibull competing risk	9.3
Model III(b)-4	Generalized competing risk	9.5
Model III(c)-1	Twofold Weibull multiplicative	10.2
Model III(c)-2	Twofold inverse Weibull multiplicative	10.3
Model III(d)-1	Weibull sectional Model-1	11.2.1
Model III(d)-2	Weibull sectional Model-2	11.2.2
Model III(d)-3	Weibull sectional Model-3	11.2.3
Model III(d)-4	Weibull sectional Model-4	11.2.4
Model III(d)-5	Weibull sectional Model-5	11.2.5
Model III(d)-6	Weibull sectional Model-6	11.2.6

the preliminary and final model selections, respectively. Following this, we illustrate the approach through three case studies in Section 16.5.

16.2 DATA-RELATED ISSUES

16.2.1 Data Size

Let n denote the size of the data. This has implications for the WPP (IWPP) plotting as indicated below.

Case A Small data set ($n < 20$). In this case one uses the complete data for the WPP plot.

Case B Medium data set ($20 \leq n \leq 50$). In this case one can use either the bootstrap or the jackknife approach. The bootstrap approach involves generating different sets of data by sampling (n data) from the initial data set with replacement in the bootstrap approach and by deleting one data at a time in the jackknife approach. The WPP plot of each sampled set results in the selection of a model from the models under consideration. A statistical analysis of the models (obtained from the different sampled sets) forms the basis for the selection of the most appropriate model.

Case C Large data set ($n > 50$). In this case the original data set (S) is divided into two disjointed subsets (S_1 and S_2). Let $n_1 (\approx 0.8n)$ and $n_2 (= n - n_1)$ denote the number of data in the two subsets. The WPP plot of the data is carried out using the data from S_1, and the data from S_2 is used for model validation.

16.2.2 Nature of Data

As mentioned in Chapter 4, the data can be uncensored or censored. In the latter case, the censoring can be either right or left censored, single or multiple censored, or interval censored. These have implications for both WPP plotting as well as for parameter estimation, as discussed in Chapters 4 and 5.

In the case of censored data, we have an additional issue—the fraction of censored data. If this is large (> 0.9), then one can have problems with the modeling of the data.

16.3 PRELIMINARY MODEL SELECTION AND PARAMETER ESTIMATION

16.3.1 Model Selection

The preliminary model selection is based on the WPP (or IWPP) plot. In Chapters 3 to 11 we discussed the WPP (or IWPP) plots for the models listed in Table 16.1. Table 16.2 gives a classification scheme for the different shapes for the WPP plots. Table 16.3 lists the different shapes for the WPP (or IWPP) plot for the models in Table 16.1.

Table 16.2 Classification of the Shapes for the WPP and IWPP Plots

Type	Description
A	Straight line
B	Concave
B1	Concave with left asymptote vertical
C	Convex
C1	Convex with right asymptote vertical
D	Single inflection point (S-shaped) with parallel asymptotes
D1	Single inflection point (S-shaped) with vertical asymptotes
E1	Bell shaped
E(n)	Multiple inflection points ($n \geq 1$ and odd)

Table 16.3 Shapes of WPP or IWPP Plots for the Models in Table 16.1

Model	Shapes	Comments
Model 1	A	WPP
Model I(a)-2	B1	WPP
Model I(b)-3	A	
Model II(b)-1	B ($\nu < 1$); C ($\nu > 1$)	WPP
Model II(b)-2	B ($\nu > 1$); C ($\nu < 1$)	WPP
Model II(b)-3	C	WPP
Model II(b)-9	D1	WPP
Model II(b)-13		WPP
Model III(a)-1	D ($\beta_1 = \beta_2$); E(3) $\beta_1 \neq \beta_2$	WPP
Model III(a)-2	Same as Model III(a)-1	IWPP
Model III(b)-1	C	WPP
Model III(b)-2	C	IWPP
Model III(b)-4	C; E1	WPP
Model III(c)-1	B	WPP
Model III(c)-2	C	IWPP
Model III(d)-1	B ($\beta_1 > \beta_2$); C ($\beta_1 < \beta_2$)	WPP
Model III(d)-2	C	WPP
Model III(d)-3	B	WPP
Model III(d)-4	E(1)	WPP
Model III(d)-5	B; C; E(1)	WPP
Model III(d)-6	B, C; E(1)	WPP

The model selection process is as follows. The WPP (or IWPP) plot of the data is carried out as discussed in Chapter 4. A smooth curve is fitted to the plotted data and its shape is assessed visually. It is compared with the WPP shapes listed in Table 16.2. If there is no match, then none of the models in Table 16.1 are suitable for the modeling of the given data. In this case, one needs to look at models derived from other distributions. On the other hand, if there is a match, then one or more of the Weibull models might be appropriate for the modeling of the given data. In this case, Table 16.3 is used to identify the models that have similar shaped WPP plots, and these are potential candidates for modeling.

16.3.2 Parameter Estimation

For the potential candidates, one needs to estimate the model parameters. In Chapters 4 to 11 we discussed the estimation based on the WPP plots. They involve fitting asymptotes, determining intersection and inflection points, and so forth.

An alternate approach is to estimate the parameters through a least-squares fit. Let x_i and y_i, $1 \leq i \leq n$, denote the Weibull transformed values in the WPP plotting of the data set. Let $y(x_i; \theta)$, $1 \leq i \leq n$, denote the Weibull transformed values for the model with parameter θ. The parameter estimate is the value of θ that minimizes the objective function given by

$$J(\theta) = \sum_{i=1}^{n} [y(t_i; \theta) - y_i]^2 \tag{16.1}$$

The optimization needs to be carried out using a computational algorithm.

16.4 FINAL MODEL SELECTION, PARAMETER ESTIMATION, AND MODEL VALIDATION

Often the preliminary model selection results in more than one model being identified as potential candidates for modeling the given data set. In this case one needs to choose between these alternate models to decide which is most appropriate. Also, the parameter estimates, based on the WPP plot, are crude. They can be used as the starting point for obtaining more refined estimates. This, however, depends to a large extent on the size of the data set. In this section we discuss these issues for the three different cases defined in Section 16.2.1.

16.4.1 Case A: Small Data Set

Model Selection

If one of the potential candidates (from the preliminary selection) has a value for $J(\theta)$ [given by (16.1)] that is considerably smaller than that for the other models, then it is selected as the most appropriate model (within the class of Weibull models) to model the given data set.

An alternate approach is to choose the model based on the principle of parsimony—given several models that fit the data reasonably well, then the model with the least number of parameters is the preferred one. This can be formalized through a function $J(\theta, K)$, involving $J(\theta)$ given by (16.1) and another function $\Phi(K_p)$ that increases with K_p (the number of parameters in the models). The model parameters are the values that minimize $J(\theta)$.

Two different forms for $\Phi(K_p)$ are the following: $\Phi(K_p) = \gamma K_p$ and $\Phi(K_p) = K_p^\gamma$ with $\gamma > 0$. As a result, the two different forms for $J(\theta, K_p)$ are given by $J(\theta, K_p) = J(\theta) + \gamma K_p$ and $J(\theta, K_p) = \ln J(\theta) + \gamma \ln K_p$. The model that yields the smallest value for $J(\theta, K_p)$ is the model that is selected as the final model. The larger the value of γ, the greater is the penalty associated with the number of parameters in the model. This implies that the value for γ and the form of $\Phi(K_p)$ have a significant impact on the final model selected.

If two or more potential candidates have nearly the same value for $J(\theta, K_p)$, then one can look at additional properties of the WPP plot to decide on the final model. This would involve comparing the error between some characteristics (such as the intersection with the x or y axis of the WPP plot based on the model and the smooth fit to the plot based on the data.

Table 16.4 Shapes of the Density Function for Different Models in Table 16.1

Model	Shapes	Comments
Model 1	Types 1, 2	
Model I(a)-2	Types 1, 2	
Model I(b)-3	Type 2	
Model II(b)-1		Not yet studied
Model II(b)-2	Types 1, 2	
Model II(b)-3		Not yet studied
Model II(b)-9	Types 1, 2	+ Increasing shape
Model II(b)-13		Not yet studied
Model III(a)-1	Types 1, 2, 3, 4	
Model III(a)-2	Types 2, 4	
Model III(b)-1	Types 1, 2, 3, 4	
Model III(b)-2	Types 2, 4	
Model III(b)-4	Types 1, 2, 3, 4	
Model III(c)-1	Types 1, 2, 4	
Model III(c)-2	Type 2	
Model III(d)-1	Types 1, 2, 3, 4	
Model III(d)-2	Types 1, 2, 3, 4	
Model III(d)-3	Types 1, 2, 3, 4	
Model III(d)-4	Types 1, 2, 3, 4	
Model III(d)-5	Types 1, 2, 3, 4, 5, 6	
Model III(d)-6	Types 1, 2, 3, 4, 5, 6	

Finally, one can also look at the properties of the model (such as the shapes for the density function or the hazard function) in deciding on the final model selection. In Chapter 3 we defined a classification scheme for the different shapes for the density and hazard functions. Tables 16.4 and 16.5 summarize the different shapes for the density and hazard functions for the models in Table 16.1.

Parameter Estimation

In Chapter 4 we discussed several different statistical methods for estimating the model parameters, and these included the method of moments, the method of percentile, the method of maximum likelihood, and the Bayesian method. Unfortunately, none of these methods (except the Bayesian) are appropriate for small data sets. Hence, the estimates obtained from the WPP plot are often the final estimates.

Model Validation

In Chapter 5 we discussed many different statistical tests for validating a model. These require data that is different from the data used for model selection and parameter estimation. For small data sets, since all the data is used for model selection and parameter estimation, there is no data left for validation.

Table 16.5 Shapes of Hazard Function for Different Models in Table 16.1

Model	Shapes
Model 1	Types 1, 2, 3
Model I(a)-2	Types 1, 2, 3
Model I(b)-3	Type 5
Model II(b)-1	Types 1, 3, 6, 7
Model II(b)-2	Types 1, 3, 4, 5
Model II(b)-3	Types 1, 4
Model II(b)-9	Types 1, 3
Model II(b)-13	Types 3, 4
Model III(a)-1	Types 1, 3, 5, 6, 7, 9
Model III(a)-2	Types 5, 9
Model III(b)-1	Types 1, 3, 4
Model III(b)-2	Types 5, 9
Model III(b)-4	Types 1, 3, 4, 6, 7, 8
Model III(c)-1	Types 1, 3, 5, 6
Model III(c)-2	Types 5, 9
Model III(d)-1	Types 1, 3, 4, 5
Model III(d)-2	Types 1, 3, 4, 5
Model III(d)-3	Types 1, 3, 4, 5, 6, 7
Model III(d)-4	Types 1, 3, 4, 5, 6, 7
Model III(d)-5	Types 1, 3, 4, 5, 6, 7
Model III(d)-6	Types 1, 3, 4, 5, 6, 7

16.4.2 Case B: Medium Data Set

Model Selection

One starts as in case A. The shape of the WPP plot (based on data) indicates which of the Weibull models are not appropriate. The remaining models are potential candidates. As before one can examine $J(\theta)$ to determine the degree of fit between the model and the data. If one of the models has a value that is considerably smaller then the rest, then it is selected as the final model.

When several models have roughly the same value of $J(\theta)$, then one needs to use some other approach to model selection. Two such approaches are the bootstrap approach and the jackknife approach. The bootstrap approach is as follows [see Efron (2000) for details]. Let $M_j, 1 \leq j \leq J$ denote the j models under consideration; n (the size of the original data set) data are sampled from the original data set with replacement. Based on the WPP plot of the sampled data, $J(\theta)$ is computed for each model based on the best fit between the data and the model. The process is repeated K times. Let M^k denote the model that yields the smallest value for $J(\theta)$ during iteration $k\,(1 \leq k \leq K)$. One carries out an analysis of the best fits for the different runs and, based on this one decides on the final model.

The jackknife approach is similar. Here the different models are fitted to n different data sets, each of size $n - 1$, obtained by deleting one data point at a time from the original data set. The best fits for each set are then analyzed to decide on the final model.

Finally, as discussed in case A, one might examine the shapes of the density and hazard functions of the different models under consideration before the final selection is made.

Parameter Estimation

In this case one can use one or more of the different statistical methods to estimate the final model parameters using either all the data available or part of the data. The method of maximum likelihood is the most preferred because of its desirable properties.

Model Validation

If the parameters are estimated using a subset of the data, then the remaining data can be used for model validation as discussed in Chapter 4. If all the data is used in model selection and parameter estimation, then one can still use the tests indicated in Chapter 4, taking into account the loss in the degree of freedom. Further details can be found in many books; see, for example, Blischke and Murthy (2000) and Meeker and Escobar (1998b).

6.4.3 Case C: Large Data Set

Model Selection

The initial selection is based on the data from subset S_1. The final model selection is based on using the approach outlined in either case A (i.e., the model with the smallest value for $J(\theta)$ [or $J(\theta, K_p)$] is the model selected) or in case B (based on the bootstrap and/or jackknife approaches).

Model Parameters

The parameters can be estimated using statistical methods as in case B using the subset of the original data set.

Model Validation

In this case the final model selected in stage 1 is viewed as the true model, and the remaining data (data not used for model selection and parameter estimation) is used to test the validity of the model using one or more of the tests discussed in Chapter 5.

16.5 CASE STUDIES*

In this section we discuss three case studies involving real data to illustrate the modeling approach discussed earlier. The data for the three cases are as follows:

*The WPP plotting and model fitting were carried out by Dr Michael Bulmer.

Case 1: Small (19 data points, comprised of failure times)

Case 2: Medium (50 data points comprised of failure and censored times)

Case 3: Large (153 data points comprised of failure and censored times)

16.5.1 Case 1: Cleaning Web Failures

Modern photocopiers are complex system involving several components. The "cleaning web" is one such component. Whenever it fails, it needs to be replaced by a new one.

The data recorded from a $4\frac{1}{2}$-year service history of a photocopier is given in Table 16.6. The first and third columns give the number of copies made at the time of replacement and the second and fourth columns give the corresponding day (subsequent to the machine being put into operation).

The difference between two successive values in a column gives the age at failure, and the number of copies produced before failure, for the different cleaning webs over the time period. The days and copies between failures of the cleaning web were strongly correlated ($r = 0.916$). The mean number of days between failures was 115.1 and the mean number of copies between failures was 72,670.

A smooth fit to the WPP plot (based on age between failures) is shown in Figure 16.1. On comparing it with the shapes indicated in Table 16.2, we see that it is shape B. From Table 16.3 we see that Models II(b)-1 (extended Weibull), II(b)-2 (exponentiated Weibull), III(c)-1 (Weibull multiplicative model), and III(d)-1 [also III(d)-3, III(d)-5, and III(d)-6] (Weibull sectional model) are potential candidates for modeling the data and that the remaining models need to be rejected.

Table 16.7 shows the parameter estimates and the corresponding $J(\theta)$ based on the WPP plot for these models and for some of the rejected models. A genetic algorithm was used in the optimization. For details, see Bulmer and Eccleston (1998). An important thing to note is that, in general, $J(\theta)$ for the rejected models are much bigger than that for the potential candidates. Among the possible models listed above, it is clearly seen that $J(\theta)$ is the smallest for Model III(c)-1. This implies that this model yields the best fit, and this can be seen from Figure 16.1 where the WPP plot for the model is shown as a continuous line and the Weibull

Table 16.6 Cleaning Web Data (Case Study 1)

Copies	Age (days)	Copies	Age (days)
60,152	29	900,362	1356
132,079	128	933,785	1412
365,075	397	938,100	1448
427,056	563	994,597	1514
501,550	722	1,045,893	1583
597,739	916	1,068,124	1609
675,841	1016	1,077,537	1640
716,636	1111		

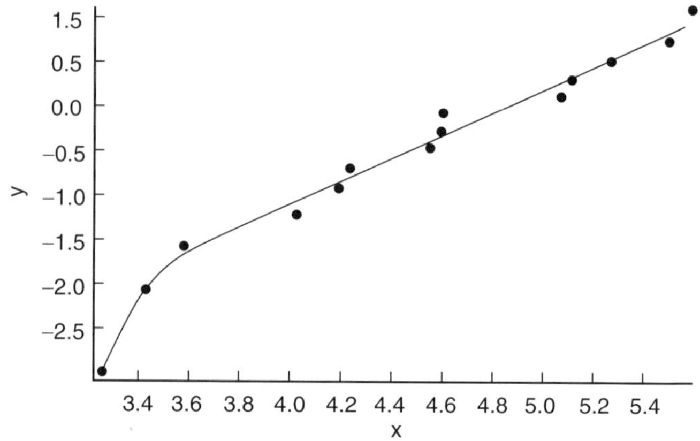

Figure 16.1 WPP plots for Weibull multiplicative model (Case Study 1); based on days between failures.

transformed data as dots. Model III(d)-1 has the second smallest value for $J(\theta)$, and the values for the remaining potential candidates are much higher. As a result, one should accept Model III(c)-1 as the most appropriate model to model cleaning web failures based on age at failure.

A similar study based on the number of copies produced before failure yields a slightly different outcome. Figure 16.2 shows the WPP plot of the data. It indicates that the shape is D (see Table 16.2) and that the only potential candidate is Model III(a)-1 (Weibull mixture model with different shape parameters), and all others need to be rejected.

Table 16.8 shows the parameter estimates obtained by best fit and the corresponding $J(\theta)$ based on the WPP plot for this model and for some of the rejected

Table 16.7 Parameter Estimates Based on Days between Failures

Model	$J(\theta)$	Parameter Estimates
Model 1	0.752	$\hat{\alpha} = 129$, $\hat{\beta} = 1.48$
Model I(a)-2	0.314	$\hat{\alpha} = 101$, $\hat{\beta} = 0.953$, $\hat{\tau} = 21.1$
Model I(b)-3	0.660	$\hat{\alpha} = 61.8$, $\hat{\beta} = 1.23$
Model II(b)-1	0.544	$\hat{\alpha} = 345$, $\hat{\beta} = 1.96$, $\hat{v} = 0.078$
Model II(b)-2	0.752	$\hat{\lambda} = 0.000744$, $\hat{\beta} = 1.48$, $\hat{v} = 0.0$
Model II(b)-3	0.480	$\hat{\alpha} = 7.74$, $\hat{\beta} = 0.449$, $\hat{v} = 13.7$
Model III(a)-1	0.424	$\hat{\alpha}_1 = 55$, $\hat{\alpha}_2 = 184$, $\hat{\beta}_1 = 2.4$, $\hat{\beta}_2 = 2.2$, $\hat{p} = 0.39$
Model III(a)-1 ($\beta_1 = \beta_2 = \beta$)	0.424	$\hat{\alpha}_1 = 56.8$, $\hat{\alpha}_2 = 189$, $\hat{\beta} = 2.34$, $\hat{p} = 0.421$
Model III(b)-1	0.752	$\hat{\alpha}_1 = 203$, $\hat{\alpha}_2 = 210$, $\hat{\beta}_1 = 1.48$, $\hat{\beta}_2 = 1.48$
Model III(c)-1	0.173	$\hat{\alpha}_1 = 28.8$, $\hat{\alpha}_2 = 128$, $\hat{\beta}_1 = 6.62$, $\hat{\beta}_2 = 1.29$
Model III(d)-1	0.314	$\hat{\alpha}_1 = 46$, $\hat{\alpha}_2 = 101$, $\hat{\beta}_1 = 5.06$, $\hat{\beta}_2 = 0.954$, $\hat{\tau} = 21.6$, $\hat{t}_1 = 26$

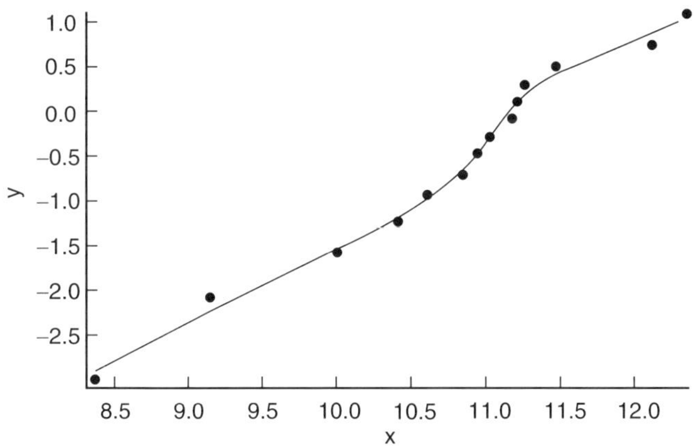

Figure 16.2 WPP plots for Weibull mixture model (Case Study 1); based on number of copies between failures.

models. As can be seen $J(\theta)$ for the all other models are very much larger than that for Model III(a)-1. This implies that Model III(a)-1 yields the best fit, and this can be seen from Figure 16.2 where the WPP plot for the model is shown as a continuous line and the Weibull transformed data as dots. As a result, one should accept Model III(a)-1 as the most appropriate model to model cleaning web failures based on number of copies produced before failure.

16.5.2 Case 2: Throttle Failures

The "throttle" is a component of a vehicle. The data presented in Table 16.9 is from Carter (1986) and gives distance traveled (in thousands of kilometers) before

Table 16.8 Parameter Estimates Based on Number of Copies between Failures

Model	$J(\theta)$	Parameter Estimates
Model 1	0.543	$\hat{\alpha} = 80000$, $\hat{\beta} = 1.06$
Model I(a)-2	0.543	$\hat{\alpha} = 80100$, $\hat{\beta} = 1.06$, $\hat{\tau} = 0.0279$
Model I(b)-3	2.32	$\hat{\alpha} = 27800$, $\hat{\beta} = 0.651$
Model II(b)-1	0.394	$\hat{\alpha} = 2820$, $\hat{\beta} = 0.442$, $\hat{\nu} = 43.2$
Model II(b)-2	0.539	$\hat{\lambda} = 0.00000803$, $\hat{\beta} = 1.03$, $\hat{\nu} = 0.0000005$
Model II(b)-3	0.527	$\hat{\alpha} = 95200$, $\hat{\beta} = 1.26$, $\hat{\nu} = 0.785$
Model III(a)-1	0.0924	$\hat{\alpha}_1 = 79400$, $\hat{\alpha}_2 = 67900$, $\hat{\beta}_1 = 0.851$, $\hat{\beta}_2 = 5.23$, $\hat{p} = 0.674$
Model III(a)-1 ($\beta_1 = \beta_2 = \beta$)	0.543	$\hat{\alpha}_1 = 80000$, $\hat{\alpha}_2 = 80300$, $\hat{\beta} = 1.06$, $\hat{p} = 0.902$
Model III(b)-1	0.502	$\hat{\alpha}_1 = 96100$, $\hat{\alpha}_2 = 1050000$, $\hat{\beta}_1 = 1.27$, $\hat{\beta}_2 = 0.63$
Model III(c)-1	0.543	$\hat{\alpha}_1 = 39100$, $\hat{\alpha}_2 = 80100$, $\hat{\beta}_1 = ??$, $\hat{\beta}_2 = 1.06$
Model III(d)-1	0.531	$\hat{\alpha}_1 = 78200$, $\hat{\alpha}_2 = 50600$, $\hat{\beta}_1 = 1.08$, $\hat{\beta}_2 = 0.753$, $\hat{\tau} = 28200$, $\hat{t}_1 = 92800$

Table 16.9 Throttle Failure Data (Case Study 2)[a]

0.478	0.959	1.847+	3.904	6.711+
0.484+	1.071+	2.400	4.443+	6.835+
0.583	1.318+	2.550+	4.829	6.947+
0.626+	1.377	2.568+	5.328	7.878+
0.753	1.472+	2.639	5.562	7.884+
0.753	1.534	2.944	5.900+	10.263+
0.801	1.579+	2.981	6.122	11.019
0.834	1.610+	3.392	6.226+	12.986
0.850+	1.729+	3.393	6.331	13.103+
0.944	1.792+	3.791+	6.531	23.245+

[a] Distance traveled before failure or censoring (denoted by +); unit, 1000 km.

failure, or the item being suspended before failure, for a preproduction general-purpose load-carrying vehicle. Hence, the independent variable is the distance traveled with 1000 km being the unit of measurement.

A smooth fit to the WPP plot (based on age between failures) is shown in Figure 16.3. On comparing with the shapes listed in Table 16.2, we see that it is type B. From Table 16.3 we see that Models II(b)-1 (extended Weibull), II(b)-2 (exponentiated Weibull), III(c)-1 (Weibull multiplicative model), and III(d)-1 [also III(d)-3, III(d)-5, and III(d)-6] (Weibull sectional model) are potential candidates for modeling the data, and the remaining models are to be rejected.

Table 16.10 shows the parameter estimates and the corresponding $J(\theta)$ based on the WPP plot for these models and for some of the rejected models. Among the possible models listed above it is clearly seen that $J(\theta)$ is the smallest for Model III(c)-1. From Table 16.10 we see that $J(\theta)$ for Models II(b)-1 and II(b)-2 are much larger than that for Model III(c)-1, and hence these models can be rejected. An

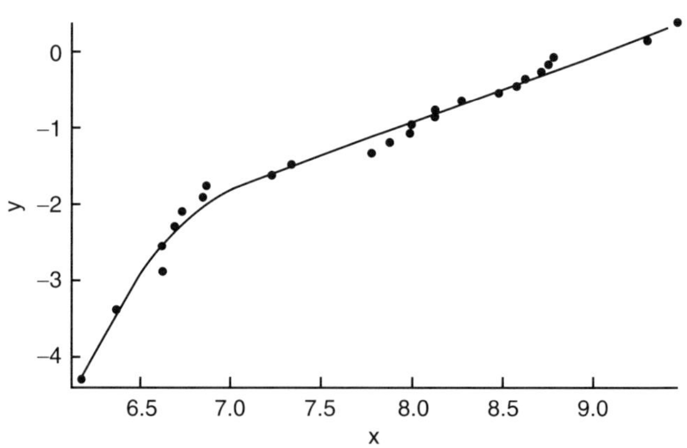

Figure 16.3 WPP plots for Weibull multiplicative model (Case Study 2).

Table 16.10 Parameter Estimates for Case Study 2

Model	$J(\theta)$	Parameter Estimates
Model 1	2.77	$\hat{\alpha} = 7470$, $\hat{\beta} = 1.13$
Model I(a)-2	0.633	$\hat{\alpha} = 8370$, $\hat{\beta} = 0.755$, $\hat{\tau} = 450$
Model I(b)-3	1.36	$\hat{\alpha} = 3110$, $\hat{\beta} = 0.706$
Model II(b)-1	2.31	$\hat{\alpha} = 80800$, $\hat{\beta} = 1.35$, $\hat{\nu} = 0.025$
Model II(b)-2	2.77	$\hat{\lambda} = 0.0000407$, $\hat{\beta} = 1.13$, $\hat{\nu} = 0.0$
Model II(b)-3	1.86	$\hat{\alpha} = 117$, $\hat{\beta} = 0.292$, $\hat{\nu} = 13.8$
Model III(a)-1	0.412	$\hat{\alpha}_1 = 865$, $\hat{\alpha}_2 = 9160$, $\hat{\beta}_1 = 6.86$, $\hat{\beta}_2 = 1.38$, $\hat{p} = 0.136$
Model III(a)-1 ($\beta_1 = \beta_2 = \beta$)	1.55	$\hat{\alpha}_1 = 1320$, $\hat{\alpha}_2 = 10500$, $\hat{\beta} = 2.46$, $\hat{p} = 0.305$
Model III(b)-1	2.77	$\hat{\alpha}_1 = 13400$, $\hat{\alpha}_2 = 14100$, $\hat{\beta}_1 = 1.13$, $\hat{\beta}_2 = 1.13$
Model III(c)-1	0.37	$\hat{\alpha}_1 = 8700$, $\hat{\alpha}_2 = 723$, $\hat{\beta}_1 = 0.87$, $\hat{\beta}_2 = 3.93$
Model III(d)-1	0.573	$\hat{\alpha}_1 = 1190$, $\hat{\alpha}_2 = 8720$, $\hat{\beta}_1 = 4.68$, $\hat{\beta}_2 = 0.699$ $\hat{\tau} = 511$, $\hat{t}_1 = 601$

interesting feature is that $J(\theta)$ for Model III(a)-1 is comparable to that for Model III(c)-1 and Models I(a)-2 and III(d)-1. Figure 16.3 shows the fit between the data and Model III(c)-1 while Figure 16.4 shows the fit with Model III(a)-1. As can be seen, both indicate a reasonable fit. This suggests that these models need to be explored further before the final decision is made.

The bootstrap approach was used to decide on the final model based on 200 independent runs (the data for each run generated through random sampling of the original data set 50 times with replacement) and yielded the following results. For 42% of the runs Model III(a)-1 had the smallest $J(\theta)$, for 24% it was Model III(d)-1, for 18% it was Model III(c)-1, and for 13% it was Model I(a)-2. This indicates that a Weibull mixture model might be the most appropriate one to model the data.

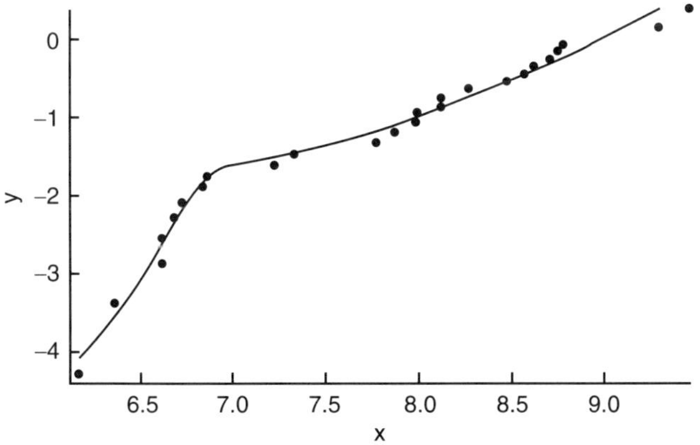

Figure 16.4 WPP plots for Weibull mixture model (Case Study 2).

The jackknife approach (based on 50 runs with 49 data in each run obtained by deleting one data at a time) yielded the following results. For 60% of the runs Model III(c)-1 had the smallest $J(\theta)$, and for the remaining 40% it was Model III(a)-1. This suggests that Model III(c)-1 be accepted as the most appropriate one to model the data.

In both the bootstrap and jackknife approaches all the models listed in Table 16.10 were fitted to the data generated for each run. An interesting thing to note is that the jackknife approach selected either Model III(a)-1 or Model III(c)-1 in each run. This is in contrast to the bootstrap approach where, for some runs, other models provided better fits. This is to be expected as sampling with replacement can produce data sets that differ significantly (in a probabilistic sense) from the original data set.

Based on the above analysis, Model III(c)-1 is the final model selected. The parameter estimates based on the method of maximum likelihood are as follows:

$$\hat{\alpha}_1 = 8920 \qquad \hat{\alpha}_2 = 779 \qquad \hat{\beta}_1 = 0.836 \qquad \hat{\beta}_2 = 2.42 \qquad \ln(\hat{\theta}) = -246.1$$

16.5.3 Case 3: Aircraft Windshield

The windshield on a large aircraft is a complex piece of equipment, comprised basically of several layers of material, including a very strong outer skin with a heated layer just beneath it, all laminated under high temperature and pressure. Failures of these items are not structural failures. Instead, they typically involve damage or delamination of the nonstructural outer ply or failure of the heating system. These failures do not result in damage to the aircraft but do result in replacement of the windshield.

Data on all windshields are routinely collected and analyzed. At any specific point in time, these data will include failures to date of a particular model as well as service times of all items that have not failed. Data of this type are incomplete in that not all failure times have as yet been observed.

Data on failure and service times for a particular model windshield are given in Table 16.11 from Blischke and Murthy (2000). The data consist of 153 observations, of which 88 are classified as failed windshields, and the remaining 65 are service times of windshields that had not failed at the time of observation. The unit for measurement is 1000 h.

A smooth fit to the WPP plot (based on all the data) is shown in Figure 16.5. On comparing with the shapes listed in Table 16.2, we see that it is type C. From Table 16.3 we see that Model II(b)-1 (extended Weibull), Model II(b)-2 (exponentiated Weibull), Model II(b)-3 (modified Weibull), Model III(b)-1 (Weibull competing risk), Model III(b)-2 (inverse Weibull competing risk), Model III(b)-4 and Model III(c)-2 (inverse Weibull multiplicative), and Model III(d)-2 have similar shapes and qualify as potential models while the remaining are rejected.

Table 16.11 Windshield Failure Data (Case Study 3)[a]

Failure Times				Service Times		
0.040	1.866	2.385	3.443	0.046	1.436	2.592
0.301	1.876	2.481	3.467	0.140	1.492	2.600
0.309	1.899	2.610	3.478	0.150	1.580	2.670
0.557	1.911	2.625	3.578	0.248	1.719	2.717
0.943	1.912	2.632	3.595	0.280	1.794	2.819
1.070	1.914	2.646	3.699	0.313	1.915	2.820
1.124	1.981	2.661	3.779	0.389	1.920	2.878
1.248	2.010	2.688	3.924	0.487	1.963	2.950
1.281	2.038	2.823	4.035	0.622	1.978	3.003
1.281	2.085	2.890	4.121	0.900	2.053	3.102
1.303	2.089	2.902	4.167	0.952	2.065	3.304
1.432	2.097	2.934	4.240	0.996	2.117	3.483
1.480	2.135	2.962	4.255	1.003	2.137	3.500
1.505	2.154	2.964	4.278	1.010	2.141	3.622
1.506	2.190	3.000	4.305	1.085	2.163	3.665
1.568	2.194	3.103	4.376	1.092	2.183	3.695
1.615	2.223	3.114	4.449	1.152	2.240	4.015
1.619	2.224	3.117	4.485	1.183	2.341	4.628
1.652	2.229	3.166	4.570	1.244	2.435	4.806
1.652	2.300	3.344	4.602	1.249	2.464	4.881
1.757	2.324	3.376	4.663	1.262	2.543	5.140

[a] Failure times and service times (unfailed items); unit, thousands hours.

Table 16.12 shows the parameter estimates and the corresponding $J(\theta)$ based on the WPP plot for some of these models (as well as for some of the rejected models). Among the possible models listed above it is clearly seen that $J(\theta)$ is the smallest for Model III(a)-1 (Weibull mixture). The WPP plot for the Weibull mixture has an

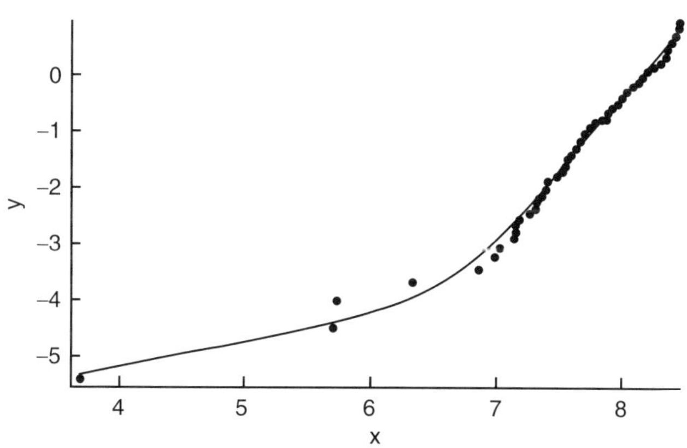

Figure 16.5 WPP plot for mixture model (windshield data).

Table 16.12 Parameter Estimates (Based on WPP Plot) for Case Study 3

Model	$J(\theta)$	Parameter Estimates
Model 1	18.0	$\hat{\alpha} = 3980$, $\hat{\beta} = 1.74$
Model I(a)-2	18.0	$\hat{\alpha} = 3980$, $\hat{\beta} = 1.74$, $\hat{\tau} = 0.00$
Model I(b)-3	62.9	$\hat{\alpha} = 3140$, $\hat{\beta} = 0.458$
Model II(b)-1	3.01	$\hat{\alpha} = 293$, $\hat{\beta} = 0.699$, $\hat{\nu} = 164$
Model II(b)-2	4.12	$\hat{\lambda} = 0.000138$, $\hat{\beta} = 0.794$, $\hat{\nu} = 0.000678$
Model II(b)-3	11.3	$\hat{\alpha} = 4910$, $\hat{\beta} = 7.54$, $\hat{\nu} = 0.206$
Model III(a)-1	1.3	$\hat{\alpha}_1 = 82300$, $\hat{\alpha}_2 = 3210$, $\hat{\beta}_1 = 0.429$, $\hat{\beta}_2 = 2.99$, $\hat{p} = 0.136$
Model III(a)-1 ($\beta_1 = \beta_2 = \beta$)	18.0	$\hat{\alpha}_1 = 4010$, $\hat{\alpha}_2 = 3980$, $\hat{\beta} = 1.74$, $\hat{p} = 0.0578$
Model III(b)-1	1.79	$\hat{\alpha}_1 = 286000$, $\hat{\alpha}_2 = 3540$, $\hat{\beta}_1 = 0.632$, $\hat{\beta}_2 = 2.9$
Model III(c)-1	18.0	$\hat{\alpha}_1 = 0.006$, $\hat{\alpha}_2 = 3980$, $\hat{\beta}_1 = 0.323$, $\hat{\beta}_2 = 1.74$
Model III(d)-1	2.05	$\hat{\alpha}_1 = 1470000$, $\hat{\alpha}_2 = 4560$, $\hat{\beta}_1 = 0.527$, $\hat{\beta}_2 = 3.67$ $\hat{\tau} = 1070$, $\hat{t}_1 = 180$

inflection point that is not to be seen from the WPP plot. This is due to most of the data coming from one subpopulation (as indicated by a small value for the mixing parameter p). From Table 16.12 we see that $J(\theta)$ for Model III(b)-1 is close to that for Model III(a)-1. This suggests that we need to look at these two models before a final decision is made.

If the smallest failure data (0.04) is viewed as an outlier, then one needs to delete it from the data set. Since the data set is large, the model selection can be done using 80% of the data so that the remaining 20% can be used for model validation. This division of data into two subsets (S_1 and S_2) was done by random selection from the original set after the outlier was deleted. Model III(a)-1 had the smallest value for $J(\theta)$ (implying the best fit among all the models considered) based on S_1. The parameter estimates (based on WPP plot) are as follows:

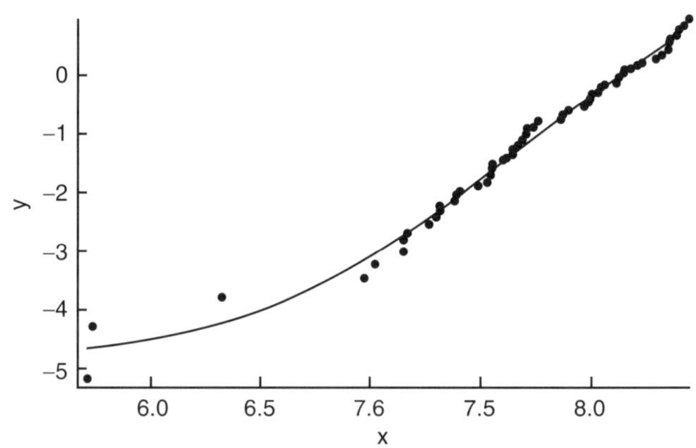

Figure 16.6 WPP plot for mixture model (Case Study 3); 80% of data, excluding outlier.

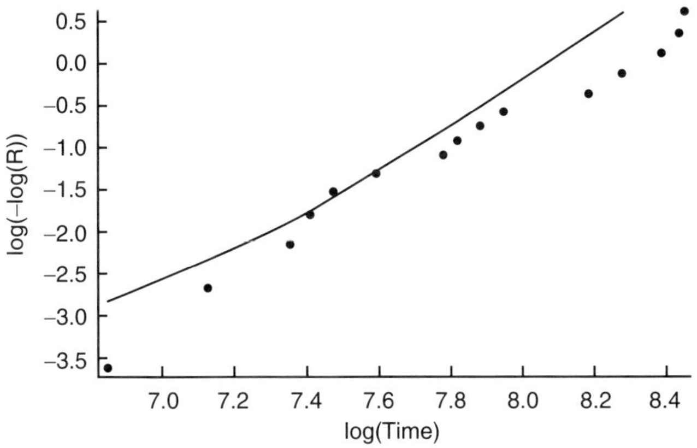

Figure 16.7 WPP plot for mixture model (Case Study 3); remaining 20% of data, excluding outlier.

$$\hat{\alpha}_1 = 1.3 \times 10^6 \qquad \hat{\alpha}_2 = 3160 \qquad \hat{\beta}_1 = 0.295 \qquad \hat{\beta}_2 = 3.15 \qquad \hat{p} = 0.113$$

Figure 16.6 shows the WPP plot of model as a continuous line and the WPP transformed data as dots. As can be see, the model fits the data reasonably well. The model validation using the data in S_2 was done as follows. Since the data comprised of both failure and censor times, it was not possible to use the various statistical tests discussed in Chapter 5. Instead, the data for validation was plotted on the WPP paper and compared with the WPP plot for Model III(a)-1.

The two plots are shown in Figure 16.7. The fit can be viewed as being acceptable in a qualitative sense. One would need to repeat stages 1 and 2 of the modeling K times. Each run would first involve dividing the data into two subsets (comprising 80 and 20% of the original data) using random sampling. This would generate different data sets for each run of the K runs. The model validation is based on comparing the fits for the K runs.

16.6 CONCLUSIONS

In this chapter we discussed model selection based solely on data. The process involves a two-stage approach. The role of WPP plots for model selection in the first stage and the different issues involved the second stage leading to a final model were highlighted. There are several topics that need further study and some of these are listed under the Exercises.

EXERCISES

Data Set 16.1: Helicopter Failures A helicopter consists of several components and some of these are critical (in the sense that any failure of such components during operation leads to a crash) and others not. Luxhoj and Shyur (1995) report on failures of three different components of a helicopter. The data includes aircraft flight hours as well as operational time to failure of each component. The data can also be found in Table 2.15 of Blischke and Murthy (2000).

Data Set 16.2: Compressor Failures For compressors located near a seacoast, salt in the air is recognized as a major cause for compressor failures. Abernathy et al. (1983) reports on 202 such units. The data consists of 10 failures (over the observation period) and the failure times are given. The data for the remaining 192 units (nonfailed items) is interval censored. The data can also be found in Table 2.16 of Blischke and Murthy (2000).

Data Set 16.3: Aircraft Air-conditioner Failures Proschan (1963) reports on the time between failures of air-conditioning units on aircraft. The data can also be found in Table 2.10 of Blischke and Murthy (2000).

Data Set 16.4: Diesel Generator Fan Failure Table 1.1 of Chapter 4 of Nelson (1982) gives data on failures for 70 fans of diesel generator sets. The fans were put into operation at different time instants. Over the observation period, 12 fans failed. The data consists of the failure times for these and the censoring (right censoring) for the remaining fans. The data can also be found in Appendix Table C.1 of Meeker and Escobar (1998).

Data Set 16.5: Vehicle Shock Absorber Failures O'Connor (1995, p. 85) reports on distance to failure and the failure modes for the failure of 38 vehicle shock absorbers. The data can also be found in Table C.2 of Meeker and Escobar (1998b).

16.1. Use the methodology outlined in Chapter 16 to determine whether one or more of the Type I to III models discussed in Chapters 3 to 11 are appropriate to model Data Set 16.1.

16.2. Repeat Exercise 16.1 for Data Set 16.2.

16.3. Repeat Exercise 16.1 for Data Set 16.3.

16.4. Repeat Exercise 16.1 for Data Set 16.4.

16.5. Repeat Exercise 16.1 for Data Set 16.5.

16.6. Failure data can be found in many books on reliability and life testing; see, for example, Kalbfleisch and Prentice (1980), Meeker and Escobar (1998b), and Blischke and Murthy (2000) to name a few. Compile a list of these data sets and categorize them based on the area of application.

16.7. Carry out a search of databases to collect data sets from the real world. Discuss critically the limitations of the data sets and what additional information would have been useful from the modeling point of view.

PART G

Applications in Reliability

CHAPTER 17

Modeling Product Failures

17.1 INTRODUCTION

All products are unreliable in the sense that they degrade with age and/or usage and ultimately fail. The reliability of a product depends on technical decisions made during the design and manufacturing of the product. Product failures, in turn, have an impact on the commercial aspect of business as frequent failures affect warranty and maintenance costs as well as new sales. This implies that product reliability decisions must be done in a framework that integrates the technical and commercial issues in an effective manner. We discuss this in the next section and the role of Weibull models to assist in the decision-making process. The models for decision making require modeling product failures, and in this chapter we focus our attention on this topic.

The failure of a product is influenced by several factors. These include the usage mode (continuous or intermittent), usage intensity (high, medium, or low), operating environment (normal or abnormal), and operator skills (for certain types of products). In general, a product is complex and comprised of many components. The failure of a product is due to the failure of one or more of its components. For a nonrepairable product, there is only one failure. In contrast, for a repairable product, there can be several failures over its useful life. These failures are influenced by maintenance (preventive and corrective) actions. As a result, the modeling can be done at either system level or component level or anywhere in between. Also, the modeling depends on the information available. The information can relate to the degradation processes at the component level, the interconnections between the different components at the system level, the usage environment, and so forth. This information can vary from very minimal (or none) to complete or anywhere in between. In this chapter we look at all these issues in the context of modeling product failures and highlight the use of the Weibull models discussed in earlier chapters.

Weibull Models, by D.N.P. Murthy, Min Xie, and Renyan Jiang.
ISBN 0-471-36092-9

The outline of the chapter is as follows. We start with a brief discussion of some basic concepts in Section 17.2. In general, most products consist of several components, and in Section 17.3 we indicate the decomposition of a product into various levels. The modeling of product failure can be done at any level, and this is discussed in Section 17.4 along with different approaches to modeling. The different approaches to modeling at the component level are discussed in Sections 17.5 to 17.7 and Sections 17.8 and 17.9 deal with modeling at the system level. We illustrate the modeling through some cases involving real data. Our presentation is terse based to a large extent on Blischke and Murthy (2000) where interested readers can get more details.

17.2 SOME BASIC CONCEPTS

17.2.1 Failures

A dictionary definition of *failure* is "falling short in something expected, attempted, or desired, or in some way deficient or lacking." From an engineering point of view, it is useful to define failure in a broader sense. Witherell (1994) elaborates as follows: "It [failure] can be any incident or condition that causes an industrial plant, manufactured product, process, material, or service to degrade or become unsuitable or unable to perform its intended function or purpose safely, reliably, and cost-effectively." Accordingly, "the definition of failure should include operations, behaviour, or product applications that lead to dissatisfaction, or undesirable, unexpected side effects." More formal definitions of failure are as follows:

- "Failure is the termination of the ability of an item to perform a required function" [International Electronic Commission, IEC 50(191)].
- "Equipment fails, if it is no longer able to carry out its intended function under the specified operational conditions for which it was designed" (Nieuwhof, 1984).
- "Failure is an event when machinery/equipment is not available to produce parts at specified conditions when scheduled or is not capable of producing parts or perform scheduled operations to specification. For every failure, an action is required" (Society of Automotive Engineers, "Reliability and Maintainability Guideline for Manufacturing Machinery and Equipment").
- "Recent developments in products-liability law has given special emphasis to expectations of those who will ultimately come in direct contact with what we will do, make or say or be indirectly affected by it. Failure, then, is any missing of the mark or falling short of achieving these goals, meeting standards, satisfying specifications, fulfilling expectations, and hitting the target" (Witherell, 1994).

17.2.2 Faults and Failure Modes

A *fault* is the state of the product characterized by its inability to perform its required function. (Note that this excludes situations arising from preventive maintenance or any other intentional shutdown period during which the product is unable to perform its required function.) A fault is hence a state resulting from a failure.

A *failure mode* is a description of a fault. It is sometimes referred to as fault mode [e.g., IEC 50(191)]. Failure modes are identified by studying the (performance) function of a product. Blache and Shrivastava (1994) suggest a classification scheme for failure modes. A brief description of the different failure modes is as follows:

1. *Intermittent failures*: Failures that last only for a short time. A good example of this is software faults that occur intermittently.
2. *Extended failures*: Failures that continue until some corrective action rectifies the failure. They can be divided into the following two categories:
   a. *Complete failure*: This results in total loss of function.
   b. *Partial failure*: This results in partial loss of function.

   Each of these can be further subdivided into the following:
   a. *Sudden failures*: Failures that occur without any warning.
   b. *Gradual failures*: Failures that occur with signals to warn of the occurrence of a failure.

A complete and sudden failure is called a *catastrophic failure*, and a gradual and partial failure is designated a *degraded failure*.

17.2.3 Failure Mechanisms

The failure of a component (of a product) occurs due to a complex set of interactions between the material properties and other physical properties of the part and the stresses that act on the part. The process through which these interact and lead to a component failure is complex and is different for different types of parts (e.g., failure mechanisms that lead to failure of mechanical parts are different from those that lead to failure of electrical parts).

According to Dasgupta and Pecht (1991), one can divide the mechanisms of failure into two broad categories: (i) overstress mechanisms and (ii) wear-out mechanisms. In the former case, an item fails only if the stress to which the item is subjected exceeds the strength of the item. If the stress is below the strength, the stress has no permanent effect on the item. In the latter case, however, the stress causes damage (e.g., crack length) that usually accumulates irreversibly. The accumulated damage does not disappear when the stress is removed, although sometimes annealing is possible. The cumulative damage does not cause any performance degradation as long as is it below the endurance limit. Once this limit is reached, the item fails. The effects of stresses are influenced by several factors—geometry of the part, constitutive and damage properties of the materials, manufacturing, and operational environment.

17.3 PRODUCT STRUCTURE

Products can vary from simple (such as light bulbs, tires, or toasters) to complex (such as bridges, power generating systems, or communications networks). A product can be viewed as a system consisting of several parts and can be decomposed into a hierarchy of levels, with the system at the top level and components at the lowest level. Various intermediate levels are as indicated follows:

Level	Characterization
1	System
2	Subsystem
3	Major assembly
4	Assembly
5	Subassembly
6	Component

17.4 MODELING FAILURES

The modeling of failures can be done at any level ranging from system to component level. We confine our attention to modeling at component and system levels. Modeling of failures depends on the kind of information available and the goal (or purpose) that the model builder has in mind. For example, if the goal is to determine the spare parts needed for nonrepairable components, then the modeling of failure needs to be done at the component level. On the other hand, if one is interested in determining the expected warranty servicing cost, one might model failures at the system level.

As mentioned earlier, the information available for modeling can vary. At the component level, a good understanding of the different mechanisms of failure at work will allow building a physics-based model. In contrast, when no such understanding exists, one might need to model the failures based solely on failure (and possibly censored) data. In this case the modeling is empirical or data driven. These are the two extreme situations and are referred to as the *white-box* and *black-box* approaches to modeling. In between, we have different degrees of understanding or information. For example, an engineer, through fault tree analysis, might identify several modes of failure so that the modeling needs to take this into account. In this case, we have a *gray-box* approach to modeling. A similar situation arises at the system level. In the white-box approach, the different components of the systems and their interrelationships are known, whereas in the black-box approach this information is not known.

17.5 COMPONENT-LEVEL MODELING (BLACK-BOX APPROACH)

Here the component is characterized as being in one of two states—working or failed.

17.5.1 Modeling First Failure

We need to consider two cases—static and dynamic. In the static case, because of manufacturing defects, a part produced can initially be in a failed state. When such a part is put into operation, its (failed) state is detected immediately. In the dynamic case, the part is in its working state to start with and fails after a certain length of time (called the time to first failure). Note that the static case can be viewed as a special case of the dynamic case with the time to first failure being zero.

Since the time to failure is uncertain, it needs to be modeled by a distribution function $F(t;\theta)$. The form of the distribution function and the appropriate parameter values are decided based on the failure and censored data. As indicated in Chapters 3 to 11, many different Weibull models have been used for modeling failure data, and Chapter 16 deals with the methodology for model selection and parameter estimation.

17.5.2 Modeling Subsequent Failures

This depends on whether the component is repairable or not.

Nonrepairable Component

In this case, the failed component needs to be replaced by either a new or used (but working) component. The modeling depends on whether the failure is detected immediately or not and on the time needed for replacement.

The simplest case is where (i) the failures are detected immediately; (ii) the replacement times are small relative to the mean life of component so that they can be ignored; (iii) the items used are statistically similar [with failure distribution $F(t;\theta)$]; and (iv) the failures are statistically independent. In this case the failures over time occur according to a renewal process (discussed in Chapter 15) associated with $F(t;\theta)$. If the replacement times are random variables with a distribution function $F_r(t)$, then failures over time occur according to an alternating renewal process (discussed in Chapter 15).

Repairable Component

In this case the modeling of subsequent failures depends on the type of repair. Various kinds of repairs have been studied. Let t_i denote the time of failure and $G(t;\theta)$, $t > t_i$ denote the time to failure for the repaired item. The following two types have been used extensively in modeling:

1. *Repair as Good as New* Here, after each repair, the condition of the repaired item is assumed to be as good as that of a new item. In other words, the failure distribution of repaired items is the same as that of a new item, so that

$$G(t;\theta) = F(t - t_i;\theta) \qquad t > t_i \tag{17.1}$$

Note that this case is identical to the earlier nonrepairable case. Also, this is seldom true in real life.

2. *Minimal Repair* When a failed item is subjected to a minimal repair (Barlow and Hunter, 1961), the hazard function of the item after repair is

the same as just before it failed. In this case

$$G(t;\theta) = \frac{F(t;\theta) - F(t_i;\theta)}{1 - F(t_i;\theta)} \qquad t > t_i \tag{17.2}$$

If the failures are repaired immediately and the repair times are negligible, then failures over time are given by a point process with intensity function (see Chapter 15), which is the same as the hazard function associated with $F(t;\theta)$ (see Chapter 3).

The details of several other kinds of repair can be found in Blischke and Murthy (2000).

17.6 COMPONENT-LEVEL MODELING (WHITE-BOX APPROACH)

Here one models component failure based on the underlying degradation mechanism. We first consider failure due to overstress, following which we look at failure due to wear.

17.6.1 Overstress Failure (Stress–Strength Models)

Static Case

Because of manufacturing variability, the strength of a component, X, may vary significantly and must be modeled as a random variable with distribution function $F_X(x)$ $[f_X(x)]$. When the component is put into use, it is subjected to a stress Y. If X is smaller than Y, then the part fails immediately (time to failure is negligible) because its strength is not sufficient to withstand the stress to which it is subjected. If Y is smaller than X, then the strength of the part is sufficient to withstand the stress, and the part is functional.

Let Y be a random variable with distribution (density) function $F_Y(y)$ $[f_Y(y)]$ and X and Y are independent. Then the reliability R that the component will not fail when put into operation can be obtained using a conditional approach. Conditional on $Y = y$, we have

$$P(X > Y|Y = y) = \int_y^{\infty} f_X(x)\,dx \tag{17.3}$$

On removing the conditioning,

$$R = P(X > Y) = \int_{-\infty}^{\infty} f_Y(y)\left[\int_y^{\infty} f_X(x)\,dx\right] dy \tag{17.4}$$

which can also be written as

$$R = P(Y < X) = \int_{-\infty}^{\infty} f_X(x)\left[\int_{-\infty}^{x} f_Y(y)\,dy\right] dx \tag{17.5}$$

Example 17.1: Weibull Stress and Strength

Let $F_X(x)$ be a three-parameter Weibull distribution [given by 1.2)] with scale parameter α_1, shape parameter β_1, and location parameter τ_1. Let $F_Y(y)$ also be a three-parameter Weibull distribution with scale parameter α_2, shape parameter β_2, and location parameter τ_2. Then we have [see Kapur and Lamberson (1997) for details]

$$R = 1 - \int_0^\infty e^{-z} \exp\left\{ - \left[\frac{\alpha_1}{\alpha_2} z^{1/\beta_1} + \left(\frac{\tau_1 - \tau_2}{\alpha_2} \right) \right]^{\beta_2} \right\} dz$$

where

$$z = \left(\frac{x - \tau_1}{\alpha_1} \right)^{\beta_1} \quad \text{and} \quad dz = \frac{\beta_1}{\alpha_1} \left(\frac{x - \tau_1}{\alpha_1} \right)^{\beta_1 - 1} dx$$

One needs to use numerical methods to evaluate the integral in (17.6). For the special case when the two location parameters are zero and the two shape parameters are 1 (so that both distributions are exponential), then we have an analytical expression for R given by $R = \alpha_1/(\alpha_1 + \alpha_2)$.

Kapur and Lamberson (1977) discuss the case where the strength and stress have different distributions that include one of them being the Weibull and the other being non-Weibull such as the normal, extreme value, and so forth.

Dynamic Case

Here strength and stress are functions of time. The strength $X(t)$ degrades with time so that it is nonincreasing. The stress $Y(t)$ can change with time in an uncertain manner (e.g., the stress on a tall structure induced by wind or stress on the legs of an offshore platform due to waves). In this case, the time to failure (T) is the first instant $X(t)$ falls below $Y(t)$. In other words,

$$T = \min[t | X(t) - Y(t) < 0] \tag{17.6}$$

Alternate characterization of $X(t)$ and $Y(t)$ lead to different models.

Example 17.2: Shock-Induced Stress

Stresses occur at random points in time due to external shocks. Let the shocks occur according to a point process $N(t)$ modeled by a Weibull intensity function (see Chapter 15) with parameters (α, β). The stresses resulting from shocks are random variables with distribution $G(y)$, and let $X(t) = x$, which is a constant. Let T denote the time to failure and $F(t)$ denote its distribution function. Then, conditional on $N(t) = n$, we have no failure only when each of n shocks causes a stress that is below the item strength x. As a result,

$$P[T > t | N(t) = n] = [G(x)]^n$$

On removing the conditioning, we have

$$\bar{F}(t) = P(T > t) = \sum_{n=0}^{\infty} [G(x)]^n P[N(t) = n]$$

One can model $N(t)$ in different ways.

1. $N(t)$ is modeled by a renewal process of the time intervals between shocks distributed according to the two-parameter Weibull distribution.
2. $N(t)$ is modeled by a Weibull intensity model given by (15.6).

Similarly $G(y)$ can be modeled by any of the different distributions from the Weibull family discussed in Chapters 3, 6, and 7.

A special case is when $N(t)$ is modeled by a Weibull intensity model with $\beta = 1$ (so that is a Poisson process) and define $\vartheta = G(x)$. Note that $\vartheta < 1$. Then we have

$$F(t) = 1 - \exp\{-[(1 - \vartheta)/\alpha]t\}$$

In other words, the time to failure is given by an exponential distribution. The mean time to failure is

$$E(T) = \frac{\alpha}{1 - \vartheta}$$

Note that as x increases (i.e., the component becomes stronger), ϑ increases; and, as a result, the mean time to failure also increases as would be expected. Similarly, as α increases (shocks occur less frequently), the mean time to failure increases, as to be expected.

17.6.2 Wear-Out Failures

Wear is a phenomenon whereby the effect of damage accumulates with time, ultimately leading to item failure. Typical examples are crack growth in a mechanical part, a tear in a conveyor belt, or bearings wearing out. These can be modeled by a variable $Z(t)$ that increases with time in an uncertain manner, and failure occurs when the value of $Z(t)$ reaches some threshold level z^*.

Let T denote the time to failure, and it is given

$$T = \min(t : Z(t) > z^*) \tag{17.7}$$

Let $F(t)$ denote the distribution function of T.

The changes in $Z(t)$ can either occur at discrete points in time as a result of some external shocks or occur continuously over time (appropriate for corrosion or fatigue-type failures). In the former case, $Z(t)$ is a jump process where the jumps coincide with the occurrence of shocks. Again, alternate formulations for $Z(t)$ lead to different models for failure.

Example 17.3: Cumulative Shock Damage Model

Here damage is done because of shocks that occur randomly over time. The damage from each shock is uncertain and the damage is cumulative. Let Z_i, $i = 1, 2, \ldots$ denote the damage caused by shock i. These are assumed to be independent and identically distributed random variables with distribution $G(z)$. Let $N(t)$ denote the number of shocks received in $[0, t)$. Then the total damage at time t (assuming that the item has not failed) is given by $Z(t) = 0$ if $N(t) = 0$ and by $Z(t) = \sum_{i=1}^{N(t)} Z_i$ for $N(t) = 1, 2, \ldots$. Note that $Z(t)$ is a cumulative process.

The item will not have failed up to time t if the total damage $Z(t) < z^*$. In other words,

$$P(T > t) = 1 - F(t) = P(Z(t) \leq z^*)$$

We obtain $P[Z(t) < z^*]$ by conditioning on $N(t)$. The result is

$$P[Z(t) \leq z^*] = \sum_{n=0}^{\infty} P[Z(t) \leq z^* | N(t) = n] \, P[N(t) = n]$$

Since $Z(t)$, conditional on $N(t) = n$, is a sum on n-independent variables, we have

$$P[Z(t) \leq z^* | N(t) = n] = G^{(n)}(z^*)$$

where $G^{(n)}(z)$ is the n-fold convolution of $G(z)$ with itself.

If the shocks occur according to a renewal process associated with a distribution function $H(t)$, then

$$P[N(t) = n] = H^{(n)}(t) - H^{(n+1)}(t)$$

where $H^{(n)}(t)$ is the n-fold convolution of $H(t)$ with itself. In this case we have

$$F(t) = 1 - \sum_{n=0}^{\infty} G^{(n)}(z^*) \, [H^{(n)}(t) - H^{(n+1)}(t)]$$

[with $G^{(0)}(x) = H^{(0)}(x) = 1$ for all x]. This can be rewritten as

$$F(t) = \sum_{n=0}^{\infty} H^{(n+1)}(t) \, [G^{(n)}(z^*) - G^{(n+1)}(z^*)]$$

For the special case $H(t) = 1 - e^{-\lambda t}$ and $G(z) = 1 - e^{-\upsilon z}$, we have from Cox (1962),

$$F(t) = 1 - e^{-\lambda t}[1 + \sqrt{\lambda \upsilon t z^*} \int_0^{z^*} e^{-\upsilon z} z^{-1/2} I_1(2\sqrt{\lambda \upsilon t z}) \, dz]$$

where $I_1(x)$ is the Bessel function of order 1 [see Abromowitz and Stegun (1964)]. The associated hazard function is given by

$$r(t) = \frac{\lambda\, e^{-\lambda t - \upsilon z^*}\, I_0(2\sqrt{\lambda \upsilon t z^*})}{1 + \sqrt{\lambda \upsilon t z^*} \int_0^{z^*} e^{-\upsilon z}\, z^{-1/2}\, I_1(2\sqrt{\lambda \upsilon t z})\, dz}$$

This model has been extended in several ways. Nakagawa and Osaki (1974) deal with two other models. In the first, each shock does damage with probability p and no damage with probability $(1 - p)$. The second looks at z^* as a random variable. Hameed and Proschan (1973) consider the case where shocks occur according to a nonstationary Poisson process.

17.7 COMPONENT-LEVEL MODELING (GRAY-BOX APPROACH)

The black-box approach and the white-box approach can be viewed as the two extremes in modeling. In the black-box approach we have no information (or understanding) regarding the mechanism of failure, and the modeling is based solely on the failure (and censored) data. In contrast, in the white-box approach we have complete understanding of the underlying failure mechanisms. Gray-box approach can be viewed as something in between the two extremes. Here we incorporate other relevant information in the selection of an appropriate model to model failures.

The kind of extra information that can be used in modeling can vary. We briefly discuss some scenarios to indicate this.

1. If there are several modes of failure at work, then one might choose a competing risk model (see Chapter 9) to model the data.
2. Often there is wide variability in manufacturing of components. In this case, one might choose a mixture model (see Chapter 8) to model the data.
3. Suppose that the components used are bought from different suppliers. If the failure data does not include the manufacturer for each component, then the components need to be treated as the pooling of all the components. In this case, the mixture model is again appropriate to model the data.
4. For consumer durables (such as cars, washing machines, etc.) the usage mode and environment can vary significantly across the consuming population. In this case, the modeling of failures under warranty need to take this into account. In this case, models with random parameters (see Chapter 12) are more appropriate to model failures.
5. During the design and development of new products, prototypes are tested under different stress (or load). Here, failure at each stress level is modeled by a suitable distribution (e.g., Weibull) and the parameters modeled as functions of the stress (see Chapter 12).

6. Finally, sectional models (see Chapter 11) might be appropriate if the degradation mechanism changes after a certain age.

17.8 SYSTEM-LEVEL MODELING (BLACK BOX APPROACH)

The failure of a system is often due to the failure of one or more of its components. At each system failure, the number of failed components that must be restored back to their working state is usually small relative to the total number of components in the system. The system is made operational by either repairing or replacing these failed components. If the time to restore the failed system to its operational state is very small relative to the mean time between failures, then it can effectively be ignored. For practical purposes, this situation is equivalent to minimal repair (see Chapter 6) and system failures can be modeled as follows.

The system failures are modeled by a point process formulation with a cumulative intensity function $\Lambda(t;\theta)$, where t represents the age of the system. $\Lambda(t;\theta)$ is an increasing function of t, reflecting the effect of age. In general, it is necessary to specify a form for $\Lambda(t;\theta)$ and estimate its parameters using failure data. One form for $\Lambda(t;\theta)$ is given by (15.11) where failures occur according to the power law process.

Systems often undergo a major overhaul (a preventive maintenance action) or a major repair (due to a major failure), which alters the failure rate of the system significantly. The failure rate after such an action is smaller than the failure rate just before failure or overhaul. In the case of an automobile, this would correspond to actions such as the reconditioning of the engine, a new coat of paint, and so on. The time instants at which these actions are carried out can be either deterministic (as in the case of preventive maintenance based on age) or random (as, e.g., in the case of a major repair subsequent to an accident). An approach to modeling this situation is as follows.

Let $t = 0$ correspond to the time when a new system is put into use. Suppose that the system is subjected to $K-1$ overhauls at time instants t_k, $k = 1 \le k \le (K-1)$, and discarded at time t_K. The failure rate of the system after the kth action, $1 \le k \le (K-1)$, is given by $\Lambda_k(t-t_k;\theta)$ for $t > t_k$. The failure rate of the system when it is new is given by $\Lambda_0(t;\theta)$. All failures between t_{k-1} and t_k, $1 \le k \le (K-1)$ (with $t_0 = 0$), are repaired minimally. The functions $\Lambda_k(x;\theta)$, $1 \le k \le (K-1)$, have the following properties:

1. For a fixed k, $\Lambda_k(x;\theta)$ is an increasing function in x for $0 \le k \le (K-1)$.
2. For a fixed x, $\Lambda_k(x;\theta) - \Lambda_{k-1}(x;\theta) > 0$ for $1 \le k \le (K-1)$.

Property 1 implies that the failure rate is always increasing with x (the time lapsed subsequent to an overhaul) and property 2 implies that after each overhaul the failure rate decreases and the decrease becomes smaller as k (the number of overhauls) increases.

It is necessary to determine the form for $\Lambda_k(x; \theta)$, $0 \le k \le (K-1)$, based on either failure data or some other basis (either theoretical or intuitive) and estimate the parameters using failure data.

17.8.1 Case 4: Photocopier Failures

The photocopier can be viewed as a system involving several components. In case 1 (see Chapter 16) we looked at the modeling of one component (cleaning web) of the photocopier. Here we focus on modeling failures at the system level. Whenever a failure occurs, it is rectified by a qualified service agent under a service contract. The motivation for the modeling is to determine the expected number of failures over the next several years so as to assist the service agent in pricing the service contact. The results presented are based on Bulmer and Eccleston (2002).

The failure data for a particular photocopier over its first $4\frac{1}{2}$ years is given in Table 17.1. For each of the 42 failures over this period, the table shows the cumulative number of copies made and the age of the photocopier (measured in days) at failure. As to be expect, the number of days (time) and the number of copies between system failures, denoted n_i and t_i, $1 \le i \le 42$, respectively, are positively correlated ($r = 0.753$). A further variable of interest is the usage rate, defined as the number of copies between failures divided by the number of days between failures. The mean usage between failures was 670.2 copies per day with a standard deviation of 409.8 copies per day.

Modeling Failures over Time

Figure 17.1 shows a plot of the cumulative number of failures against time and against the total copies made. It suggests that a power law process with $\Lambda(t)$ given by (15.11) is an appropriate model to model failures over time. One can either use

Table 17.1 Photocopier Failure Data

Copies	Days	Copies	Days	Copies	Days
60,152	29	597,739	916	936,597	1,436
132,079	128	624,578	956	938,100	1,448
220,832	227	660,958	996	944,235	1,460
252,491	276	675,841	1,016	984,244	1,493
365,075	397	684,186	1,074	994,597	1,514
370,070	468	716,636	1,111	1,005,842	1,551
378,223	492	769,384	1,165	1,014,550	1,560
390,459	516	787,106	1,217	1,045,893	1,583
427,056	563	840,494	1,266	1,057,844	1,597
449,928	609	851,657	1,281	1,068,124	1,609
472,320	677	872,523	1,312	1,072,760	1,625
501,550	722	900,362	1,356	1,077,537	1,640
533,634	810	933,637	1,410	1,099,369	1,650
583,981	853	933,785	1,412	1,099,369	1,650

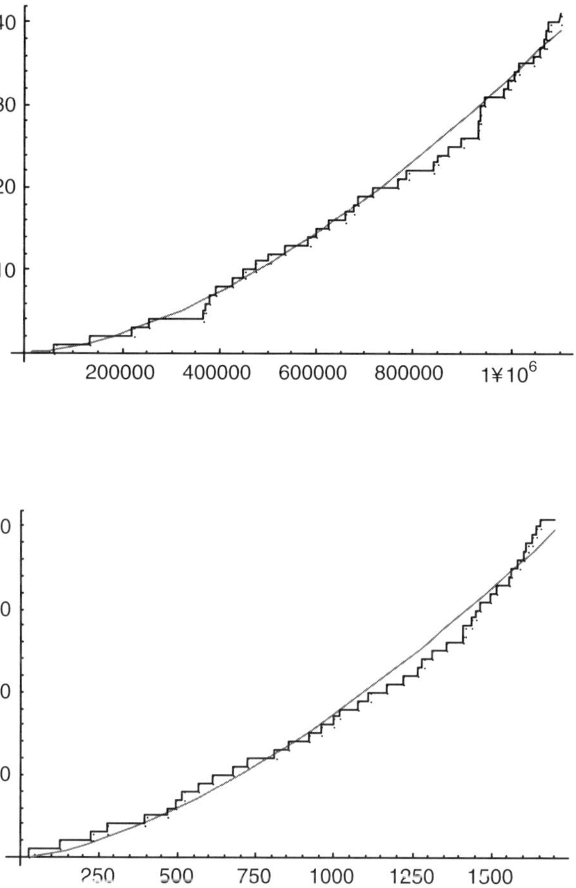

Figure 17.1 Cumulative failures (number of copies, above, and days, below) with Weibull power law fits.

the age or the number of copies made as the independent variable (t) of the formulation. The two parameters can be determined using least-squares fitting or by maximum-likelihood estimation. Figure 17.1 shows the fitted power functions obtained by the method of least squares. The parameter estimates are as follows:

- Based on age: $\hat{\beta} = 1.55$ and $\hat{\alpha} = 157.5$ (days)
- Based on number of copies: $\hat{\beta} = 1.64$ and $\hat{\alpha} = 118400$

As can be seen, the fits look reasonable and one can view the model as an adequate model based on this.

We confine our attention to the case where the independent variable is the age of the photocopier. The expected number of failures in year $j, j = 1, 2, \ldots$, is given by the difference between $\Lambda(j365)$ and $\Lambda[(j-1)365]$. The values, based on the above

Table 17.2 Estimated Service Calls for 10 Years

Year	1	2	3	4	5	6	7	8	9	10
Expected number of failures	3.7	7.1	9.4	11.3	13.0	14.6	16.0	17.3	18.5	19.7

estimates, are given in Table 17.2. Note that the estimated number of service calls is increasing each year, there will come a time where it will be cheaper to replace the copier rather than paying the servicing costs. Correspondingly, the age of the machine will be an important factor when entering into a service contract. The precise details will depend on the particular service costs.

17.9 SYSTEM-LEVEL MODELING (WHITE-BOX APPROACH)

In the white-box approach, system failure is modeled in terms of the failures of the components of the system. We discussed the modeling of component failures in Section 17.5. The linking of component failures to system failures can be done using two different approaches. The first is called the *forward* (or bottom-up) approach and the second is called the *backward* (top-down) approach.

In the forward approach, one starts with failure events at the part level and then proceeds forward to the system level to evaluate the consequences of such failures on system performance. *Failure mode and effects analysis* (FMEA) uses this approach. In the backward approach, one starts at the system level and then proceeds downward to the part level to link system performance to failures at the part level. *Fault tree analysis* (FTA) uses this approach.

Instead of the fault tree, one can use a network representation to model the links between different components, and the system failure is obtained in terms of component failures based on the linking between components. In either case, the state of the system (a binary-valued variable) can be expressed in terms of the component states (each of which is also binary valued) through the structure function.

17.9.1 Structure Function

Let $X_i(t)$, $1 \leq i \leq n$, denote the state of component i, at time t, with

$$X_i(t) = \begin{cases} 1 & \text{if component } i \text{ is in working state at time } t \\ 0 & \text{if component } i \text{ is in failed state at time } t \end{cases} \tag{17.8}$$

Let $X(t) = [X_1(t), X_2(t), \ldots, X_n(t)]$ denote the state of the n components at time t. The state of the system, $X_S(t)$, at time t is given by a function $\phi[X(t)]$, which is called the structure function with

$$\phi[X(t)] = \begin{cases} 1 & \text{if the system is in working state at time } t \\ 0 & \text{if the system is in a failed state at time } t \end{cases} \tag{17.9}$$

The form of $\phi(X)$ depends on the fault tree structure (or the equivalent network structure). Most systems can be represented as a network involving series and parallel connections, and in this case it is not too difficult to obtain $\phi[X(t)]$. The details can be found in most books on reliability; see, for example, Blischke and Murthy (2000) and Hoyland and Rausand (1994).

1. *System with Series Structure* Here the system fails whenever a component fails. As a result,

$$\phi(X) = X_1, X_2, \ldots, X_n = \prod_{i=1}^{n} X_i \tag{17.10}$$

2. *System with Parallel Structure* Here the system fails only when all the components fail. As a result,

$$\phi(X) = 1 - \prod_{i=1}^{n}(1 - X_i) = \coprod_{i=1}^{n} X_i \tag{17.11}$$

3. *k-out-of-n System* Here the system is functioning if at least k of the n components are functioning. The structure function is given by

$$\phi(X) = \begin{cases} 1 & \text{if } \sum_{i=1}^{n} X_i \geq k \\ 0 & \text{if } \sum_{i=1}^{n} X_i < k \end{cases} \tag{17.12}$$

For $n = 3$ and $k = 2$ we have

$$\phi(X) = X_1X_2 + X_1X_3 + X_2X_3 - 2X_1X_2X_3 \tag{17.13}$$

Finally, a component is said to be irrelevant if the system state is not affected by the state of the component. A system is said to be coherent if it has no irrelevant components.

17.9.2 System Reliability

We assume the following:

1. Component failures are statistically independent.
2. Components are new and in working state at $t = 0$.
3. Components and the system are nonrepairable.

Let

$$p_i(t) = P[X_i(t) = 1] \tag{17.14}$$

for $1 \leq i \leq n$ and

$$p_S(t) = P[X_S(t) = 1] \tag{17.15}$$

Note these represent the survivor functions for the n components and for the system. In other words, $p_S(t) = 1 - F_S(t)$ and $p_i(t) = 1 - F_i(t)$ where $F_S(t)$ and $F_i(t)$ are the failure distributions for the system and component i. Since the component and system states are binary valued, we have

$$p_S(t) = P\{\phi[X(t)] = 1\} = E\{\phi[X(t)]\} \tag{17.16}$$

This can be written as

$$p_S(t) = E\{\phi[X(t)]\} = \phi\{E[X(t)]\} = \phi[p(t)] \tag{17.17}$$

where $p(t)$ is the vector $[p_1(t), p_2(t), \ldots, p_n(t)]$. As a result, we have

$$F_S(t) = 1 - \phi[p(t)] \tag{17.18}$$

with $p_i(t) = 1 - F_i(t)$, $1 \leq i \leq n$.

1. *System with Series Structure* From (17.18) and (17.10) we have

$$\bar{F}_S(t) = \prod_{i=1}^{n} \bar{F}_i(t) \tag{17.19}$$

This is the competing risk model discussed in Chapter 9.

2. *System with Parallel Structure* From (17.18) and (17.11) we have

$$F_S(t) = \prod_{i=1}^{n} F_i(t) \tag{17.20}$$

This is the multiplicative model discussed in Chapter 10.

17.9.3 Case 5: Dragline Optimization

A dragline is a moving crane with a bucket at the end of a boom. It is used primarily in coal mining for removing the dirt to expose the coal. The bucket volume varies around 90 to 120 m^3 and the dragline is operated continuously (24 h/day and 365 days/year) except when it is down undergoing either corrective or preventive maintenance actions. A performance indicator of great importance to a mining business is the yield (annual output) of a dragline. This is a function of the dragline (bucket) load, speed of operation, and availability. Availability depends on two factors: (i) degradation of the components over time and (ii) maintenance (corrective and preventive) actions used. Degradation depends on the stresses on different components, and these in turn are functions of the dragline load. As a result, availability is a function of the dragline load and the maintenance effort. Availability decreases as the dragline load increases and increases as the maintenance effort increases. This implies that the annual total output is a complex function of dragline load. The problem of interest to the mine operators is the optimum dragline load to maximize the yield.

Decomposition of Dragline

A dragline is a complex system. It can be broken down into many subsystems. Each subsystem in turn can be decomposed into assemblies, and these in turn can be broken down still further into modules and modules into parts. Townson et al. (2002) model the dragline as a series structure comprising of 25 components, as indicated in Table 17.3.

Table 17.3 Decomposition of Dragline[a]

i	Component
1	Generator (hoist)
2	Generator (drag)
3	Generator (swing)
4	Generator (synchronous)
5	Motor (hoist)
6	Motor (drag)
7	Motor (swing)
8	Motor (propel)
9	Machinery (hoist)
10	Machinery (drag)
11	Machinery (swing)
12	Machinery (propel)
13	Ropes (hoist)
14	Ropes (drag)
15	Ropes (dump)
16	Ropes (suspension)
17	Ropes (fairleads)
18	Ropes (deflection sheaves)
19	Bucket and rigging
20	Frame (boom)
21	Frame (tub)
22	Frame (revolving)
23	Frame (A frame)
24	Frame (mast)
25	Others

[a] "Others" include the following: lube system, air system, transformers, brakes, blowers, fans, cranes, winches, communication equipment, air conditioning, and safety equipment.

Table 17.4 Estimates of Failure Model Parameters

Component i	Component	Scale	Shape	Method
1	Hoist generator	1.1076×10^3	1.1904	Maximum likelihood
2	Drag generator	1.4852×10^3	1.3232	Maximum likelihood
3	Swing generator	5.3328×10^3	2.1095	Maximum likelihood
4	Synchronous motor	9.2290×10^3	1.5852	Maximum likelihood
5	Hoist motor	2.0674×10^3	1.1365	Maximum likelihood
6	Drag motor	3.8743×10^3	1.2330	Least squares
7	Swing motor	1.2466×10^3	1.1367	Least squares
8	Propel motor	6.2218×10^3	1.3211	Least squares
9	Hoist machinery	565.7741	1.1010	Least squares
10	Drag machinery	769.7037	1.1153	Least squares
11	Swing machinery	565.7506	1.1010	Least squares
12	Propel machinery	1.9862×10^3	1.5889	Maximum likelihood
13	Hoist rope	423.3406	1.0455	Maximum likelihood
14	Drag rope	302.2390	1.3213	Maximum likelihood
15	Dump rope	188.0395	1.0455	Not fitted (used hoist rope shape)
16	Suspension ropes	8.9572×10^3	3.4450	Maximum likelihood
17	Fairlead	2.8608×10^3	1.3660	Maximum likelihood
18	Deflection sheaves	3.0781×10^3	1.6706	Maximum likelihood
19	Bucket and rigging	64.6218	1.0661	Least squares
20	Boom	789.9312	1.1117	Least squares
21	Tub	3.4554×10^3	1.4647	Maximum likelihood
22	Revolving frame	372.5864	1.0938	Least squares
23	A frame	3.0607×10^3	1.2628	Maximum likelihood
24	Mast	2.7577×10^3	1.1827	Least squares
25	Other	47.5795	1.0625	Least squares

Data for Modeling

The dragline is subject to a major shutdown once every 5 years and completely reassembled so that it can be viewed as being as good as new. The failure data was collected over a 2-year period subsequent to a major shutdown. For further details of the failure data, see Townson (2002).

Modeling of Component Failures

For each component, failures over time were modeled by the Weibull power law process given by (15.11). This is justifiable as failures are rectified through minimal repair, and the time to rectify a failure is relatively small in relation to the mean time between failures for a component. As a result, the model for each component involves two parameters—a scale parameter α and a shape parameter β.

Estimation of Model Parameters

The estimates for the parameters ($\alpha_i, \beta_i, 1 \leq i \leq 25$) were obtained using the field data and the method of maximum likelihood or the method of least squares. The final estimates for $\alpha_i, \beta_i, 1 \leq i \leq 25$ are as shown in Table 17.4.

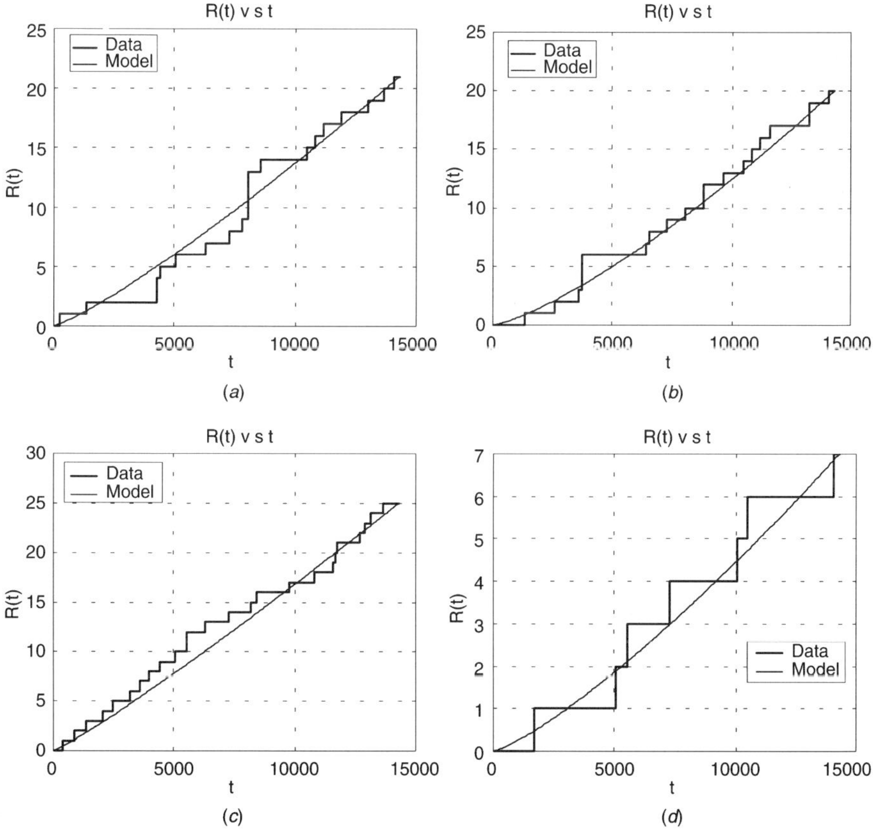

Figure 17.2 Cumulative failure plots of data and fitted model: (*a*) hoist generator, (*b*) drag generator, (*c*) boom, and (*d*) A frame.

Table 17.5 Acceleration Factors γ_i for Different Components

i	Component	γ_i
1	Generator (hoist)	7.0
2	Generator (drag)	1.0
3	Generator (swing)	1.0
4	Generator (synchronous)	1.0
5	Motor (hoist)	7.0
6	Motor (drag)	1.0
7	Motor (swing)	1.0
8	Motor (propel)	1.0
9	Machinery (hoist)	9.0
10	Machinery (drag)	1.0
11	Machinery (swing)	1.0
12	Machinery (propel)	1.0
13	Ropes (hoist)	5.7
14	Ropes (drag)	1.0
15	Ropes (dump)	5.7
16	Ropes (suspension)	5.7
17	Ropes (Fairleads)	1.0
18	Ropes (Deflection sheaves)	1.0
19	Bucket and rigging	1.0
20	Frame (boom)	3.1
21	Frame (tub)	3.1
22	Frame (revolving)	3.1
23	Frame (A frame)	3.1
24	Frame (mast)	3.1
25	Others	1.0

Model Validation

Proper validation of the model requires data different from that used in parameter estimation to compare model output with data. This was not possible since the sample size for each component was small, and all of the data were required for parameter estimation. Instead, a plot of the expected number of failures (based on the model) versus time for the model was prepared and visually compared with the failure data. For some of the components, the two plots were in reasonable agreement as shown in Figure 17.2, and hence one can accept with some moderate degree of confidence that the Weibull model is a reasonable model. For the remaining, the match between the two plots was poor. These could be either due to lack of data and/or the Weibull model not being appropriate.

Modeling the Effect of Bucket Load

The effect of bucket load deviating from the nominal load was modeled through an accelerated failure model (see Section 12.3.1) that affects the scale parameters for some of the components. As a result, for these components, the scale parameter

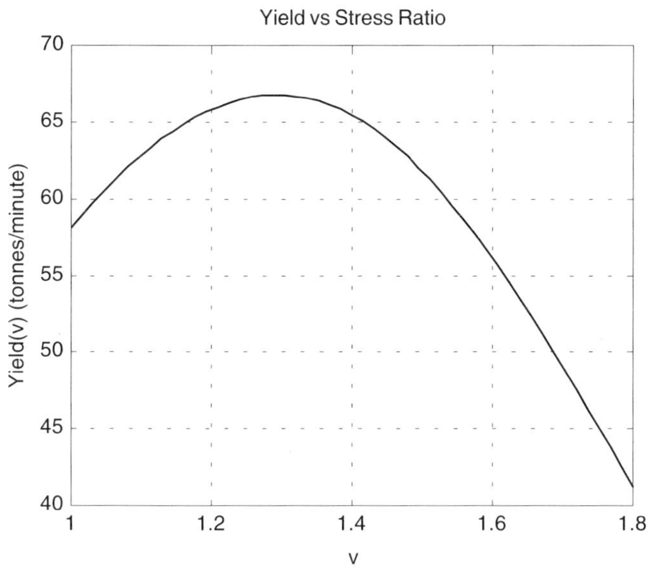

Figure 17.3 Yield versus v.

$\alpha(V)$ at bucket load V is related to the scale parameter $\alpha(V_0)$ at the base load V_0 by the relationship

$$\alpha(V) = \left(\frac{V_0}{V}\right)^{\gamma} \alpha(V_0) \tag{17.21}$$

Typical values of γ are indicated in Table 17.5. Note that $\gamma = 1$ implies that the corresponding component is not affected by bucket load whereas $\gamma > 1$ implies the reverse.

Optimal Bucket Load

Define $v = V/V_0$. The yield is a complicated function of the load and the preventive and corrective maintenance actions employed. We omit giving the details and they can be found in Townson et al. (2002). A plot of the yield versus v is shown in Figure 17.3. As can be seen from the plot of yield versus v, the maximum yield occurs at $v \approx 1.3$.

CHAPTER 18

Product Reliability and Weibull Models

18.1 INTRODUCTION

In Chapter 1 we defined the reliability of a product as the probability that the product will perform its intended function for a specified time period when operating under normal environmental conditions. The inherent reliability of a product is determined by the decisions made during the design and development and in the manufacturing stages of the product development. Better design and quality control will ensure higher reliability, and this in turn results in higher unit manufacturing cost. The actual reliability, however, depends on the inherent reliability and on the environment and the mode of usage.

Product reliability is an important issue in the purchase decisions of customers as it impacts product performance. This in turn affects customer satisfaction in the case of consumer durables (such as washing machines, televisions, cars, etc.) and profits in the case of commercial and industrial products (such as trucks, pumps, computers, etc.). Manufacturers can signal reliability of a product through warranty. In the simplest form, a *warranty* is a contract that requires the manufacturer to either rectify (through repair or replacement) or compensate through refund should the purchased item fail within the warranty period specified in the contract. Longer warranty periods signal greater reliability. Hence, warranties not only serve as an assurance to customers but also serve as a promotional tool for manufacturers to promote their products. This implies that product reliability is important in the context of marketing a new product.

The cost of operating the product subsequent to its sale is of interest to both the manufacturer and the customer. For the manufacturer it is the cost of servicing the warranty offered with the product. The warranty-servicing cost depends on product reliability. For the customer it is the cost associated with the maintenance of an item

Weibull Models, by D.N.P. Murthy, Min Xie, and Renyan Jiang.
ISBN 0-471-36092-9

over its useful life. This cost is comprised of two elements—preventive and corrective maintenance cost—and they in turn depend on reliability of the product.

As a result, effective management of product reliability is very important for manufacturers. This requires making proper reliability-related decisions using a product life-cycle framework discussed in Chapter 1. It involves the following three phases: (i) premanufacturing phase, (ii) manufacturing phase, and (iii) post-sale phase. Decisions at each of these phases must take into account the interaction between product reliability and many other variables such as costs, performance, and the like. In this chapter our focus is on the use of models to solve a variety of reliability-related decision problems at each of the three phases. The solutions to these problems require building appropriate models, and this, in turn, requires failure models (from component through to product levels) discussed in the last chapter.

The outline of the chapter is as follows. Sections 18.2 to 18.4 discuss a range of topics for each of the above three phases. These lead to a variety of reliability-related decision problems. These are discussed in Section 18.5 where we review the literature dealing with such decision problems for failures modeled by Weibull models.

18.2 PREMANUFACTURING PHASE

The premanufacturing phase deals with the design and development activities that result in a prototype of the product. The process of design involves two stages (conceptual and detail). These two and the development stages are sequentially linked with a number of reliability-related decision problems that need to be solved at each stage.

18.2.1 Design Process

During the conceptual design stage, the capabilities of different design approaches and technologies are evaluated. The goal is to ensure that a product that embodies these design features is capable of meeting all the stated requirements with regard to product performance (which includes different reliability measures) taking into account manufacturing implications and cost constraints. The objective is to decide on the best design approach in fairly general terms—the structure for the product, the materials and technologies to be used, and so on.

An important factor is cost analysis, an important role in optimal decision making. Two commonly used analyses are *design to cost* (DTC) and *life-cycle cost* (LCC). In the design-to-cost methodology, the aim is to minimize the unit manufacturing cost. This cost includes the cost of design and development, testing, and manufacturing. DTC is used to achieve the business strategy of a higher market share through increased sales. It is used for most consumer durables and many industrial and commercial products. In the life-cycle cost methodology, the cost under consideration includes the total cost of acquisition, operation, and maintenance over the life of the item as well as the cost associated with discarding the item at the end of its useful life. LCC is used for expensive defense and industrial products. Buyers of such products often require a cost analysis from the manufacturer as a part of the acquisition process.

The detailed design stage begins with an initial design that is subjected to detailed analysis. Based on this analysis, the design is improved and the process is repeated until analysis of the design indicates that the performance requirements are met.

A product can be viewed as a system comprising of several components and can be decomposed into a hierarchy of levels, with the product (or system) at the top level and components at the lowest level. There are many ways of describing this hierarchy, and Blischke and Murthy (2000) describe an eight-level decomposition where the system is decomposed into subsystems that in turn are decomposed into assemblies, subassemblies, and the like. In the remainder of the chapter we shall use the concept of "system" to indicate the highest level and component to indicate the lowest level in the decomposition.

Reliability Allocation

The process of determining subsystem and component reliability goals based on the system reliability goal is called *reliability allocation* or *reliability apportionment*. The reliability allocation to each subsystem must be compatible with its current state of development, expected improvement, and the amount of testing effort (in terms of time and money) needed for the development process.

Suppose that the system is composed of n components, indexed by subscript i, $1 \le i \le n$, and the focus is on satisfactory performance of the product for a period of time T. This is given by a reliability goal that $R(T) \ge R^*$, where R^* is the specified minimum target value. Let $R_i(T)$ denote the reliability allocated to component i. Then, depending on the configuration of the product (i.e., the linking between the different components), the state of the final product (working or failed) can be described in terms of the states of the n components through a structure function as discussed in Section 17.9. Using this one can express the system reliability in terms of the component reliabilities.

The cost of a component depends on its reliability and increases in a highly nonlinear manner as the reliability increases. Let $C_i[R_i(t)]$ denote the cost (design and development) of component i with reliability $R_i(T)$. The objective is to find the optimal allocation so as to minimize the total cost to achieve the desired reliability. This results in the following optimization problem:

$$\min_{R_i(T), 1 \le i \le n} \sum_{i=1}^{n} C_i[R_i(T)] \tag{18.1}$$

subject to

$$R(T) = \phi(R_1(T), R_2(T), \ldots R_n(T)) \ge R^* \tag{18.2}$$

Note: A related optimization problem is to maximize the system reliability subject to a cost constraint.

Reliability Prediction

Reliability prediction may be interpreted in two ways. First, it may mean prediction of system reliability in the design stage, where little or no "hard" information is available for many of the parts and components of the system. Second, predictions may be made later in the development stage in situations where test data and other information are available on components and on the system as a whole.

Reliability prediction at the design stage is concerned with systems not yet built and is heavily model oriented using nominal values of parameters as inputs. It is important to use all available information. Reliability prediction provides a basis for evaluating reliability growth during the development stage.

Typical examples of predictions are:

1. Prediction of the reliability of a system for a given design and selected set of components. This involves linking component reliability to system reliability through the structure function as mentioned earlier.
2. Prediction of the reliability of a system in a different environment from those for which data are available (especially environmental factors).
3. Prediction of the reliability of the system at the end of the development program. This topic is related to reliability growth and is discussed later in this section.

18.2.2 Development Process

The development process involves transforming the design into a physical product. It starts at component level and proceeds through a series of operations (such as fabrication, assembly) to yield the prototype.

Reliability Assessment

Reliability assessment is basically concerned with evaluation of the current reliability during the development process. It can be at any level—system down to component. The assessment is based on the data obtained from the tests carried out. From a statistical point of view, this involves estimation based on available information and data at the point in time when the assessment is being made.

Development Tests

Developmental tests are carried out during the development phase to assess and improve product reliability. A test can be defined as the application of some form of stimulation to a unit (system, component, or any other intermediate level) so that the resulting performance can be measured and compared to design requirements. Some of the development tests are as follows.

Testing to Failure Tests to failure can be performed at any level (component to system). The test involves subjecting the item to stress until a failure occurs. Each failure is analyzed and fixed.

Tests Involving Progressive Censoring Here a fraction of the surviving items are removed at several time instants to carry out detailed studies relating to the degradation mechanisms responsible for causing failure.

Progressive Type I Right Censoring Scheme This involves putting n items on life test at time zero. Then m censoring times, $T_1, T_2, \ldots, T_{m-1}$, are fixed such that at these times $K_1, K_2, \ldots, K_{m-1}$ surviving units are randomly removed (censored) from the test provided the number of surviving items is greater than the number to be removed. The test terminates at T_m if there are still some items that have survived. This is an extension of the conventional type I censoring discussed in Chapter 4.

Progressive Type II Censoring Scheme This involves putting n items on life test at time zero. Immediately following the first failure, K_1 surviving units are removed (censored) from the test at random. This process is repeated with K_i surviving units removed at the ith failure. The test terminates at the mth failure. This is an extension of the conventional type II censoring discussed in Chapter 4.

For more on progressing censoring see, Mann (1969, 1971), Yuen and Tse (1969), Gibbons and Vance (1983), Cacciari and Montannari (1987), Balakrishnan and Aggarwala (2000), Balasooriya (2001) and Aggarwala (2001).

Environmental and Design Limit Testing This can be done at any level and should include worst-case operating conditions, including operations at the maximum and minimum specified limits. Test conditions can include temperature, shock, vibration, and so forth. All failures resulting from the test are analyzed (through root cause analysis) and fixed through design changes. These tests are to assure that the product performs at the extreme conditions of its operating environment.

Accelerated Life Testing When a product is very reliable, it is necessary to use accelerated life tests to reduce the time required for testing. This involves putting items on test under environmental conditions that are far more severe than those normally encountered. Models for accelerated life testing are discussed in Chapter 12.

Reliability assessment requires the following:

1. Test data at one or more levels from carefully designed experiments.
2. Statistical estimation and hypothesis testing (discussed in Chapters 4 and 5).
3. This in turn may require model selection and validation (discussed in Chapter 5).

Both classical and Bayesian statistical analysis have been used in reliability assessment. If prior information is available, the Bayesian approach is preferable because it is a methodology for incorporating this information into the assessment process and provides a well-developed means for updating as more information, primarily from tests, becomes available. Problems are that modeling of information from many sources may be difficult, subjective information may not be reliable or may not be agreed upon, and analytical difficulties are encountered. The classical method would lead to assessments based primarily on the likelihood function and hence would be "objective," although many decisions that are at least partially subjective, for example, choice of the life distribution of a part or item, are involved here as well. It is important, whichever approach is used, to provide interval estimates of system reliability (confidence intervals or Bayesian intervals) at each stage.

Testing involves additional costs that depend on the type of tests, number of items tested and the duration of the tests and results in better estimates of reliability. This in turn leads to better decision making. As a result, the optimal testing effort must be based on a trade-off between the testing costs and the benefits derived through better assessment of reliability.

For more on accelerated testing, see Meeker and Nelson (1975), Nelson and Meeker (1978), Glasser (1984), Meeker (1984), Khamis and Higgins (1996), Meeker and Escobar (1993, 1998a).

Reliability Improvement

Achieving a desired reliability at the component or subsystem level often requires further development to improve reliability at that level. This is also known as reliability growth. Development involves time and money, and analysis of the process requires models that relate reliability improvement to the time and effort expended. A complicating factor is that the outcome of the development process is often uncertain.

There are basically two approaches to improving product reliability. These are (i) using redundancy and (ii) reliability growth through a development program.

Redundancy

Here one or more components are replicated to improve the reliability of the product. As such, for a component that is replicated, we use a module of k (≥ 2) components instead of a single component. The three different types of redundancy that have been used are (i) hot standby, (ii) cold standby, and (iii) warm standby. Let $H(x)$ and $F(x)$ denote the failure distribution for the module and component, respectively.

Hot Standby In hot standby all k replicates are connected in parallel and in use so that the module failure time is the largest of the k-component failure times. As a result, $H(x)$ is given by

$$H(x) = [F(x)]^k \tag{18.3}$$

Note that this is a special case of the multiplicative model discussed in Chapter 10.

Cold Standby In cold standby, only one component is in use at any given time. When it fails, it is replaced by a working component (if available) through a switching mechanism. If the switch is perfect and components do not degrade when not in use, then the module failure time is the sum of the k-component failure times. As a result, $H(x)$ is given by

$$H(x) = F(x)^* F(x)^* \ldots {}^* F(x) \tag{18.4}$$

where * is the convolution operator.

Warm Standby In warm standby, the component in use has a failure distribution $F(x)$ and the remaining nonfailed components (in partially energized state as opposed to fully energized state when put into operation) have a failure distribution $F_s(x)$ [$< F(x)$ for $x \geq 0$]. As a result, a component can fail before it is put into use. This implies that the failure distribution of a component put into use is a complicated function of $F(x)$, $F_s(x)$, and the duration for which it was partially loaded before being put into use. In the case $k = 2$, the distribution function for the module is given by

$$H(x) = F(x) - \int_0^x \bar{F}\{x - y + \bar{F}^{-1}[\bar{F}_s(y)]\} f(y)\, dy \tag{18.5}$$

For general k, the expressions for $H(x)$ are more complex and can be found in Hussain (1997).

For all three types of redundancy, $H(x) < F(x)$, implying that a module is more reliable than a single component. The reliability increases with k, the number of replicates in a module. If the switch is not perfect, then $H(x)$ [given by (18.4) for cold standby and by (18.5) for warm standby] gets modified. For further details, see Hussain (1997).

Note that cost of a module depends on k (the number of replication) and the type of redundancy (as both cold and warm standby redundancies involve a switching mechanism). The cost of a module is kC_m in the case of hot standby and $C_m(k + \psi)$, $0 \leq \psi \leq 1$, in the case of cold and warm standby, where ψC_m is the additional cost of the switching mechanism.

Reliability Growth

This involves research and development (R&D) effort where the product is subjected to an iterative process of test, analyze, and fix cycles. During this process, an item is tested for a certain period of time or until a failure occurs. Based on the analysis of the test run and failure mode, design and/or engineering modifications are made to improve the reliability. The process is repeated, resulting in reliability improvement (or growth) of the product. The reliability growth models can be broadly categorized into two groups: continuous and discrete models. Fries and Sen (1996) give a comprehensive review of the discrete models, and a discussion of continuous time models can be found in Hussain (1997).

In continuous time models, the improvement in reliability can be modeled through a parameter (λ) of the distribution, which decreases with development. Usually, λ is the inverse of the scale parameter so that in the case where $F(x)$ is exponential, it is the failure rate associated with $F(x)$.

In one of the well-known continuous time models, developed by Duane (1964), the parameter corresponds to the failure rate for an exponential distribution. After development for a period τ, the failure rate $\lambda(\tau)$ is given by

$$\lambda(\tau) = \lambda_0 \beta \tau^{\beta - 1} \tag{18.6}$$

where λ_0 is the failure rate before the development program is initiated, and $0 < \beta < 1$. Note that this is a deterministic function that decreases with τ increasing. In the case of the Weibull distribution one often assumes that reliability improvement affects the scale parameter but not the shape parameter. In this case the change in the scale parameter can be modeled by $\alpha(\tau) = 1/\lambda(\tau)$ with $\lambda(\tau)$ given by (18.6).

The number of modifications, $N(t)$, carried out during the development period is, in general, a random variable. Crow (1974) modeled it as a nonhomogeneous Poisson process with the intensity function given by (18.6) and τ replaced by t. The failure rate $\lambda(\tau)$, after development for a period τ, is given by (18.6), and the expected number of design modification is given by

$$E[N(\tau)] = \int_0^{\tau} \lambda(t)\, dt \tag{18.7}$$

In real life, the reliability improvement achieved at the end of the development period is uncertain. Murthy and Nguyen (1987) build on the Crow model and treat changes in the failure rate after each modification as random variables. As a result, the failure rate at the end of the development period is uncertain.

Hussain and Murthy (1999) propose another model where the outcome of the development process is uncertain. Their model is as follows. As the development program continues, λ decreases, thereby improving the reliability of the product. However, the outcome $\tilde{\lambda}(\tau)$, after a development (testing) period of τ, is uncertain. Let λ_0 denote the initial value of λ, that is, the value before the development program is initiated, and let λ_m be the limiting minimum value after development for an infinite time; $\tilde{\lambda}(\tau)$ is given by

$$\tilde{\lambda}(\tau) = \lambda_0 - (\lambda_0 - \lambda_m) Z_\tau \tag{18.8}$$

where Z_τ is a random variable that is a function of the development time τ and is modeled as

$$Z_\tau = [1 - \exp(-\rho\tau)]Y \tag{18.9}$$

where $Y \in (0, 1)$ is a random variable distributed according to a beta distribution and $\rho > 0$. Note that conditional on Y, $E(Z_\tau)$ increases and $E[\tilde{\lambda}(\tau)]$ decreases as τ increases.

The cost of development, $C_d(\tau)$, is an increasing function of the development period τ, and this increases the unit manufacturing cost. Let S denote the total sales. Then the effect of reliability growth is to increase the unit manufacturing cost given by $C_d(\tau)/S$.

Reliability improvement results in an increase in the unit manufacturing cost and is worthwhile only if the benefits derived from improvement (such as lower warranty costs, increased revenue through greater sales, etc.) exceed the cost of improvement.

18.3 MANUFACTURING PHASE

The manufacturing phase deals with the process of producing items in bulk in such a way that all items conform to the stated design performance specifications and doing so in the most economical manner.

The process used for manufacturing a product depends on the demand for the product and is determined by economic considerations. If the demand is high, then it is economical to use a continuous production process. If the demand is low to medium, then it is more economical to use a batch production process, where items are produced in lots (or batches).

18.3.1 Quality Variations

Due to variability in the manufacturing process, some of the items do not conform to design specifications, and these are termed *nonconforming* in contrast to the remaining, which are termed *conforming*. The characteristics (e.g., reliability) of a nonconforming item are inferior to a conforming item. We call this *manufacturing quality*, and higher quality implies fewer nonconforming items being produced in a probabilistic sense.

Let $F_c(x)$ and $F_n(x)$ denote the failure distributions of conforming and nonconforming items, where $F_c(x) < F_n(x)$ implying that the reliability of a nonconforming item is smaller than that for a conforming item. One needs to differentiate two types—types A and B—of nonconforming items.

A type A nonconforming item occurs where $F_n(x) = 1$ for $x > 0$. This implies that the item is nonfunctional and is detected immediately after it is put in use. This type of nonconformance is usually due to defects in assembly (e.g., a dry solder joint). For a type B nonconforming item, the mean to first failure (μ) is greater than zero and hence cannot be detected easily as a type A nonconforming item.

The probability that an item produced is nonconforming depends on the state of the manufacturing process. In the simplest characterization, the state can be modeled as being either in control or out of control. When the state is in control, all the assignable causes are under control, and, although nonconformance cannot be avoided entirely, the probability that an item produced is nonconforming is very small. When the state changes to out of control, this probability increases significantly. The process starts in control and changes to out of control after a certain (random) number of items are produced, and it stays there until some action is initiated to change it back to in control.

Modeling Occurrence of Nonconforming Items

Let p_i (p_0) denote the probability that an item produced is conforming when the process is in control (out of control) and $p_i \gg p_0$. In the extreme cases, $p_i = 1$, implying that all items produced are conforming when the state is in control, and $p_0 = 0$, implying that all items produced are nonconforming when the process is out of control.

Continuous Production Here the manufacturing process starts in control, and, after a random length of time, it changes to out of control. Since the failure distributions of the two types of items are $F_c(x)$ and $F_n(x)$, respectively, the failure distribution of an item produced, with the process in control, is given by

$$F_i(x) = p_i F_c(x) + (1 - p_i) F_n(x) \tag{18.10}$$

Note that this mixture model was discussed in Chapter 8. Similarly, given that the process is out of control, the item failure distribution is given by the mixture distribution

$$F_0(x) = p_0 F_c(x) + (1 - p_0) F_n(x) \tag{18.11}$$

Once the process state changes from in control to out of control, it remains in that state until it is brought back to an in control state through some corrective action.

Batch Production Here the items are produced in lots of size Q. At the start of each lot production, the state of the process is checked to ensure that it is in control. If the process is in control at the start of the production of an item, it can change to out of control with probability $1 - q$ or continue to be in control with probability q. Once the state changes to out of control, it remains there until completion of the lot. As mentioned previously, an item produced with the state in control (out of control) is conforming with probability $p_i(p_0)$ where $p_i \gg p_0$.

Let $\tilde{N}_c$ denote the number of conforming items in a lot. The expected value of this [for details see Djamaludin (1993)] is given by

$$E(\tilde{N}_c) = \frac{q(p_i - p_0)(1 - q^Q)}{1 - q} + p_0 Q \tag{18.12}$$

and the expected fraction of conforming items in a lot of size Q, $\phi(Q)$, is given by

$$\phi(Q) = \frac{q(p_i - p_0)(1 - q^Q)}{(1 - q)Q} + p_0 \tag{18.13}$$

It is easily seen that $\phi(Q)$ is a decreasing sequence in Q. This implies that the expected fraction of conforming items in a batch decreases as the batch size Q increases.

18.3.2 Quality Control

Quality control refers to the set of actions to ensure that the items produced conform to the specification.

Acceptance Sampling

The quality of input material (raw material, components, subassemblies, etc.) has a significant impact on the conformance quality of items produced. The input material is obtained from external suppliers in batches. The quality of input material can vary from batch to batch. One way of ensuring high input quality is to test for the quality and based on the outcome of the test a batch is either accepted or rejected. The test is based on a small sample from the batch. This is known as acceptance sampling. Many different acceptance sampling schemes have been proposed and studied in detail and these can be found in most books on quality control; for example, see Schilling (1982).

The cost associated with testing depends on the sample size and the type and duration of tests.

Output Inspection and Testing

The aim of output inspection and testing is to detect nonconforming items and to weed them out before items are released for sale. For type A nonconformance, testing takes very little time since a nonconforming item is detected immediately after it is put into operation. In contrast, the detection of nonconforming items for type B nonconformance involves testing for a significant length of time. In either case, testing involves additional costs.

Another issue that needs to be addressed is the level of testing. One can either test 100% or less than 100%. Again, cost becomes an important factor in deciding on the level of inspection. The quality of testing is still another variable that must be considered. If testing is perfect, then every nonconforming item tested is detected. With imperfect testing, not only are nonconforming items not detected but also a conforming item may be classified as nonconforming. As a result, the outgoing quality (the fraction of conforming items) depends on the level of testing and the quality of testing.

The cost of testing per unit is an increasing function of the testing period.

Control Charts

As mentioned earlier, the occurrence of nonconforming items increases when the state of the process changes from in control to out of control. The reason for the change is due to one or more of the controllable factors deviating significantly from its nominal values.

In the case of continuous production, the process starts in control and changes to out of control with the passage of time. If the change is detected immediately, then it can be brought back into control at once, and the high occurrence of nonconforming items, when the state is out of control, would thereby be avoided. This is achieved through the use of control charts. This involves taking small samples of the output periodically and plotting the sample statistics (such as the mean, the spread, the number or fraction of defective items) on a chart. A significant deviation in the statistics is more likely to be the result of a change in the state of the process. When this occurs, the process is stopped and controllable factors that have deviated are restored back to their nominal values before the process is put into operation. A variety of charts have been proposed and studied in detail and these can be found in most books on quality control; see, for example, Grant and Leavenworth (1988) and Del Castillo (2002). The cost of quality control depends on the frequency of sampling, the number of items sampled each time, and the nature and duration of the testing involved.

Lot Sizing

In the case of batch production, the process can change from in control to out of control during the production of a lot. As mentioned earlier, this affects the fraction of nonconforming items in a lot, and the expected fraction of nonconforming items in a lot increases with the lot size Q. This implies that the outgoing quality increases as Q decreases. However, this is achieved at a cost as smaller batch size implies more frequent setups, and this increases the unit manufacturing cost.

18.3.3 Quality Improvement

The design of the manufacturing process has a significant impact on p_i, the probability that an item is conforming when the process is in control. Ideally, one would like to have this probability equal to one so that no item produced is nonconforming. One way of ensuring a high value for p_i is through optimal choice for the nominal values for the different controllable factors. This is achieved through proper design of experiments. Details can be found in many books; see, for example, Dehnad (1989) and Moen et al. (1991).

Burn-In

Due to quality variations, the failure distribution of an item produced is given by a mixture distribution

$$F(x) = pF_c(x) + (1 - p)F_n(x) \tag{18.14}$$

where p is the probability that the item is conforming and $F_c(x)$ and $F_n(x)$ are the failure distributions of conforming and nonconforming items, respectively, with $\bar{F}_n(x) < \bar{F}_c(x)$. Burn-in is used to improve the reliability of the item released for sale.

Burn-in involves testing items for a period τ. Those that fail during testing are scrapped so as to get rid of the manufacturing defect. The probability that a conforming (nonconforming) item will fail during testing for a period τ is given by $F_c(\tau)[F_n(\tau)]$. As a result, the probability that an item that survives the test is conforming is given by

$$p_1 = \frac{p\bar{F}_c(\tau)}{p\bar{F}_c(\tau) + (1 - p)\bar{F}_n(\tau)} \tag{18.15}$$

Since $\bar{F}_n(\tau) < \bar{F}_c(\tau)$, we have $p_1 > p$. The failure distribution of an item that survives the test and is released for sale is given by

$$G(x) = p_1 F_{c1}(x) + (1 - p_1)F_{n1}(x) \tag{18.16}$$

where $F_{c1}(x)$ and $F_{n1}(x)$ are given by

$$F_{c1}(x) = \frac{F_c(x+\tau) - F_c(\tau)}{1 - F_c(\tau)} \tag{18.17}$$

and

$$F_{n1}(x) = \frac{F_n(x+\tau) - F_n(\tau)}{1 - F_n(\tau)} \tag{18.18}$$

for $x \geq 0$. Note that as τ increases, p_1 (the probability that an item released is conforming) increases, and, hence, the reliability of the items is improved. However, this is achieved at the expense of the useful life of conforming items released being reduced by an amount τ and the burn-in cost.

18.4 POSTSALE PHASE

As the name implies this phase deals with topics related to postsale. The two topics are warranties and maintenance.

18.4.1 Warranties

Warranty Concept

A warranty is a manufacturer's assurance to a buyer that a product or service is or shall be as represented. It may be considered to be a contractual agreement between buyer and manufacturer (or seller), which is entered into upon sale of the product or service. A warranty may be implicit or it may be explicitly stated.

The purpose of a warranty is to establish liability of the manufacturer in the event that an item fails or is unable to perform its intended function when properly used. The contract specifies both the performance that is to be expected and the redress available to the buyer if a failure occurs or the performance is unsatisfactory. The warranty is intended to assure the buyer that the product will perform its intended function under normal conditions of use for a specified period of time. In very simple terms, the warranty assures the buyer that the manufacturer will either repair or replace items that do not perform satisfactorily or refund a fraction or the whole of the sale price.

Warranty Policies

Many different warranty policies have been proposed and studied in detail. The type of warranty offered depends on the product (consumer durable, industrial or commercial, and special custom-built products). For a classification of the different types of warranty policies, see Blischke and Murthy (1994, 1996). Two warranties that are commonly offered with consumer durables are the following:

Free Replacement Warranty (FRW) Policy

Under a FRW policy the manufacturer agrees to repair or provide replacements for failed items free of charge up to a time W from the time of the initial purchase. The warranty expires at time W (the warranty period) after purchase. In the case of nonrepairable items, should a failure occur at age X (with $X < W$), under this policy the replaced item has a warranty for a period $(W - X)$, the remaining duration of the original warranty. Should additional failures occur, this process is repeated until the total service time of the original item and its replacements is at least W. In the case of repairable items, repairs are made free of charge until the total service time of the item is at least W.

Pro-Rata Warranty (PRW) Policy

Under a PRW policy the manufacturer agrees to refund a fraction of the purchase price should the item fail before time W (warranty period) from the time of the initial purchase. The buyer is not constrained to buy a replacement item. The refund depends on the age of the item at failure (X), and it can be either linear or a nonlinear function of $W - X$, the remaining time in the warranty period. Let $\varsigma(x)$ denote this function. This defines a family of pro-rata policies characterized by the form of the refund function. Two forms commonly offered are as follows:

1. Linear function: $\varsigma(x) = [(W - x)/W]C_b$.
2. Proportional function: $\varsigma(x) = [\gamma(W - x)/W]C_b$, where $0 < \gamma < 1$.

where C_b is the sale price.

Warranty Cost per Unit Sale

Whenever an item is returned for rectification action under warranty, the manufacturer incurs various costs (handling, material, labor, facilities, etc.). These costs can also be random variables. The total warranty cost (i.e., the cost of servicing all warranty claims for an item over the warranty period) is thus a sum of a random number of such individual costs, since the number of claims over the warranty period is also a random variable.

Cost Analysis of FRW Policy

Nonrepairable Product In this case any failures during warranty require the replacement of the failed item by a new item. Let C_s denote the cost of replacing a failed item by a new item. If failures result in instantaneous claims and the time to replace is small relative to the time between failures, then failures over the warranty period occur according to a renewal process discussed in Section 15.6. As a result, the cost $C_m(W)$ to the manufacturer over the warranty period W is a random variable given by

$$C_m(W) = C_s N(W) \tag{18.19}$$

where $N(W)$ is the number of failures over the warranty period. Let $F(x)$ denote the failure distribution, then the expected number of failures during warranty, $E[N(W)]$

is given by

$$E[N(W)] = M(W) \tag{18.20}$$

where $M(\cdot)$ is the renewal function given by

$$M(t) = F(t) + \int_0^t M(t-x)f(x)\,dx \tag{18.21}$$

As a result, the expected warranty cost per unit to the manufacturer is given by

$$E[C_m(W)] = C_s M(W) \tag{18.22}$$

Repairable Product In this case, the cost depends on the nature of the repair. If all failures over the warranty period are minimally repaired and the repair times are small in relation to the mean time between failures, then failures over the warranty period occur according to a nonhomogeneous Poisson process with intensity function equal to the hazard function $r(x)$ associated with $F(x)$. Let C_r denote the cost of each repair. Then the expected warranty cost to the manufacturer is given by

$$E[C_m(W)] = C_r \int_0^W r(x)\,dx \tag{18.23}$$

Cost Analysis of PRW Policy
Let the rebate function be given by

$$\varsigma(x) = \begin{cases} (1 - x/W)C_b & 0 \le x < W \\ 0 & \text{otherwise} \end{cases} \tag{18.24}$$

where C_b is the sale price.

The cost per unit to the manufacturer is given by $C_m(W) = \varsigma(X)$, where X is the lifetime of the item supplied and is a random variable with distribution function $F(x)$. As a result, the expected cost per unit to manufacturer is given by

$$E[C_m(W)] = \int_0^W \varsigma(x)f(x)\,dx \tag{18.25}$$

Using (18.24) in (18.25) and carrying out the integration, we have

$$E[C_m(W)] = C_b\left[F(W) - \frac{\mu_W}{W}\right] \tag{18.26}$$

where

$$\mu_W = \int_0^W xf(x)\,dx \tag{18.27}$$

is the partial expectation of X.

The expected cost per unit to the buyer is given by

$$C_b(W) = C_b - \varsigma(X) \tag{18.28}$$

where X is the item failure time and $\varsigma(\cdot)$ is given by (18.24). As a result, the expected per unit cost to the buyer is given by

$$E[C_b(W)] = C_b\left[\frac{\mu_W}{W} + 1 - F(W)\right] \tag{18.29}$$

18.4.2 Maintenance

Maintenance can be defined as actions to (i) control the deterioration process leading to failure of a system and (ii) restore the system to its operational state through corrective actions after a failure. The former is called *preventive* maintenance (PM) and the latter *corrective* maintenance (CM). Carrying out maintenance involves additional costs to the buyer and is worthwhile only if the benefits derived from such actions exceed the costs. From the buyer's viewpoint, this implies that maintenance must be examined in terms of its impact on the system performance.

Preventive maintenance actions are divided into the following categories:

1. Clock-based maintenance: Here PM actions are carried out at set times. An example of this is the *block replacement* policy.
2. Age-based maintenance: Here PM actions are based on the age of the component. An example of this is the *age replacement* policy.
3. Usage-based maintenance: Here PM actions are based on usage of the product.
4. Condition-based maintenance: Here PM actions are based on the condition of the component being maintained. This involves monitoring of one or more variables characterizing the wear process (e.g., crack growth in a mechanical component).
5. Opportunity-based maintenance: This is applicable for multicomponent systems, where maintenance actions (PM or CM) for a component provide an opportunity for carrying out PM actions on one or more of the remaining components of the system.
6. Design-out maintenance: This involves carrying out modifications through redesign of the component. As a result, the new component has better reliability characteristics.

Component-Level Maintenance

When a component fails in operation, it can have serious implications with a high cost of rectifying the failure. Many different preventive maintenance policies have been proposed where the item is replaced before failure at a much cheaper cost. Preventive maintenance involves sacrificing the useful life of an item. Frequent

PM actions reduce the likelihood of failure (and the resulting costs) but result in increased PM costs. This implies that the parameters characterizing the preventive maintenance policy needs to be selected properly to achieve a sensible trade-off between preventive and corrective maintenance costs.

The optimal PM depends on the item failure distribution $[F(x)]$, the costs of each corrective maintenance replacement (C_f) and each preventive maintenance replacement $C_p(\ll C_f)$, the time horizon, and the objective function for optimization. Often the time horizon is very large relative to the mean time to failure for the item. In this case, the time horizon can be treated as being infinite. The objective function can be the some performance measure such as the asymptotic availability or the asymptotic cost per unit time. Expressions for the objective can be obtained using the renewal reward theorem (see Section 15.6).

Two policies that have been used extensively are the age and block policies. We present the results for these policies without presenting the details. They can be found in many books on reliability. See, for example, Blischke and Murthy (2000).

Age Policy

Under this policy a component is replaced (preventive maintenance) when it reaches age T or on failure (corrective maintenance) should it fail earlier. The policy is characterized by a single parameter T that needs to be selected optimally.

The asymptotic cost per unit time is given by

$$J_c(T) = \frac{C_f F(T) + C_p \bar{F}(T)}{\int_0^T x f(x)\, dx + T\bar{F}(T)} \tag{18.30}$$

where T^*, the optimal T to minimize the asymptotic cost per unit time, is the value that yields a minimum for $J_c(T)$.

Block Policy

Under this policy, the component is replaced (preventive maintenance) by a new one at set times $t = kT, k = 1, 2, \ldots$. Any failures in between these times are rectified through replacement (corrective maintenance) of the failed item by a new one.

The asymptotic cost per unit time is given by

$$J_c(T) = \frac{C_p + C_f M(T)}{T} \tag{18.31}$$

where $M(\cdot)$ is the renewal function associated with $F(\cdot)$. As with the age policy, T^*, the optimal T, is the value that yields a minimum for $J_c(T)$ given by (18.31).

System-Level Maintenance

Here the system is subjected to major preventive maintenance (called *shutdown* maintenance), which rejuvenates the system. Any failures in between two consecutive preventive maintenance actions are rectified through minimal repair. If the time for minimal repair is small relative to the mean time between failures, then it can be ignored. In this case, failures between two preventive maintenance actions occur according to a point process. Let the intensity function, for the occurrence of failures subsequent to the jth PM action, be given by $\Lambda_j(t)$, $j \geq 1$, with $t = 0$ corresponding to instant the system was put into operation subsequent to the PM action. Let $\Lambda_0(t)$ denote the intensity function for a new system. Then we have

1. For a given j, $j \geq 0$, $\Lambda_j(t)$ is an increasing function of t. This implies that the rate of occurrence of failures increases with t.
2. For a given t, $\Lambda_j(t) < \Lambda_{j+1}(t)$ for $j \geq 0$. This implies that PM actions rejuvenate the system, but the performance decreases as the number of PM actions carried out increases.

The parameters of the policy are (i) the number of times PM actions should be carried out and (ii) the time interval between successive PM actions.

18.5 DECISION MODELS INVOLVING WEIBULL FAILURE MODELS

In the last three sections we discussed a variety of topics at the premanufacturing, manufacturing, and postsale phases for a product. Associated with each topic are several decision problems. The solutions to these problems involve building suitable models. In this section we carry out a limited review of the literature (as the literature is vast) where the models for decision making involve Weibull failure models.

18.5.1 Premanufacturing Phase

Reliability Allocation

Tillman et al. (1980) deal with a range of reliability allocation issues with failures given by a general distribution function. More recent references are Aggarwal and Shaswati (1993) dealing with a practical approach to reliability allocation, Majety et al. (1999), who deal with a series–parallel structure for the system, and Mettas (2000) dealing with reliability allocation with component failures given by Weibull distribution.

Reliability Prediction

Lawless (1973) examines predicting the smallest value for a future sample of k observations based on an observed sample and presents conditional confidence interval. Hsieh (1996) considers the case where n items are put on test and the first r failures observed (type II censoring). He derives prediction intervals for the life of the remaining components. Nelson (2000) derives simple prediction limits for the number of failures that will be observed in a future inspection of a sample of items based on past data, which is the cumulative number of failures in a previous inspection of the same sample of units with the shape parameter known.

Usher (1996) deals with the prediction of reliability of a component based on masked data. The component under consideration is an element of a system with a serial configuration. The exact cause (which component failed) is often not known or known partially and hence the masked data. The prediction involves maximum-likelihood estimates and confidence intervals for component life distribution based on masked data. Lin et al. (1996) deal with Bayes's estimation of component reliability based on masked data.

The prediction of the future number of failures in an interval is discussed in Nordman and Meeker (2002).

Singpurwalla and Song (1988) deal with a model where the prediction involves the incorporation of expert opinion. A Bayesian framework is used for the elicitation of expert opinion on the median life and the codification of the expert's opinion on the Weibull shape parameter.

Reliability Assessment

Nelson (1985) considers the case where the shape parameter is known and derives confidence limits for the scale parameter, percentiles, and reliabilities based on few or no failures.

McCool (1970a) looks at several inference problems in the context of reliability assessment. More specifically, he looks at tests for the following hypothesis:

1. Weibull shape parameter is equal to a specified value.
2. Weibull percentile is equal to a specified value.
3. The shape parameters of two Weibull populations are equal.
4. The specified percentile of two Weibull distributions are equal given that the shape parameters are different.

McCool (1970b) deals with the sudden death test, which is a special case of multiply censored life test wherein an equal number of randomly selected surviving items are removed from testing following the occurrence of each failure. See also Ferdous et al. (1995).

Tseng and Chang (1989) deal with selecting the best design based on type II censored data. This problem is similar to the subset selection problem discussed in the next section.

Testing

Meeker and Escobar (1993) review accelerated testing and research needed to improve accelerated test planning and methods. Meeker and Escobar (1998) deal with the pitfalls in accelerated testing. Nelson (1990) and Meeker and Escobar (1998b) deal with accelerated testing in depth. The test plan involves the number to be tested, the termination of the test, and the stress levels in the case of accelerated testing.

A small sample of the more recent literature on accelerated testing in the context of Weibull failure models is as follows:

- Seo and Yum (1991): Accelerated life test plans under intermittent inspection and type I censoring

- Bai et al. (1992): Simple ramp tests with type I censoring
- Achar and Louzada-Neto (1992): Bayesian approach to accelerated life tests
- Yang and Jin (1994): Three-level constant-stress accelerated life test plans
- Tang et al. (1996): Step-stress accelerated life test with linear cumulative exposure model and multiple censoring
- Khamis (1997b): Comparison between constant and step-stress tests
- McSorely et al. (2002): Accelerated life tests with various sample plans

Most models assume that the shape parameter is constant and the scale parameter is a function of the stress level. Wang and Kececioglu (2000) consider models where both the shape and scale parameters are functions of stress level. As mentioned earlier, most of the test plans assume standard Weibull failure model. Klein and Basu (1981, 1982) deal with accelerated life tests with competing causes of failures.

Reliability Improvement
Murthy and Hussain (1994) and Murthy (1996) examine optimal hot and cold standby redundancy to achieve an optimal trade-off between manufacturing cost and expected warranty cost. Hussain and Murthy (1998, 1999, 2000) deal with redundancy decisions when there are variations in the quality of components and examine testing at component and module level for weeding out nonconforming items. Pan (1998) examines reliability prediction with imperfect switching. Pham and Phan (1992) look at the mean life of k-out-of-n system. Hasanuddin (1985) looks at cold redundancy.

Yamada and Osaki (1983) review nonhomogeneous Poisson process models in the context of reliability growth for both hardware and software. Yamada et al. (1993) deal with software reliability growth models with Weibull test effort. Miller (1984) deals with predicting future reliability parameters for a Weibull process model.

Lakey and Rigdon (1993) use experimental design approach to improve reliability with the scale parameter of the failure distribution being a function of the independent variables under the control of the designer.

18.5.2 Manufacturing Phase

Quality Variations and Control
For items produced using a continuous production line, the process can switch from in control to out of control. When this occurs, the process needs to be brought back to in control. The detection of a switch from in control to out of control is done using control charts. This involves testing a sample of items at regular intervals to determine the quality. The sample statistics are plotted using some $\bar{X}$ (mean), R (range), p (fraction nonconforming), or some other chart to determine if a switch has occurred or not. The design of charts requires modeling the in-control time intervals. Rahim (1993) deals with economic design of $\bar{X}$ charts based on Weibull distribution for in-control times. Costa and Rahim (2000) consider both $\bar{X}$ and R

charts with the process changing from in control to out of control modeled by a Weibull shock model. Chen and Yang (2002) consider the case with multiple assignable causes for the change in the state of the process and in-control times are modeled by a Weibull distribution.

Nelson (1979) looks at the case where the observed samples are the lifetimes modeled by a Weibull distribution. The sample statistics used in the plotting of the charts are the sample median, sample range, and first-order statistics. Here, the change from in control to out of control results in a change in scale parameter for the item failure distribution.

Murthy et al. (1993) assume that the process is in steady state with the occurrence of nonconforming items purely due to variations in the uncontrollable factors. The items produced are tested for a period T, and any item that fails is scrapped (as it is more likely to be a nonconforming item). The optimal T^* is obtained by minimizing the total asymptotic cost per item sold, which is the sum of the asymptotic manufacturing and inspection costs and the asymptotic warranty cost per item sold. They consider three different warranty policies and determine conditions to determine whether testing is the optimal strategy or not.

For items produced using batch production, Djamaludin et al. (1994) looks at quality control through lot sizing. The optimal lot size is obtained by minimizing the total cost given by the sum of the asymptotic manufacturing cost and the asymptotic warranty cost. Djamaludin et al. (1995) deal with a model similar to that in Djamaludin et al. (1994), except that it also involves testing a fraction of items in some lots. At the end of each lot production of Q items, the state of the process is assessed. If it is found to be in-control, then the lot is released without testing. However, if the state is found to be out of control, the last K items in the lot are tested for a period T. Items that fail during testing are scrapped, and the others that are not tested and that survive the test are released for sale. An item released for sale can be one of the four types: type A (conforming and not tested), type B (nonconforming and not tested), type C (conforming and tested), and type D (nonconforming and tested). The optimal values for Q, K, and T are obtained by minimizing the asymptotic total cost per item, which is the sum of the asymptotic manufacturing cost per item and the asymptotic warranty cost per item. Djamaludin et al. (1997) deal with a model similar to that in Djamaludin et al. (1995), except that at the end of production of a lot, the process state is not known. The testing scheme is as follows. Items are numbered sequentially (1 through Q) in the order in which they are produced in a lot. Item Q (last item produced in a lot) is life tested for a period of time T. If it survives, then it and the remaining items in the lot are released with no further testing. If it fails, then it is scrapped and item K is tested for a period of time T. If item K survives, then it and the remaining items in the batch are released with no further testing. However, if it fails, then items $(K + 1)$ to $(Q - 1)$ are life tested for a period of time T. Those that survive the test are released along with the first $(K - 1)$ items, which are released with no testing. An item released for sale can be one of the four types (types A, B, C, and D) as Djamaludin et al. (1995). They consider three different warranty policies (i) FRW for repairable product, (ii) PRW for nonrepairable product, and (iii) FRW for nonrepairable product. The optimal Q, K, and T are obtained by minimizing the asymptotic total cost per item released, which is the sum of the asymptotic manufacturing cost per item and the asymptotic warranty servicing cost per item.

Yeh and Lo (1998) use both lot size and burn-in to control quality for products sold with free replacement policy. The optimal burn-in and lot size achieve a balance between quality control and warranty costs. Yeh et al. (2000) extend the results of Djamaludin et al. (1994) to include inventory holding costs under the assumption that the demand rate and production rates are constants. For a different model formulation, see Chen et al. (1998).

Acceptance Sampling

In manufacturing, components are often bought from external suppliers. Acceptance sampling involves testing a small number of items from each lot to determine whether the quality of items is acceptable or not. If the quality is acceptable, the lot is accepted; if not, it is rejected.

Schneider (1989) looks at acceptance sampling where the quality is determined by the lifetimes of items in a lot. An item with lifetime $Y < L$ (a prespecified value) is deemed as nonconforming and conforming otherwise. A sample of n items are drawn from each lot and tested until r fail (type II censoring). Based on data from the test, a decision is made either to accept or reject the lot. Under the transformation the Weibull model is transformed into the log Weibull model with location parameter μ and scale parameter σ. Let $\hat{\mu}$ and $\hat{\sigma}$ be the estimates based on the test data, and the test statistics $\hat{t}$ (given by $\hat{t} = \hat{\mu} - k\hat{\sigma}$) are compared with $\ln(L)$ where k is the acceptability constant. If $\hat{t} \geq \ln(L)$, then the lot is accepted and rejected if not. The optimal n and k are derived taking into account the buyer and seller risks.

Kwon (1996) deals with a method to find the optimal sampling plans that minimize the expected average cost per lot. The accepted lots are sold under pro-rata warranty (PRW). The item failure distribution is Weibull with a known shape parameter and an unknown scale parameter. A sample of n items is drawn at random and life tested simultaneously. The test is terminated when r ($r \leq n$) items have failed. The optimal sampling plan is the plan that minimizes the expected average cost per lot of size Q.

In most acceptance sampling schemes one uses point estimates for the failure distribution. Hisada and Arizino (2002) consider the acceptance sampling schemes with shape parameters specified as interval estimates. Finally, Huei (1999) deals with sampling plans for verification of vehicle component reliability. Fertwig and Mann (1980) deals with sampling plan for life testing.

Subset Selection

Often, a manufacturer has a choice to select the component supplier from several component manufacturers. The reliability of the components differs across component manufacturers, and the problem facing the manufacturer is to select the best component supplier. This problem can be posed as selecting the best population from a collection of populations and is called the subset selection problem. In order to do this, one needs to define more precisely the notion of "best," and several different notions have been proposed and studied.

The earliest study is by Rademaker and Antle (1975), who looked at the optimal sample size to decide on which of the two populations has the larger reliability. Kingston and Patel (1980a, 1980b) focus on selecting the population with the largest reliability based on type II censoring. The shape parameter can be the same or different and needs to be estimated. Hsu (1982) assumes that the shape parameter is the same for all populations and ranks them based on the scale parameters. This is equivalent to ranking based on the mean life of components. Sivanci (1986) also uses the scale parameter for comparing two populations with random censoring. Gupta and Miescke (1987) also rank the population based on the scale parameter but based on a two-stage test involving censoring. Tseng and Wu (1990) look at the case where the shape parameters are known as well as the case where they are unknown. In the latter case, they assume a common beta prior distribution for the shape parameter. Gill and Mehta (1994) examine selecting the populations whose shape parameters are at least that of the control population. See also, Kingston and Patel (1980b) and Tseng and Wu (1990).

Burn-In

When the failure rate of a product exhibits a bathtub shape, then burn-in is an effective way to improve product reliability. Burn-in involves testing the item for a period τ, called the burn-in period. Any item that survives the burn-in has a higher reliability than before burn in. If the product is repairable, then failures during burn-in are rectified minimally so that reliability after burn-in is effectively the same as those that survive the burn-in.

Burn-in costs money and the optimal burn-in achieves a trade-off between the burn-in cost and the resulting reduction in the expected warranty cost. Blischke and Murthy (1994), Murthy (1996), and Mi (1997) discuss the optimal burn-in period for different warranty policies. Kar and Nachlas (1997) study a net-profit model to optimally select the warranty period, burn-in, and stress variables to maximize the expected net-profit subject to constraints that reflect the extent to which the system components can endure various stress variables.

18.5.3 Postsale Phase

Warranties

The warranty-servicing cost for an item sold with a warranty is a random variable that depends on the reliability of the product and the warranty-servicing strategy. A detailed study of warranty cost analysis can be found in Blischke and Murthy (1994, 1996), where they derive expressions for various expected costs. Sahin and Polatoglu (1998) and Polatoglu and Sahin (1998) derive the probability distribution for the warranty cost and some related variables. When a repairable item is returned to the manufacturer for repair under free replacement warranty, the manufacturer has the option of either repairing it or replacing it by a new one. The optimal strategy is one that minimizes the expected cost of servicing the warranty over the warranty period. Blischke and Murthy (1994) and Murthy (1996) discuss two suboptimal strategies for one-dimensional warranties, and Iskandar and Murthy (1999) look at the two-dimensional case. See also Yeh and Lo (2000).

Jack and Schouten (2000) deal with the optimal repair–replace strategies. Here the decision to repair or replace is based on the age of the item at failure. Jack and Murthy (2001) examine a suboptimal policy that is very close to the optimal strategy and involves at most one replacement over the warranty period. Iskandar et al. (2003) study a similar strategy in the context of two-dimensional warranties.

In general, the cost to repair a failed item is a random variable that can be characterized by a distribution function $H(z)$. Analogous to the notion of a failure rate, one can define a repair cost rate given by $h(z)/\bar{H}(z)$, where $h(z)$ is the derivative of $H(z)$. Depending on the form of $H(z)$, the repair cost rate can increase, decrease, or remain constant with z. A decreasing repair cost rate is usually an appropriate characterization for the repair cost distribution (Mahon and Bailey, 1975). Optimal repair limit strategy is discussed in Blischke and Murthy (1994), Chung (1994), Murthy (1996), and Zuo et al. (2000).

Maintenance

The literature on maintenance is very large. There are several review studies that have appeared over the last 30 years. These include McCall (1965), Pierskalla and Voelker (1976), Monahan (1982), Jardine and Buzzacot (1985), Sherif and Smith (1986), Thomas (1986), Gits (1986), Valdez-Flores and Feldman (1989), Pintelton and Gelders (1992), and Scarf (1997). Cho and Parlar (1991) and Dekker et al. (1997) and Archibald and Dekker (1996) deal with the maintenance of mutlicomponent systems. Many of the works cited deal with failures modeled by Weibull distribution. Tadikmalla (1980) deals with age policy at the component level, and a similar study for the block and periodic policies can be found in Blischke and Murthy (2000). See also Huang et al. (1995). Mine and Nakagawa (1978) consider the case where the failure distribition involves a mixture model. Dagpunar (1996) examines optimal maintenance policies with opportunities and interrupt replacement options. Love and Guo (1996) examines repair limit policies. Mazzuchi and Soyer (1996), Percy and Kobbacy (1996), and Percy et al. (1998) deal with the Bayesian approach to preventive maintenance.

References

Aarset, M.V. (1987). How to identify bathtub hazard rate, *IEEE Transactions on Reliability*, **36**, 106–108.

Abernathy, R.B., Breneman, J.E., Medlin, C.H., and Reinman, G.L. (1983). *Weibull Analysis Handbook*, Report AFWAL-TR-83-2079, Air Force Wright Aeronautical Laboratories, Ohio.

Abramowitz, M. and Stegun, I.A. (1964). *Handbook of Mathematical Functions with Formulas, Graphs, and Mathematical Tables*, Dover, New York.

Achar, J.A. and Louzada-Neto, F. (1992). A Bayesian approach for accelerated life tests considering the Weibull distribution, *Computational Statistics*, **7**, 355–369.

Agarwal, S.K. and Al-Saleh, J.A. (2001). Generalized Gamma type distribution and its hazard rate function, *Communications in Statistics—Series A: Theory and Methods*, **30**, 309–318.

Agarwal, S.K. and Kalla, S.L. (1996). A generalized gamma distribution and its application in reliability. *Communications in Statistics—Series A: Theory and Methods*, **25**, 201–210.

Aggarwal, S.K. and Shaswati, G. (1993). Reliability allocation in a general system with non-identical components—a practical approach, *Microelectronics and Reliability*, **33**, 1089–1093.

Aggarwala, R. (2001). Progressive censoring: A review, in *Advances in Reliability, Handbook of Statistics*, Vol. 20, N. Balakrishnan and C.R. Rao (eds.), Elsevier, Amsterdam.

Ahmad, K.E. (1994). Modified weighted least-squares estimators for the three-parameter Weibull distribution, *Applied Mathematics Letters*, **7**, 53–56.

Ahmad, K.E. and Abdelrahman, A.M. (1994). Updating a nonlinear discrimant function estimated from a mixture of 2-Weibull distributions, *Mathematics and Computer Modelling*, **18**, 41–51.

Ahmad, K.E., Moustafa, H.M., and AbdElrahman, A.M. (1997). Approximate Bayes estimation for mixtures of two Weibull distributions under type-2 censoring, *Journal of Statistical Computation and Simulation*, **58**, 269–285.

Akersten, P-A., Klefsjö, B., and Bergman, B. (2001). Graphical techniques for analysis of data from repairable systems, in *Handbook of Statistics—Advances in Reliability*, N. Balakrishnan and C.R. Rao (eds.), Elsevier, Amsterdam, pp. 469–484.

Weibull Models, by D.N.P. Murthy, Min Xie, and Renyan Jiang.
ISBN 0-471-36092-9

Ali, M.A. (1997). Characterization of Weibull and exponential distributions, *Journal of Statistical Studies*, **17**, 103–106.

Almeida, J.B. (1999). Application of Weilbull statistics to the failure of coatings, *Journal of Material Processing Technology*, **93**, 257–263.

Ananda, M.M. and Singh, A.K. (1993). Estimation of a composite hazard rate model, *Microelectronics and Reliability*, **33**, 2013–2019.

Archibald, Y.W. and Dekker, R. (1996). Modified block replacement for multiple-component systems, *IEEE Transactions on Reliability*, **45**, 75–83.

Aroian, L.A. and Robinson, D.E. (1966). Sequential life tests for the exponential distribution with changing parameters, *Technometrics*, **8**, 217–227.

Ascher, H. and Feingold, H. (1984). *Repairable System Reliability*, Marcel Dekker, New York.

Ashour, S.K. (1987). Multicensored sampling in mixed Weibull distribution. *IAPQR Transactions*, **12**, 51–56.

Attia, A.F. (1993). On estimation for mixtures of two Rayleigh distribution with censoring. *Microelectronics and Reliability*, **33**, 859–867.

Badarinathi, R. and Tiwari, R. (1992). Hierarchical Bayesian approach to reliability estimation under competing risk, *Microelectronics and Reliability*, **32**, 249–258.

Bai, D.S., Cha, M.S., and Chung, S.W. (1992). Optimum simple ramp-tests for the Weibull distribution and Type-I censoring, *IEEE Transactions on Reliability*, **41**, 407–413.

Bain, L.J. (1978). *Statistical Analysis of Reliability and Life Testing Models*, Marcel Dekker, New York.

Bain, L.J. and Engelhardt, M. (1980). Inference on the parameters of current system reliability for a time truncated Weibull process, *Technometrics*, **22**, 421–426.

Bain, L.J. and Engelhardt, M. (1986). On the asymptotic behaviour of the mean time between failures for repairable systems, in *Reliability and Quality Control*, A.P. Basu (ed.), North-Holland, Amsterdam, pp. 1–7.

Bain, L.J. and Engelhardt, M. (1991). *Statistical Analysis of Reliability and Life-Testing Models*, Marcel Dekker, New York.

Bain, L.J. and Weeks, D.L. (1965). Tolerance limits for the generalized Gamma distribution, *Journal of the American Statistical Association*, **60**, 1142–1152.

Balakrishnan, N. and Aggarwala, R. (2000). *Progressive Censoring Theory, Methods and Applications*, Birkhauser, Boston.

Balakrishnan, N. and Basu, A.P. (1996). *The Exponential Distribution: Theory, Methods and Applications*, Gordon and Breach, Singapore.

Balakrishnan, N. and Joshi, P.C. (1981). A note on order statistics from Weibull distribution, *Scandinavian Actuarial Journal*, **2**, 121–122.

Balakrishnan, N. and Kocherlakota, S. (1985). On the double Weibull distribution: Order statistics and estimation, *Sankhya*, **47**, 161–176.

Balakrishnan, N. and Rao, C.R. (1998). *Order Statistics: Theory & Methods*, North-Holland, Amsterdam.

Balasooriya, U. (2001). Progressively censored variable-sampling plans for life testing, in *Advances in Reliability, Handbook of Statistics*, Vol. 20, N. Balakrishnan and C.R. Rao (eds), Elsevier, Amsterdam.

Barakat, H.M. and Abdelkader, Y.H. (2000). Computing the moments of order statistics from nonidentically distributed Weibull variables, *Journal of Computational and Applied Mathematics*, **117**, 85–90.

Bar-Lev, S., Lavi, I., and Reiser, B. (1992). Bayesian inference for the power law process, *Annals of the Institute of Statistical Mathematics*, **44**, 632–639.

Barlow, R.E. and Campo, R. (1975). Total time on test processes and applications to failure data analysis, in *Reliability and Fault Tree Analysis*, R.E. Barlow, J. Fussell, and N.D. Singpurwalla (eds), SIAM, Philadelphia, pp. 451–491.

Barlow, R.E. and Hunter, L. (1961). Optimum preventive maintenance policies, *Operations Research*, **8**, 90–100.

Bartolucci, A.A., Singh, K.P., Bartolucci, A.D., and Bae, S. (1999). Applying medical survival data to estimate the three-parameter Weibull distribution by the method of probability-weighted moments, *Mathematics and Computers in Simulation*, **48**, 385–392.

Bassin, W.M. (1973). A Bayesian optimal overhaul model for the Weibull restoration process, *Journal of the American Statistical Association*, **68**, 575–578.

Basu, A.P. and Klein, J.P. (1982). Some recent results in competing risk theory, in Survival Analysis, vol. 2, Proceedings of the Special Topics Meeting sponsored by the Institute of Mathematical Statistics, October 26–28, Columbus, Ohio, pp. 216–229.

Basu, S., Basu, A.P., and Mukhopadhyay, C. (1999). Bayesian analysis for masked system failure data using non-identical Weibull models, *Journal of Statistical Planning and Inference*, **78**, 255–275.

Baxter, L.A., Scheuer, E.M., McConalogue D.J., and Blischke, W.R. (1982). On the tabulation of the renewal function, *Technometrics*, **24**, 151–156.

Beetz, C.P., Jr. (1982). Analysis of carbon fibre strength distributions exhibiting multiple modes of failure, *Fibre Science and Technology*, **16**, 45–59.

Bell, B.M. (1988). Generalized gamma parameter estimation and moment evaluation. *Communications in Statistics—Series A: Theory and Methods*, **17**, 507–517.

Berger, J.O. and Sun, D.C. (1993). Bayesian analysis for the Poly-Weibull distribution, *Journal of the American Statistical Association*, **88**, 1412–1418.

Bergman, B. and Klefsjo, B. (1984). The total time on test and its use in reliability theory, *Operations Research*, **32**, 596–606.

Berman, M. (1981). Inhomogeneous and modulated gamma processes, *Biometrika*, **68**, 143–152.

Bjarnason, H. and Hougaard, P. (2000). Fisher information for two gamma frailty bivariate Weibull models, *Lifetime Data Analysis*, **6**, 59–71.

Blache, K.M. and Shrivastava, A.B. (1994). Defining failure of manufacturing machinery and equipment, *Proceedings of the Annual Reliability and Maintainability Symposium*, 69–75.

Black, S.E. and Rigdon, S.E. (1996). Statistical inference for a modulated power law process. *Journal of Quality Technology*, **28**, 81–90.

Blischke, W.R. (1974). On nonregular estimation. II. Estimation of the location parameter of the gamma and Weibull distributions, *Communications in Statistics*, **3**, 1109–1129.

Blischke, W.R. and Murthy, D.N.P. (1994). *Product Warranty Handbook*, Marcel Dekker, New York.

Blischke, W.R. and Murthy, D.N.P. (1996). *Warranty Cost Analysis*, Marcel Dekker, New York.

Blischke, W.R. and Murthy, D.N.P. (2000). *Reliability*, Wiley, New York.

Block, H.W. and Savits, T.H. (1981). Multivariate classes in reliability theory. *Mathematics of Operations Research*, **6**, 453–461.

Boorla, R. and Rotenberger, K. (1997). Load variability of a two-bladed helicopter, *Journal of American Helicopters*, **42**, 15–26.

Broido, A. and Yow, H. (1977a). Application of the Weibull distribution function to the molecular weight distribution of cellulose, *Journal of Applied Polymer Science*, **21**, 1667–1676.

Broido, A. and Yow, H. (1977b). Resolution of molecular weight distributions in slightly pyrolyzed cellulose using the Weibull, *Journal of Applied Polymer Science*, **21**, 1677–1685.

Bulmer, M.B. and Eccleston, J.A. (1998). Automated reliability modeling with a genetic algorithm, *Proceedings of the 6th International Applied Statistics in Industry Conference*, Melbourne, 1–8.

Bulmer, M. and Eccleston, J.A. (2002). Photo copier reliability modelling using evolutionary algorithms, in *Case Studies in Reliability and Maintenance*, W.R. Blischke and D.N.P. Murthy (eds.), Wiley, New York.

Cacciari, M. and Montannari, G.C. (1987). A method to estimate the Weibull parameters for progressive censored tests, *IEEE Transactions on Reliability*, **R-36**, 87–93.

Cacciari, M., Contin, A., Rabach, G., and Montannari, G.C. (1993). Risk functions for partial discharge inference in insulation systems, *Proceedings of the International Conference on Partial Discharge*, 64–65.

Cacciari, M., Contin, A., and Montanari, G.C. (1995). Use of mixed-Weibull distribution for the identification of PD phenomena, *IEEE Transactions on Dielectrics and Electrical Insulation*, **2**, 1166–1179.

Calabria, R. and Pulcini, G. (1989). Confidence limits for reliability and tolerance limits in the inverse Weibull distribution, *Reliability Engineering and Systems Safety*, **24**, 77–85.

Calabria, R. and Pulcini, G. (1994). Bayesian 2-sample prediction for the inverse Weibull distribution, *Communications in Statistics – A: Theory and Methods*, **23**, 1811–1824.

Calabria, R. and Pulcini, G. (2000). Inference and test in modelling the failure/repair process of repairable mechanical equipment, *Reliability Engineering and System Safety*, **67**, 41–53.

Calabria, R., Guida, M., and Pulcini, G. (1990). Bayes estimation of prediction intervals for a power law process, *Communications in Statistics—Series A: Theory and Methods*, **19**, 3023–3035.

Canavos, G. C. and Tsokos, C.P. (1973). Bayesian estimation of life parameters in the Weibull distribution, *Operations Research*, **21**, 755–763.

Cancho, V.G., Bolfarine, H., and Achcar, J.A. (1999). A Bayesian analysis for the exponentiated-Weibull distribution. *Journal of Applied Statistical Science*, **8**, 227–242.

Carter, A.D.S. (1986). *Mechanical Reliability*, 2nd ed., Wiley, New York.

Chan, V. and Meeker, W.Q. (1999). A failure-time model for infant-mortality and wearout failure modes, *IEEE Transactions on Reliability*, **48**, 377–387.

Chandra, M., Singpurwalla, N.D., and Stephens, M.A. (1981). Kolmogorov statistics for tests of fit for the extreme value and Weibull distributions, *Journal of the American Statistical Association*, **76**, 729–731.

Chang, S.-C. (1998). Using parametric statistical models to estimate mortality structure: The case of Taiwan, *Journal of Actuarial Practice*, **6** (1 & 2).

Chapman, T.G. (1997). Stochastic models for daily rainfall in the Western Pacific, *Mathematics and Computers in Simulation*, **43**, 351–358.

Chaudhuri, A. and Chandra, N.K. (1990). A characterization of the Weibull distribution, *IAPQR Transactions*, **15**, 69–72.

Chen, Z. (2000). A new two-parameter lifetime distribution with bathtub shape or increasing failure rate function, *Statistics & Probability Letters*, **49**, 155–161.

Chen, Y.S. and Yang, Y.M. (2002). Economic design of x-control charts with Weibull in-control times and when there are multiple assignable causes, *International Journal of Production Economics*, **77**, 17–23.

Chen, K., Papadopoulos, A.S., and Tamer, P. (1989). On Bayes estimation for mixtures of two Weibull distributions under type I censoring, *Microelectronics and Reliability*, **29**, 609–617.

Chen, J., Yao, D.D., and Zheng, S.H. (1998). Quality control for products supplied with warranty. *Operations Research*, **46**, 107–115.

Cheng, K.F. and Chen, C.H. (1988). Estimation of the Weibull parameters with grouped data, *Communication in Statistics—Theory and Methods*, **17**, 325–341.

Cheng, S.W. and Fu, J.C. (1982). Estimation of mixed Weibull parameters in life testing, *IEEE Transactions on Reliability*, **31**, 377–381.

Chin, H.C., Quek, S.T., and Cheu, R.L. (1991). Traffic conflicts in expressway merging, *Journal of Transport Engineering*, **117**, 633–643.

Cho, D.I. and Parlar, M. (1991). A survey of maintenance models for multiunit systems. *European Journal of Operational Research*, **51**, 1–23.

Chou, K. and Tang, K. (1992). Burn-in time and estimation of change-point with Weibull-exponential mixture distribution, *Decision Sciences*, **23**, 973–990.

Chung, M. K. (1986). Use of a mixed Weibull model in occupational injury analysis, *Journal of Occupational Accidents*, **7**, 239–250.

Chung, K.J. (1994). An approximate calculation of optimal repair-cost limits under minimal repair. *Engineering Optimization*, **23**, 119–123.

Christopeit, N. (1994). Estimating parameters of an extreme value distribution by the method of moments. *Journal of Statistical Planning and Inference*, **41**, 173–186.

Cinlar, E. (1975). *Introduction to Stochastic Processes*, Prentice-Hall, Englewood Cliffs, NJ.

Cohen, A.C. (1973). The reflected Weibull distribution, *Technometrics*, **15**, 867–873.

Cohen, A.C. and Whitten, B.J. (1982). Modified moment and maximum likelihood estimators for parameters of the three-parameter gamma distribution, *Communications in Statistics—Series B: Simulation and Computation*, **11**, 197–216.

Cohen, A.C., Whitten, B.J., and Ding, Y. (1984). Modified moment estimation for the three-parameter Weibull distribution, *Journal of Quality Technology*, **16**, 159–167.

Coles, S.G. (1989). On goodness-of-fit tests for the two-parameter Weibull distribution derived from the stabilized probability plot. *Biometrika*, **76**, 593–598.

Colvert, R.E. and Boardman, T.J. (1976). Estimation in the piece-wise constant hazard rate model, *Communications in Statistics—Series A: Theory and Methods*, **1**, 1013–1029.

Constantine, A.G. and Robinson, N.I. (1997). The Weibull renewal function for moderate to large arguments, *Computational Statistics & Data Analysis*, **24**, 9–27.

Contin, A., Cacciari, M., and Montanari, G.C. (1994). Estimation of Weibull distribution parameters for partial discharge inference, *IEEE 1994 Annual Report*, 71–78.

Contin, A., Fenu, G.F., and Parisini, T. (1996). Diagnosis of HV stator bars insulation in the presence of multi partial-discharge phenomena, *Conference on Electrical Insulation and Dielectric Phenomena (CEIDP), Annual Report*, **2**, 488–491.

Costa, A.F.B. and Rahim, M.A. (2000). Economic design of X and R charts under Weibull shock models, *Quality and Reliability Engineering International*, **16**, 143–156.

Cox, D.R. (1962). *Renewal Theory*, Methuen and Co. Ltd., London.

Cox, D.R. (1975). Partial likelihood, *Biometrika*, **62**, 269–276.

Cox, D.R. and Miller, H.D. (1965). *The Theory of Stochastic Processes*, Methuen, London.

Cox, D.R. and Oakes, D. (1984). *Analysis of Survival Data, Chapman and Hall, New York.*

Cran, G.W. (1976). Graphical estimation methods for Weibull distributions, *Microelectronics and Reliability*, **15**, 47–52.

Cran, G.W. (1988). Moment estimators for the 3-parameter Weibull distribution, *IEEE Transactions on Reliability*, **37**, 360–363.

Crow, L.H. (1974). Reliability analysis for complex systems, in *Reliability and Biometry*, F. Proschan and R.J. Serfling (eds.), Philadelphia, SIAM, pp. 379–410.

Crowder, M. (1989). A multivariate distribution with Weibull connections, *Journal of the Royal Statistical Society—B*, **51**, 93–107.

D'Agostino, R.B. and Stephens, M.A. (1986). *Goodness-of-Fit Techniques*, Marcel Dekker, New York.

Dagpunar, J.S. (1996). A maintenance model with opportunities and interrupt replacement policies, *Journal of Operational Research Society*, **47**, 1406–1409.

Dasgupta, A. and Pecht, M. (1991). Material failure mechanisms and damage models, *IEEE Transactions on Reliability*, **40**, 531–536.

David, H.A. (1974). Parametric approaches to the theory of competing risks, in *Reliability and Biometry: Statistical Analysis of Life Length*, F. Proschan and R.J. Serfling (eds.), SIAM, Philadelphia.

Davis, D.J. (1952). An analysis of some failure data. *Journal of the American Statistical Association*, **47**, 113–150.

Davison, A.C. and Louzada-Neto, F. (2000). Inference for the poly-Weibull model. *Journal of the Royal Statistical Society—Series D: The Statistician*, **49**, 189–196.

de Rijk, W.G. (1988). Applications of the Weibull method to statistical analysis of strength parameters of dental materials, *Polymeric Materials Science and Engineering, Proceedings of the ACS Division of Polymeric Materials Science and Engineering*, **59**, 407–412.

Dehnad, K. (1989). *Quality Control, Robust Design and the Taguchi Method*, Wadsworth & Brooks/Cole, Pacific Grove, CA.

Dekker, R., Wildeman, R.E., and Schouten, F.A.V.D. (1997). A review of multi-component maintenance models with economic dependence. *Mathematical Methods in Operations Research*, **45**, 411–435.

Dekkers, A.L.M., Einmahl, J.H.J., and de Haan, L. (1989). A moment estimator for the index of an extreme-value distribution. *Annals of Statistics* **17**, 1833–1855.

Del Castillo, E. (2002). *Statistical Process Adjustment for Quality Control*, Wiley, New York.

Dellaportas, P. and Wright, D.E. (1991). Numerical prediction for the 2-parameter Weibull distribution, *Statistician*, **40**, 365–372.

De Souza, D.I. and Lamberson, L.R. (1995). Bayesian Weibull reliability estimation, *IIE Transactions*, **27**, 311–320.

Dickey, J.M. (1991). The renewal function for an alternating renewal process, which has a Weibull failure distribution and a constant repair time, *Reliability Engineering and Systems Safety*, **31**, 321–343.

Djamaludin, I. (1993). Warranty and Quality Control, Ph.D. Thesis, The University of Queensland, Brisbane, Australia.

Djamaludin, I., Murthy, D.N.P., and Wilson, R.J. (1994). Quality-control through lot-sizing for items sold with warranty, *International Journal of Production Economy*, **33**, 97–107.

Djamaludin, I., Murthy, D.N.P., and Wilson, R.J. (1995). Lot-sizing and testing for items with uncertain quality, *Mathematical and Computer Modelling*, **22**, 35–44.

Djamaludin, I., Wilson, R.J., and Murthy, D.N.P. (1997). An optimal quality control scheme for product sold with warranty, in K.S. Al-Sultan, M.A. Rahim (eds.), *Optimization in Quality Control*, Kluwer Academic, New York, Chapter 13.

Dodson, B. (1994). *Weibull Analysis*, ASQ Quality Press, Milwaukee, WI.

Doganaksoy, N. (1991). Interval estimation from censored and masked system-failure data, *IEEE Transactions on Reliability*, **40**, 280–286.

Drapella, A. (1993). Complementary Weibull distribution: Unknown or just forgotten, *Quality and Reliability Engineering International*, **9**, 383–385.

Duan, J., Selker, J., and Grant, G.E. (1998). Evaluation of probability density functions in precipitation models for the Pacific Northwest, *Journal of the American Water Resource Association*, **34**, 617–627.

Duane, J.T. (1964). Learning curve approach to reliability monitoring, *IEEE Transactions on Aerospace*, **40**, 563–566.

Dubey, S.D. (1968). A compound Weibull distribution, *Naval Research Logistics Quarterly*, **15**, 179–188.

Dubey, S.D. (1967a). Some percentile estimators for Weibull parameters, *Technometrics*, **9**, 119–129.

Dubey, S.D. (1967b). On some permissible estimators of the location parameter of the Weibull and certain other distributions, *Technometrics*, **9**, 293–307.

Durham, S.D. and Padgett, W.J. (1997). Cumulative damage model for system failure with application to carbon fibers and composites, *Technometrics*, **39**, 34–44.

Efron, B. (2000). The bootstrap and modern statistics. *Journal of American Statistical Association*, **95**, 1293–1296.

Elandt-Johnson, R.C. and Johnson, N.L. (1980). *Survival Model and Data Analysis*, Wiley, New York.

El-Arishy, S.M. (1993). Characterizations of Weibull, Rayleigh and Maxwell distributions using truncated moments and order statistics, *Egyptian Statistical Journal*, **37**, 198–212.

Eldin, M., Mahmoud, M., and Youssef, S. (1991). Moments of order statistics from parabolic and skewed distributions and a characterization of Weibull distribution, *Communications in Statistics—Series B: Simulation and Computation*, **20**, 639–645.

English, J.R., Yan, L., and Landers, T.L. (1995). A modified bathtub curve with latent failures, *Proc. of the Annual Reliability and Maintainability Symposium*, 217–222.

Erto, P. and Rapone, M. (1984). Non-informative and practical Bayesian confidence bounds for reliable life in the Weibull model, *Reliability Engineering*, **7**, 181–191.

Evans, J.W., Johnson, R.A., and Green, D.W. (1997). Goodness-of-fit tests for two-parameter and three-parameter Weibull distributions, *Advances in the Theory and Practice of Statistics*, Wiley, New York, pp. 159–178.

Everitt, B. S. and Hand, D. J. (1981). *Finite Mixture Distributions, Chapman and Hall, London.*

Fahrmeir, L. and Klinger, A. (1998). A nonparametric multiplicative hazard model for event history analysis, *Biometrika*, **85**, 581–592.

Falls, L.W. (1970). Estimation of parameters in compound Weibull distributions, *Technometrics*, **12**, 399–407.

Fang, Z., Patterson, B.R., and Turner, M.E. (1993). Modelling particle size distributions by Weibull distribution function, *Materials Characterization*, **31**, 177–182.

Fei, H.L. and Kong, F.H. (1995). Estimation for 2-parameter Weibull distribution and extreme-value distribution under multiply Type II censoring, *Communications in Statistics—Series A: Theory and Methods*, **24**, 2087–2104.

Ferdous, J., Borhon, U.D., and Pandey, M. (1995). Reliability estimation with Weibull interfailure times, *Reliability Engineering and System safety*, **50**, 285–296.

Fertig, K.W., Meyer, M.E., and Mann, N.R. (1980). On constructing prediction intervals for samples from a Weibull or extreme value distribution, *Technometrics* **22**, 567–573.

Fertwig, K.W. and Mann, N.R. (1980). Life-tests sampling plans for two-parameter Weibull populations, *Technometrics*, **22**, 165–177.

Finkelstein, J.M. (1976). Confidence bounds for the parameters of the Weibull process, *Technometrics*, **18**, 115–117.

Fisher, R.A. and Tippett, L.M.C. (1928). Limiting forms of frequency distribution of the largest or smallest member of a sample, *Proceedings of the Cambridge Philosophical Society*, **24**, 180–190.

Fok, S.L., Mitchell, B.C., Smart, J., and Marsden, B.J. (2001). A numerical study on the application of the Weibull theory to brittle materials, *Engineering Fracture Mechanics*, **68**, 1171–1179.

Fraser, R.P. and Eisenklam, P. (1956). Liquid atomisation and the drop size of sprays, *Transactions of the Institute of Chemical Engineers*, **34**, 294–319.

Freimer, M., Kollia, G., Mudholkar, G.S., and Lin, C.T. (1989). Extremes, extreme spacings and outliers in the Tukey and Weibull families, *Communications in Statistics—Series A: Theory and Methods*, **18**, 4261–4274.

Friedman, L. and Gertsbakh, I.B. (1980). Maximum likelihood estimation in a minimum-type model with exponential and Weibull failure modes, *Journal of the American Statistical Association*, **75**, 460–465.

Fries, A. and Sen, A. (1996). A survey of discrete reliability-growth models, *IEEE Transactions on Reliability*, **45**, 582–604.

Gallangher, M.A. and Moor, A.H. (1990). Robust minimum-distance estimation using the 3-parameter Weibull distribution, *IEEE Transactions on Reliability*, **39**, 575–581.

Geist, R., Smotherman, M., and Talley, R. (1990). Modeling recovery time distributions in ultrareliable fault-tolerant systems, *Digest of Papers—FTCS (Fault-tolerant Computing Symposium)* (IEEE cat n 90CH2887-9), pp. 499–504.

Gera, A.E. (1995). Lifetime modelling with aid of a modified complementary Weibull distribution. *Quality and Reliability Engineering International*, **11**, 379–388.

Gertsbakh., I.B. (1989). *Statistical Reliability Theory*, Marcel Dekker, New York.

Gertsbakh, I. and Kagan, A. (1999). Characterization of the Weibull distribution by properties of the Fisher information under type-1 censoring, *Statistics and Probability Letters*, **42**, 99–195.

Ghitany, M.E. (1998). On a recent generalization of gamma distribution. *Communications in Statistics—Series A: Theory and Methods*, **27**, 223–233.

Gibbons, D.I. and Vance, L.C. (1983). Estimators for the two-parameter Weibull distribution with progressively censored samples, *IEEE Transactions on Reliability*, **R-32**, 95–99.

Gill, A.N. and Mehta, G.P. (1994). A class of subset selection procedures for Weibull populations, *IEEE Transactions on Reliability*, **43**, 65–70.

Gits, C.W. (1986). On the maintenance concept for a technical system: II. Literature review, *Maintenance Management International*, **6**, 181–196.

Glaser, R.E. (1980). Bathtub and related failure rate characterizations. *Journal of the American Statistical Association*, **75**, 667–672.

Glaser, R.E. (1984). Estimation for a Weibull accelerated life testing model, *Naval Research Logistics Quarterly*, **31**, 559–570.

Goda, K. and Hamada, J. (1995). Creep-rupture lifetime distribution of boron fibers, *Zairyo/ Journal of the Society of Materials Science, Japan*, **44**, 1066–1074.

Gouno, E. and Balakrishnan, N. (2001). Step-stress accelerated life test, in *Handbook of Statistics*—Vol. 20, N. Balakrishnan and C.R. Rao (eds.), Elsevier, Amsterdam, pp. 623–639.

Goto, M. et al. (1989). Statistical property on the initiation and propagation of microcracks (rotating bending fatigue of heat-treated 0.45% C steel plain specimens), *Nippon Kikai Gakkai Ronbunshu, A Hen/Transactions of the Japan Society of Mechanical Engineers, Part A*, **55**, 1493–1498.

Goto, M. (1991). Statistical investigation of the behavior of microcracks in carbon steels, *Fatigue and Fracture of Engineering Materials & Structures*, **14**, 833–845.

Gourdin, E., Hansen, P., and Jaumard, B. (1994). Finding maximum likelihood estimators for the 3-parameter Weibull distribution, *Journal of Global Optimization*, **5**, 373–397.

Grant, E.L. and Leavensworth, R.S. (1988). *Statistical Quality Control*, McGraw-Hill, New York.

Griffith, W.S. (1982). Representation of distributions having monotone or bathtub-shaped failure rates, *IEEE Transactions on Reliability*, **31**, 95–96.

Groeneveld, R.A. (1986). Skewness for the Weibull family, *Statistica Neerlandica*, **40**, 135–140.

Grzybowski, S., Abu-Al-Feilat, E.A., Knight, P., and Doriott, L. (1998). Comparison of aging behavior of two-layer polymeric dielectrics aged at AC and DC voltages, *Conference on Electrical Insulation and Dielectric Phenomena (CEIDP), Annual Report*, **2**, 678–681.

Guida, M., Calabria, R., and Pulcini, G. (1989). Bayes inference a non-homogeneous Poisson process with a power law intensity, *IEEE Transactions on Reliability*, **38**, 603–609.

Gumbel, E.J. (1958). *Statistics of Extremes*, Columbia University Press, New York.

Gupta. S.S. and Miescke, K.J. (1987). Optimum two-stage selection procedures for Weibull populations, *Journal of Statistical Planning and Inference*, **15**, 147–156.

Hager, H.W. and Bain, L.J. (1970). Inferential procedures for the generalized gamma distribution, *Journal of the American Statistical Association*, **65**, 1601–1609.

Hameed, M.S. and Proschan, F. (1973). Non-stationary shock models, *Stochastic Processes and Their Applications*, **1**, 383–404.

Harris, R. (1970). A multivariate definition for increasing hazard rate distribution functions. *Annals of Mathematical Statistics*, **41**, 713–717.

Harris, C.M. (1985). Note on mixed exponential approximations for $G_i/G/1$ waiting times, *Computers & Operations Research*, **12**, 285–289.

Harris, C.M. and Singpurwalla, N.D. (1968). Life distributions derived from stochastic hazard function. *IEEE Transactions on Reliability*, **17**, 70–79.

Harris, C.M. and Singpurwalla, N.D. (1969). On estimation in Weibull distributions with random scale parameters. *Naval Research Logistics Quarterly*, **16**, 405–410.

Harter, H.L. (1967). Maximum-likelihood estimation of the parameters of a four-parameter generalized gamma population from complete and censored samples, *Technometrics*, **9**, 159–165.

Harter, H.L. (1978). A bibliography of extreme value theory, *International Statistical Review*, **46**, 279–306.

Hasanuddin, A.S. (1985). Reliability of a 3-unit cold redundant system with Weibull failure, *Microelectronics and Reliability*, **25**, 325–330.

Hassanein, K.M. (1971). Percentile estimators for the parameters of the Weibull distribution, *Biometrika*, **58**, 673–676.

Hayakawa, N., Wakita, M., Hikita, M., and Okubo, H. (1997a). Statistical analysis of scattering in quench current of AC superconducting wires, *Proceedings of the Sixteenth International Cryogenic Engineering Conference,* T. Haruyama, T. Mitsui, and K. Yamafuji (eds.) **3**, 1837–1840.

Hayakawa, N., Wakita, M., Hikita, M., and Okubo, H. (1997b). Weibull statistical analysis of quench current of A.C. superconducting wires and coils, *Cryogenics,* **37**, 91–95.

Henderson, G. and Webber, N. (1978). Waves in severe storms in the Cental English Channel, *Coastal Engineering*, **2**, 95–110.

Heo, J.H., Boes, D.C., and Salas, J.D. (2001). Regional flood frequency analysis based on a Weibull model: Part 1, Estimation and asymptotic variances, *Journal of Hydrology*, **242**, 157–170.

Herman, R.J. and Patell, R.K.N. (1971). Maximum likelihood estimation for multi-risk model, *Technometrics*, **13**, 385–396.

Hirao, K. et al. (1988). Statistical study on applied stress dependence of failure time in stress corrosion cracking of zircaloy-4 alloy, *Nippon Genshiryoku Gakkaishi/Journal of the Atomic Energy Society of Japan,* **30**, 1038–1044.

Hirose, H. (1997). Lifetime assessment by intermittent inspection under the mixture Weibull power law model with application to XLPE cables, *Lifetime Data Analysis*, **3**, 179–189.

Hirose, H. (1999). Bias correction for the maximum likelihood estimates in the two-parameter Weibull distribution, *IEEE Transactions on Dielectrics and Electrical Insulation*, **6**, 66–68.

Hirose, H. (2000). Maximum likelihood parameter estimation by model augmentation with applications to the extended four-parameter generalized gamma distribution. *Mathematics and Computers in Simulation*, **54**, 81–97.

Hirose, H. and Lai, T.L. (1997). Inference from grouped data in three-parameter Weibull models with applications to breakdown-voltage experiments, *Technometrics*, **39**, 199–210.

Hisada, K. and Arizino, I. (2002). Reliability tests for Weibull distribution with varying shape-parameter, based on complete data, *IEEE Transactions on Reliability*, **51**, 331–336.

Hoel, D.G. (1972). A representation of mortality data by competing risks, *Biometrics*, **28**, 475–488.

Hougaard, P. (1986). Survival models for heterogeneous populations derived from stable distributions, *Biometrika*, **73**, 387–396.

Hoyland, A. and Rausand, M. (1994). *System Reliability Theory*, Wiley, New York.

Hsieh, H.K. (1996). Prediction interval for Weibull observations, based on early-failure data, *IEEE Transactions on Reliability*, **45**, 666–670.

Hsu, T.A. (1982). On some optimal selection procedures for Weibull populations, *Communications in Statistics—Series A: Theory and Methods*, **11**, 2657–2668.

Huang, J., Miller, C.R., and Okogbaa, O.G. (1995). Optimal preventive replacement intervals for Weibull life distribution: Solutions and applications, *Proceedings of the Annual Reliability and Maintainability Symposium*, IEEE, New York.

Huei, Y.K. (1999). Sampling plans for vehicle component reliability verification, *Quality and Reliability Engineering International*, **15**, 363–368.

Huillet, T. and Raynaud, H.F. (1999). Rare events in a log-Weibull scenario—Application to earthquake magnitude data, *European Physical Journal B*, **12**, 457–469.

Hunter, J.J. (1974). Renewal theory in two-dimensions: Basic results, *Advances in Applied Probability*, **6**, 376–391.

Hussain, A.Z.M.O. (1997). Warranty and Product Reliability, Unpublished doctoral thesis, University of Queensland, Brisbane, Australia.

Hussain, A.Z.M.O. and Murthy, D.N.P. (1998). Warranty and redundancy design with uncertain quality, *IIE Transactions*, **30**, 1191–1199.

Hussain, A.Z.M.O. and Murthy, D.N.P. (1999). Warranty and reliability improvement through product development, *Proceedings of the First Western Pacific and Third Australia–Japan Workshop, Stochastic Models in Engineering, Technology and Management*, September, Christchurch, pp. 166–175.

Hussain, A.Z.M.O. and Murthy, D.N.P. (2000). Warranty and optimal redundancy with uncertain quality, *Mathematical and Computer Modelling*, **31**, 175–182.

Hutchinson, T.P. and Lai, C.D. (1990). *Continuous Bivariate Distributions, Emphasising Application*, Rumsby Scientific, Adelaide.

Hwang, H.S. (1996). A reliability prediction model for missile systems based on truncated Weibull distribution. *Computers and Industrial Engineering*, **31**, 245–248.

IEC 50 (191): *International Electrotechnical Vocabulary* (IEV), Chapter 191 – Dependability and Quality of Service, International Electrotechnical Commission, Geneva,1990.

Ishioka, T. and Nonaka, Y. (1991). Maximum likelihood estimation of Weibull parameters for two independent competing risks, *IEEE Transactions on Reliability*, **40**, 71–74.

Iskandar, B.P., Murthy, D.N.P., and Jack, N. (2003). A new repair-replace strategy for items sold with two dimensional warranties, *Computers and Operational Research*, Accepted for publication.

Jack, N. and Murthy, D.N.P. (2001). A servicing strategy for items sold under warranty. *Journal of Operational Research Society*, **52**, 1284–1288.

Jack, N. and Schouten, F.V.D. (2000). Optimal repair-replace strategies for a warranted product. *International Journal of Production Economy*, **67**, 95–100.

Jaisingh, L.R., Dey, D.K., and Griffith, W.S. (1993). Properties of a multivariate survival distribution generated by a Weibull and inverse-Gaussian mixture, *IEEE Transactions on Reliability*, **42**, 618–622.

Janardan, K.G. and Schaeffer, D.J. (1978). Another characterization of the Weibull distribution, *Canadian Journal of Statistics*, **6**, 77–78

Janardan, K.G. and Taneja, V.S. (1979a). Characterization of the Weibull distribution by properties of ordered statistics, *Biometrical Journal*, **21**, 3–9.

Janardan, K.G. and Taneja, V.S. (1979b). Some theorems concerning characterization of the Weibull distribution, *Biometrical Journal*, **21**, 139–144.

Jardine, A.K.S. and Buzacott, J.A. (1985). Equipment reliability and maintenance, *European J. Operational Research*, **19**, 285–296.

Jensen, F. (1995). *Electronic Component Reliability*, Wiley, New York.

Jensen, J. and Petersen, N.E. (1982). *Burn-in: An Engineering Approach to the Design and Analysis of Burn-in Procedures*, Wilcy, New York.

Jiang R. (1996). Failure Models Involving Two Weibull Distributions, Ph.D. Thesis, University of Queensland, Australia.

Jiang, R. and Ji, P. (2001). Characterization of the general limited failure population model, Submitted for publication.

Jiang, S.Y. and Kececioglu, D. (1992a). Graphical representation of 2 mixed-Weibull distribution. *IEEE Transactions on Reliability*, **41**, 241–247.

Jiang, S.Y. and Kececioglu, D. (1992b). Maximum-likelihood-estimated, from censored-data, mixed-Weibull distribution, *IEEE Transactions on Reliability*, **41**, 248–255.

Jiang, R. and Murthy, D.N.P. (1995a). Modeling failure data by mixture of two Weibull distributions: A graphical approach, *IEEE Transactions on Reliability*, **44**, 477–488.

Jiang, R. and Murthy, D.N.P. (1995b). Reliability modeling involving two Weibull distributions, *Reliability Engineering and System Safety*, **47**, 187–198.

Jiang R. and Murthy, D.N.P. (1996). A mixture model involving three Weibull distributions, Proceedings of the Second Australia–Japan Workshop on Stochastic Models in Engineering, Technology and Management (Gold Coast, Australia), pp. 260–270.

Jiang, R. and Murthy, D.N.P. (1997a). Comments on: A general linear-regression analysis applied to the 3-parameter Weibull distribution, *IEEE Transactions on reliability*, **46**, 389–393.

Jiang, R. and Murthy, D.N.P. (1997b). Two sectional models involving three Weibull distributions, *Quality and Reliability Engineering International*, **13**, 83–96.

Jiang, R. and Murthy, D.N.P. (1997c). Parametric study of competing risk model involving two Weibull distributions, *International Journal of Reliability, Quality and Safety Engineering*, **4**, 17–34.

Jiang, R. and Murthy, D.N.P. (1997d). Parametric study of multiplicative model involving two Weibull distributions, *Reliability Engineering and System Safety*, **55**, 217–226.

Jiang, R. and Murthy, D.N.P. (1998). Mixture of Weibull distributions—parametric characterization of failure rate function, *Stochastic Models and Data Analysis*, **14**, 47–65.

Jiang, R. and Murthy, D.N.P. (1999a), The exponentiated Weibull family: A graphical approach. *IEEE Transactions on Reliability*, **48**, 68–72.

Jiang, R. and Murthy, D.N.P. (1999b). Study of n-fold competing risk model, *Proceedings of the First Western Pacific and Third Australia–Japan Workshop on Stochastic Models in Engineering, Technology and Management*, R.J. Wilson, S. Osaki, and M.J. Faddy (eds.), Christchurch, September 23–25, New Zealand.

Jiang, H., Sano, M., and Sekine, M. (1997). Weibull raindrop-size distribution and its application to rain attenuation. *IEE Proceedings—Microwaves, Antennas and Propagation*, **144**, 197–200.

Jiang, R., Zuo, M.J., and Li, H.X. (1999a). Weibull and inverse Weibull mixture models allowing negative weights, *Reliability Engineering and System Safety*, **66**, 227–234.

Jiang, R., Zuo, M., and Murthy, D.N.P. (1999b). Two sectional models involving two Weibull distributions, *International Journal of Reliability, Quality and Safety Engineering*, **6**, 103–122.

Jiang, R., Murthy D.N.P., and Ji, P. (2001a). Models involving two inverse Weibull distributions, *Reliability Engineering and System Safety*, **73**, 73–81.

Jiang, R., Murthy, D.N.P., and Ji, P. (2001b). *n*-fold Weibull Multiplicative Model, *Reliability Engineering and System safety*, **74**, 211–219.

Johnson, N.L. and Kotz, S. (1969a). *Distributions in Statistics: Discrete Distributions*, Vol. 1, Wiley, New York.

Johnson, N.L. and Kotz, S. (1969b). *Distributions in Statistics: Continuous Univariate Distributions*, Vol. I, Wiley, New York.

Johnson, N.L. and Kotz, S. (1970a). *Distributions in Statistics: Continuous Univariate Distributions*, Vol. II, Wiley, New York.

Johnson, N.L. and Kotz, S. (1970b). *Distributions in Statistics: Multivariate Distributions*, Wiley, New York.

Johnson, N.L., Kotz, S., and Balakrishnan, N. (1995). *Continuous Univariate Distributions*. Wiley, New York.

Kabir, A.B.M.Z. (1998). Estimation of Weibull distribution parameters for irregular interval group failure data with unknown failure times, *Journal of Applied Statistics*, **25**, 207–219.

Kaio, N. and Osaki, S. (1982). Optimum repair limit policies with a time constraint, *International Journal of Systems Science*, **13**, 1345–1350.

Kalbfleisch, J.D. and Prentice, R.L. (1980). *The Statistical Analysis of Failure Time Data*, Wiley, New York.

Kalla, S.L., Al-Saqabi, B.N., and Khajah, H.G. (2001). A unified form of gamma-type distributions. *Applied Mathematics and Computation*, **118**, 175–187.

Kanie, H. and Nonaka, Y. (1985). Estimation of Weibull shape-parameters for two independent competing risks, *IEEE Transactions on Reliability*, **34**, 53–56.

Kao, J.H.K. (1959). A graphical estimation of mixed Weibull parameters in life-testing of electron tubes, *Technometrics*, **1**, 389–407.

Kappenman, R.F. (1982). On a method for selecting a distribution model, *Communication in Statistics—Theory and Methods*, **11**, 663–672.

Kappenman, R.F. (1989). A simple method for choosing between the lognormal and Weibull models, *Statistics & Probability Letters*, **7**, 123–126.

Kar, T.R. and Nachlas, J.A. (1997). Coordinated warranty and burn-in strategies, *IEEE Transactions on Reliability*, **46**, 512–518.

Kapur, K.C. and Lamberson, L.R. (1977). *Reliability in Engineering Design*, Wiley, New York.

Keats, J.B., Lawrence, F.P., and Wang, F.K. (1997). Weibull maximum likelihood parameter estimates with censored data, *Journal of Quality Technology*, **29**, 105–110.

Kececioglu, D. (1991). *Reliability Engineering Handbook*, Prentice-Hall, Englewood Cliffs, NJ.

Kececioglu, D. and Sun, F.B. (1994). Mixed-Weibull parameter-estimation for burn-in data using the Bayesian-approach, *Microelectronics and Reliability*, **34**, 1657–1679.

Kececioglu, D.B. and Wang, W. (1998). Parameter estimation for mixed-Weibull distribution, *Annual Reliability and Maintainability Symposium 1998 Proceedings*, pp. 247–252.

Kerscher III, W.J., Lin, T., Stephenson, H., Vannoy, E.H., and Wioskowski, J. (1989). A reliability model for total field incidents, *Proceedings of the Annual Reliability and Maintainability Symposium*, 22–28.

Keshevan, K., Sargent, G., and Conrad, H. (1980). Statistical analysis of the Hertzian fracture of Pyrex glass using the Weibull distribution function, *Journal Material Science*, **15**, 839–844.

Khamis, H.J. (1997a). The corrected Kolmogorov-Smirnov test for the two-parameter Weibull distribution, *Journal of Applied Statistics*, **24**, 301–317.

Khamis, I.H. (1997b). Comparison between constant and step-stress tests for Weibull models, *International Journal of Quality and Reliability Management*, **14**, 74–81.

Khamis, I.H. and Higgins, J.J. (1996). Optimum 3-step step-stress tests, *IEEE Transactions on Reliability*, **45**, 341–345.

Khan, A.H. and Abu-Salih, M.S. (1988). Characterization of the Weibull and the inverse Weibull distributions through conditional moments, *Journal of Information & Optimization Sciences*, **9**, 355–362.

Khan, A.H. and Beg, M.I. (1987). Characterization of the Weibull distribution by conditional variance, *Sankhya*, **49**, 268–271.

Khan, M.S.A., Khalique, A., and Abouammoh, A.M. (1989). On estimating parameters in a discrete Weibull distribution, *IEEE Transactions on Reliability*, **38**, 348–350.

Kies, J.A. (1958). The strength of glass, *Naval Research Lab. Report No. 5093*, Washington D.C.

Kim, K.N. (1998). Optimal burn-in for minimizing cost and multiobjectives, *Microelectronics Reliability*, **38**, 1577–1583.

Kimber, A.C. (1985). Tests for the exponential, Weibull and Gumbel distributions based on stabilised probability plot, *Biometrika*, **72**, 661–663.

King, J.R. (1971). *Probability Charts for Decision Making*, Industrial Press, New York.

Kingston, J.V. and Patel, J.K. (1980a). Selecting the best one of several Weibull populations, *Communications in Statistics—Series A: Theory and Methods*, **A9**, 383–398.

Kingston, J.V. and Patel, J.K. (1980b). A restricted subset selection procedure for Weibull populations, *Communications in Statistics—Series A: Theory and Methods*, **A9**, 1371–1383.

Kingston, J.V. and Patel, J.K. (1981). Interval estimation of the largest reliability of *k* Weibull populations, *Communications in Statistics—Series A: Theory and Methods*, **A10**, 2279–2298.

Klefsjo, B. and Kumar, U. (1992). Goodness-of-fit tests for power-law process based on the TTT-plot, *IEEE Transactions on Reliability*, **41**, 593–598.

Klein, J.P. and Basu, A.P. (1981). Accelerated life tests when there are competing Weibull causes of failure, *Communications in Statistics—Series A: Theory and Methods*, **10**, 2073–2100.

Klein, J.P. and Basu, A.P. (1982). Accelerated life tests under competing Weibull causes of failure, *Communications in Statistics—Series A. Theory and Methods*, **11**, 2271 2286.

Kotani, K., Ishikawa, T., and Tamiya, T. (1997). Simultaneous point and interval predictions in the Weibull distribution. *Statistica*, **7**, 221–235.

Kotz, S. and Nadarajah, S. (2000). *Extreme Value Distributions: Theory and Applications*. Imperial College Press, River Edge, NJ.

Krohn, C.A. (1969). Hazard versus renewal rate of electronic items, *IEEE Transactions on Reliability*, **18**, 64–73.

Kudlayev E.M. (1987). Estimation of the parameters of a Weibull-Gnedenko distribution (a survey), *Soviet Journal of Computer and Systems Sciences*, **25**, 44–57.

Kulasekera, K.B. (1994). Approximate MLE's of the parameters of a discrete Weibull distribution with Type I censored data, *Microelectronics and Reliability*, **34**, 1185–1188.

Kulkarni, A., Satyanarayan, K., and Rohtagi, P. (1973). Weibull analysis of strength of coir fibres, *Fibre Science Technology*, **19**, 56–76.

Kumar, D. and Klefsjo, B. (1994). Proportional hazards model—a review. *Reliability Engineering and Systems Safety*, 44, 177–188.

Kundu, D. and Basu, S. (2000). Analysis of incomplete data in presence of competing risks, *Journal of Statistical Planning and Inference*, **87**, 221–239.

Kunitz, H. (1989). A new class of bathtub-shaped hazard rates and its application in a comparison of two test-statistics, *IEEE Transactions on Reliability*, **38**, 351–353.

Kuo, L. and Tae, Y.Y. (2000). Bayesian reliability modeling for masked system lifetime data, *Statistics and Probability Letters*, **47**, 229–241.

Kuo, L. and Yiannoutsos, C. (1993). Empitical Bayes risk evaluation with Type II censored data, *Journal of Statistical Computation and Simulation*, **48**, 195–206.

Kwon, Y.I. (1996). A Bayesian life test sampling plan for products with Weibull lifetime distribution sold under warranty, *Reliability Engineering and System Safety*, **53**, 61–66.

Kwon, Y.W. and Berner, J. (1994). Analysis of matrix damage evolution in laminated composites, *Engineering Fracture Mechanics*, **48**, 811–817.

Lai, C.D., Moore, T., and Xie, M. (1998). The beta integrated model, *Proceedings of the International Workshop on Reliability Modeling and Analysis – From Theory to Practice*, National University of Singapore, 153–159.

Lai, C.D., Xie, M., and Murthy, D.N.P. (2003). Modified Weibull model, *IEEE Transactions on Reliability*, **52**, 33–37.

Lakey, M.J. and Rigdon, S.E. (1993). Reliability improvement using experimental design, *1993—ASQC Quality Congress, Transactions*, Boston, pp. 559–563.

Lamon, J. (1990). Ceramics reliability. Statistical analysis of multiaxial failure using the Weibull approach and the multiaxial elemental strength model, *Journal of the American Ceramic Society*, **73**, 2204–2212.

Landes, J.D. (1993). Two criteria statistical model for transition fracture toughness, *Fatigue and Fracture of Engineering Materials & Structures*, **16**, 1161–1174.

Landers, T.L. and Soroudi, H.E. (1991). Robustness of a semi-parametric proportional intensity model. *IEEE Transactions on Reliability*, **40**, 161–164.

Lawless, J.F. (1973). On the estimation of safe life when the underlying life distribution is Weibull, *Technometrics*, **15**, 857–865.

Lawless, J. F. (1978). Confidence interval estimation for the Weibull and extreme value distributions (with a discussion by R. G. Easterling), *Technometrics*, **20**, 355–365.

Lawless, J.F. (1982). *Statistical Models and Methods for Lifetime Data*, Wiley, New York.

Lawless, J.F. (1987). Regression methods for Poisson process data. *Journal of the American Statistical Association*, **82**, 808–815.

Lawless, J.F. and Thiagarajah, K. (1996). A point-process model incorporating renewals and time trends, with application to repairable systems. *Technometrics*, **38**, 131–138.

Lee, L. (1979). Multivariate distributions having Weibull properties, *Journal of Multivariate Analysis*, **9**, 267–277.

Lee, L. and Thompson, W.A. (1974). Results on failure time and pattern for the series system, In *Reliability and Biometry, Statistical Analysis of Life Length*, SIAM, Philadelphia, pp. 291–302.

Leemis, L.M. (1995). *Reliability: Probabilistic Models and Statistical Methods*, Prentice-Hall, Englewood Cliffs, NJ.

Lemon, G.H. (1975). Maximum likelihood estimation for the three parameter Weibull distribution based on censored samples, *Technometrics*, **17**, 247–254.

Leonard, T., Hsu, J.S.J., and Tsui, K.W. (1994). Bayesian and likelihood inference from equally weighted mictures, *Annals of the Institute of Statistical Mathematics*, **46**, 203–220.

Li, Y.M. (1994). A general linear-regression analysis applied to the 3-parameter Weibull distribution, *IEEE Transactions on Reliability*, **43**, 255–263.

Li, G. and Doss, H. (1993). Generalized Pearson-Fisher chi-Square goodness-of-fit tests, with applications to models with life history data, *Annals of Statistics*, **21**, 772–797.

Liao, M. and Shimokawa, T. (1999). A new goodness-of-fit test for type-I extreme-value and 2-parameter Weibull distributions with estimated parameters, *Journal of Statistical Computation and Simulation*, **64**, 23–48.

Lieblein, J. (1955). On moments of ordered statistics from the Weibull distribution, *Annals of the Mathematical Statistics*, **34**, 633–651.

Lieblein, J. and Zelen, M. (1956). Statistical investigation of the fatigue life of ball bearings, *Journal of Research of National Bureau of Standards*, **57**, 273–315.

Liertz, H. and Oestreich, U. (1977). Mechanical properties of optical waveguides, *NTG – Fachberichte*, **59**, 93–95.

Lin, D.K.J., Usher, J.J.S., and Guess, F.M. (1996). Bayes estimation of component-reliability from masked system-life data, *IEEE Transactions on Reliability*, **45**, 233–237.

Ling, J. and Pan, J. (1998). An optimization method of parameter estimation for mixed Weibull distributions, Reliability and Quality in Design, Proceedings of the Fourth ISSAT International Conference, August 12–14, 1998, Seattle, Washington, pp. 237–243.

Littell, R.C., McClave, J.T., and Offen, W.W. (1979). Goodness-of-fit tests for the two parameter Weibull distribution, *Communication in Statistics—Computation and Simulation*, **8**, 257–269.

Lochner, R.H., Basu, A.P., and DiPonzio, M. (1974). Covariance of Weibull order statistics, *Naval Research Logistics Quarterly*, **21**, 543–547.

Lockhart, R.A. and Stephens, M.A. (1994). Estimation and tests of fit for the 3-parameter Weibull distribution. *Journal of the Royal Statistical Society—B*, **56**, 491–500.

Lomnicki, Z.A. (1966). A note on the Weibull renewal process, *Biometrika*, **53**, 375–381.

Love, C.E. and Guo, R. (1991). Application of Weibull proportional hazards modelling to bad-as-old failure data, *Quality and Reliability Engineering International*, **7**, 149–157.

Love, C.E. and Guo, R. (1996). Utilizing Weibull failure rates in repair limit analysis for equipment replacement/preventive maintenance decisions, *Journal of the Operational Research Society*, **47**, 1366–1376.

Lu, J.C. (1989). Weibull extensions of the Freund and Marshall-Olkin bivariate exponential models, *IEEE Transactions on Reliability*, **38**, 615–619.

Lu, J.C. and Bhattacharyya, C.K. (1990). Some new constructions of bivariate Weibull models, *Annals of the Institute of Statistical Mathematics*, **42**, 543–559.

Lun, I.Y.F. and Lam, J.C. (2000). A study of Weibull parameters using long-term wind observations, *Renewable Energy*, **20**, 145–153.

Luxhoj, J. T. and Shyur, H. J. (1995). Reliability curve fitting for aging helicopter components, *Reliability Engineering and System Safety*, **48**, 229–234.

Majeske, K.D and Herrin, G.D. (1995). Assessing mixture-model goodness of fit with an application to automobile warranty data, *Proc. Annual Reliability and Maintainability Symposium*, 378–383.

Majety V.S., Dawande, M., and Rajagopal, J. (1999). Optimal reliability allocation with discrete cost-reliability data for components, *Operations Research*, **47**, 899–906.

Makino, T. (1984). Mean hazard rate and its application to the normal approximation of the Weibull distribution, *Naval Research Logistics Quarterly*, **31**, 1–8.

Malik, H.J. and Trudel, R. (1982). Probability density function of quotient of order statistics from the Pareto, power and Weibull distributions, *Communications in Statistics—Series A: Theory and Methods*, **11**, 801–814.

Mann, N.R (1968). Point and interval estimation procedures for the two-parameter Weibull and extreme-value distributions, *Technometrics*, **10**, 231–256.

Mann, N.R. (1969). Exact third-order-statistics confidence bounds on reliable life for a Weibull model with progressive censoring, *Journal of the American Statistical Association*, **64**, 306–315.

Mann, N.R. (1971). Best linear invariant estimation for Weibull parameters under progressive censoring, *Technometrics*, **13**, 521–533.

Mann, N.R. and Fertig, K.W. (1975). Simplified efficient point and interval estimators for Weibull parameters, *Technometrics*, **17**, 361–368.

Mann, N.R., Schafer, R.E., and Singpurwalla, N.D. (1974). *Methods for Statistical Analysis of Reliability and Life Data*, Wiley, New York.

Marohn, F. (1994). Asymptotic sufficiency of order statistics for almost regular Weibull type densities, *Statist. Decisions*, **12**, 385–393.

Marshall, A.W. and Olkin, I. (1967). A multivariate exponential distribution, *Journal of the American Statistical Association*, **62**, 30–44.

Marshall, A.W. and Olkin, I. (1997). A new method for adding a parameter to a family of distributions with application to the exponential and Weibull families, *Biometrika*, **84**, 641–652.

Martinez, S. and Quintana, F. (1991). On a test for generalized upper truncated Weibull distribution. *Statistics and Probability Letters*, **12**, 273–279.

Martz, H.F. and Waller, R.A. (1982). *Bayesian Reliability Analysis*, Wiley, New York.

Massey, F.J. (1951). The Kolmogorov-Smirnov test for goodness of fit, *Journal of the American Statistical Association*, **46**, 68–78.

Mau, S.T. and Chow, T. (1980). Simple model for typhoon risk analysis of a site, *Proceedings of the Southeast Asian Conference on Soil Engineering*, 247–258.

Mazzuchi, T.A. and Soyer, R. (1996). A Bayesian perspective on some replacement strategies, *Reliability Engineering and System Safety*, **51**, 295–303.

McCall, J.J. (1965). Maintenance policies for stochastically failing equipment: A survey, *Management Science*, **11**, 493–524.

McCool, J.I. (1970a). Inference on Weibull percentiles and shape parameter from maximum likelihood estimates, *IEEE Transactions on Reliability*, **19**, 2–9.

McCool, J.I. (1970b). Inference on Weibull percentiles from sudden death tests using maximum likelihood, *IEEE Transactions on Reliability*, **19**, 177–179.

McCool, J.I. (1976). Estimation of Weibull parameters with competing-mode censoring, *IEEE Transactions on Reliability*, **25**, 25–31.

McCool, J.I (1978). Competing risk and multiple comparison analysis for bearing fatigue tests, *ASLE Transactions*, **21**, 271–284.

McCool, J.I. (1998). Inference on the Weibull location parameter, *Journal of Quality Technology*, **30**, 119–126.

McEwen, R.P. and Parresol, B.R. (1991). Moment expressions and summary statistics for the complete and truncated Weibull distribution, *Communications in Statistics—Series A: Theory and Methods*, **20**, 1361–1372.

McNolty, F. (1980). Properties of the mixed exponential failure process, *Technometrics*, **22**, 555–565.

McSorely, E.O., Lu, J.C., and Li, C.S. (2002). Performance of parameter estimates in step-stress accelerated life tests with various sampling sizes, *IEEE Transactions on Reliability*, **51**, 271–278.

Meeker, W.Q. (1984). A comparison of accelerated life test plans for Weibull and lognormal distributions and type I censoring, *Technometrics*, **26**, 157–171.

Meeker, W.Q. and Escobar, L.A. (1993). A review of recent research and current issues in accelerated testing, *International Statistical Review*, **61**, 147–168.

Meeker, W.Q. and Escobar, L.A. (1998a). Pitfalls of accelerated testing, *IEEE Transactions on Reliability*, **47**, 114–118.

Meeker, W.Q. and Escobar, L.A. (1998b). *Statistical Methods for Reliability Data*, Wiley, New York.

Meeker, W.Q. and Nelson, W. (1975). Optimum accelerated life-tests for Weibull and extreme value distributions, *IEEE Transactions on Reliability*, **24**, 321–332.

Mendenhall, W. and Hader, R.J. (1958). Estimation of parameters of mixed exponentially distributed failure time distributions from censored life test data. *Biometrika*, **45**, 504–520.

Menon, M.V. (1963). Estimation of the shape and scale parameters of the Weibull distribution, *Technometrics*, **5**, 175–182.

Menon, A.Z. and Daghel, M.M. (1987). Some studies of order statistics from a Weibull distribution. *Pakistan Journal of Statistics*, **3**, 123–130.

Mettas, A. (2000). Reliability allocation and optimization for complex systems, *Proceedings of the Annual Reliability and Maintainability Symposium*, IEEE, New York, pp. 216–221.

Mi, J. (1997). Warranty policies and burn-in, *Naval Research Logistics Quarterly*, **44**, 199–209.

Michael, J. R. (1983). The stabilized probability plot. *Biometrika*, **70**, 11–17.

Miller, G. (1984). Inference on a future reliability parameter with the Weibull process model, *Naval Research Logistics Quarterly*, **31**, 91–96.

Mine, H. and Nakagawa, T. (1978). Age replacement model with mixed failure times, *IEEE Transactions on Reliability*, **27**, 173.

Mita, N. (1995). Improving the reliability of space-use traveling wave tubes and confirming the improvement, *Electronics & Communications in Japan, Part II: Electronics (English translation of Denshi Tsushin Gakkai Ronbunshi)*, **78**, 100–105.

Mittal, M.M. and Dahiya, R.C. (1989). Estimating the parameters of a truncated Weibull distribution. *Communications in Statistics—Series A: Theory and Methods*, **18**, 2027–2042.

Moen, R.D., Nolan, T.W., and Provost, L.P. (1991). *Improving Quality Through Planned Experimentation*, McGraw Hill, New York.

Moeschberger, M.L. (1974). Life tests under dependent competing causes of failure, *Technometrics*, **16**, 39–47.

Mohie El-Din, M.M., Mahmoud, M.A.W., and Abo Youssef, S.E. (1991). Moments of order statistics from parabolic and skewed distributions and a characterization of Weibull distribution, *Communications in Statistics—Series B: Simulation and Computation*, **20**, 639–645.

Molitor, J.T. and Rigdon, S.E. (1993). A comparison of estimators of the shape parameter of the power law process, *Proceedings of the Section on Physical and Engineering Sciences*, American Statistical Association, Alexandria, VA, pp. 107–115.

Moller, S.K. (1976). The Rasch-Weibull process, *Scandinavian Journal of Statistics*, **3**, 107–115.

Moltoft, J. (1983). Behind the "bathtub" curve, a new model and its consequences, *Microelectronics & Reliability*, **23**, 489–500.

Monahan, G.E. (1982). A survey of partially observable Markov decision processes: Theory, models and algorithms, *Management Science*, **28**, 1–16.

Montanari, G.C., Mazzanti, G., Cacciari, M., and Fothergill, J.C. (1997). In search of convenient techniques for reducing bias in the estimation of Weibull parameters for uncensored tests, *IEEE Transactions on Dielectrics and Electrical Insulation*, **4**, 306–313.

Mood, A.M., Graybill, F.A., and Boes, D.C. (1974). *Introduction to the Theory of Statistics*, 3rd ed., McGraw-Hill, New York.

Mu, F.C., Tan, C.H., and Xu, M.Z. (2000). Proportional difference estimate method of determining the characteristic parameters of monomodal and multimodal Weibull distributions of time-dependent dielectric breakdown, *Solid-State Electronics*, **44**, 1419–1424.

Mudholkar, G.S. and Hutson, A.D. (1996). The exponentiated Weibull family: Some properties and a flood data application, *Communications in Statistics—Series A: Theory and Methods*, **25**, 3059–3083.

Mudholkar, G.S. and Kollia, G.D. (1994). Generalized Weibull family: A structural analysis, *Communications in Statistics—Series A: Theory and Methods*, **23**, 1149–1171.

Mudholkar, G.S. and Srivastava, D.K. (1993). Exponentiated Weibull family for analyzing bathtub failure-rate data, *IEEE Transactions on Reliability*, **42**, 299–302.

Mudholkar, G.S., Srivastava, D.K., and Freimer, M. (1995). The exponentiated Weibull family—a reanalysis of the bus-motor-failure data, *Technometrics*, **37**, 436–445.

Mudholkar, G.S., Srivastava, D.K., and Kollia, G.D. (1996). A generalization of the Weibull distribution with application to the analysis of survival data, *Journal of the American Statistical Association*, **91**, 1575–1583.

Mukherjee, S.P. and Roy, D. (1993). A versatile failure time distribution, *Microelectronics and Reliability*, **33**, 1053–1056.

Murthy, D.N.P. (1996). Warranty and design, in *Product Warranty Handbook*, W.R. Blischke and D.N.P. Murthy (eds.), Marcel Dekker, New York.

Murthy, D.N.P. and Hussain, A.Z.M.O. (1994). Warranty and optimal redundancy design, *Engineering Optimization*, **23**, 301–314.

Murthy, D.N.P. and Jiang, R. (1997). Parametric study of sectional models involving two Weibull distributions, *Reliability Engineering and System Safety*, **56**, 151–159.

Murthy, D.N.P. and Nguyen, D.G. (1987). Optimal development testing policies for products sold under warranty. *Reliability Engineering*, **19**, 113–123.

Murthy, D.N.P., Djamaludin, I., and Wilson, R.J. (1993). Product warranty and quality control, *Quality and Reliability Engineering*, **9**, 431–443.

Nagode, M. and Fajdiga, M. (1999). The influence of variable operating conditions upon the general multi-modal Weibull distribution, *Reliability Engineering and System Safety*, **64**, 383–389.

Nagode, M. and Fajdiga, M. (2000). An improved algorithm for parameter estimation suitable for mixed Weibull distributions, *International Journal of Fatigue*, **22**, 75–80.

Nakagawa, T. and Osaki, S. (1974). Some aspects of damage models, *Microelectronics and reliability*, **13**, 253–257.

Nakagawa, T. and Osaki, S. (1975). The discrete Weibull distribution, *IEEE Transactions on Reliability*, **24**, 300–301.

Natesan, J. and Jardine, A.K.S. (1986). Graphical estimation of mixed Weibull parameters for ungrouped multicensored data: Its application to failure data, *Maintenance Management International*, **6**, 115–127.

Nelson, P.R. (1979). Control chart for Weibull process, *IEEE Transactions on Reliability*, **28**, 283–288.

Nelson, W. (1982). *Applied Life Data Analysis*, Wiley, New York.

Nelson, W. (1985). Weibull analysis of reliability data with few or no failures, *Journal of Quality Technology*, **17**, 140–146.

Nelson, W. (1990). *Accelerated Life Testing: Statistical Models, Test Plans and Data Analyses*, Wiley, New York.

Nelson, W. (2000). Weibull prediction of a future number of failures, *Quality and Reliability Engineering International*, **16**, 23–26.

Nelson, W. and Meeker, W.Q. (1978). Theory of optimum accelerated life tests for the Weibull and extreme value distributions, *Technometrics*, **20**, 171–177.

Neshyba, S. (1980). On the size distribution of Antarctic icebergs, *Cold Regions Science and Technology*, **1**, 241–248.

Newby, M. (1994). Perspective on Weibull proportional-hazards models, *IEEE Transactions on Reliability*, **43**, 217–223.

Nieuwhof, G.W.E. (1984). The concept of failure in reliability engineering, *Reliability Engineering*, **7**, 53–59.

Nigm, A. (1990). Prediction bounds using type I censoring for the Weibull distribution, *Statistics*, **21**, 227–237.

Nigm, E.M. and El-Hawary, H.M. (1996). On order statistics from the Weibull distribution, *Egyptian Statistical Journal*, **40**, 80–92.

Nordman, D.J. and Meeker, W.Q. (2002). Weibull prediction intervals for a future number of failures, *Technometrics*, **44**, 15–23.

Nossier, A., El-Debeiky, S., Hashad, I., and A-Bary, H. (1980). Effect of temperature on the breakdown property of liquid dielectrics, *IEEE Transactions on Electrical Insulation*, **EI-15**, 502–505.

Ochiai, S. and Osamura, K. (1987). Tensile strength of fibres containing two types of flaws and its influence on the tensile behavior of metal matrix composites, *Zeitschrift fuer Metallkunde*, **78**, 525–532.

O'Connor, D.T.P. (1995). *Practical Reliability Engineering*, Heyden & Son, London.

O'Connor, P.D.T. (1995). *Practical Reliability Engineering*, Wiley, New York.

Padgett, W.J. and Spurrier, J.D. (1985). Discrete failure models, *IEEE Transactions on Reliability*, **34**, 253–256.

Padgett, W.J., Durham, S.D., and Mason, A.M. (1995). Weibull analysis of the strength of carbon fibers using linear and power law models for the length effects, *Journal of Composite Materials*, **29**, 1873–1884.

Pagonis, V., Mian, S.M., and Kitis, G. (2001). Fit of first order thermoluminescence glow peaks using the Weibull distribution function, *Radiation Protection Dosimetry*, **93**, 11–17.

Pan, J.N. (1998). Reliability prediction of imperfect switching systems subject to Weibull failures, *Computers and Industrial Engineering*, **34**, 481–492.

Pandey, M. and Upadhyay, S.K. (1988). Bayesian inference in mixtures of two exponentials, *Microelectronics and Reliability*, **28**, 217–221.

Papadopoulos, A.S. and Hoekstra, R.M. (1995). Some robustness issues in the Bayesian analysis of a Weibull failure model, *Journal of Statistical Computation and Simulation*, **52**, 85–94.

Papadopoulos, A.S. and Tsokos, C.P. (1975a). Bayesian confidence bounds for the Weibull failure model, *IEEE Transactions on Reliability*, **24**, 21–26.

Papadopoulos, A.S. and Tsokos, C. P. (1975b). Bayesian analysis of the Weibull failure model with unknown scale and shape parameters, *Statistica*, **36**, 547–560.

Park, W.J. and Seoh, M. (1994). More goodness of fit tests for the power law process, *IEEE Transactions on Reliability*, **43**, 275–278.

Parker, R.A. (1977). Optimizing vehicle inspection resources using a stochastic model of vehicle degradation, *Journal of Safety Research*, **9**, 59–67.

Parr, V.B. and Webster, J.T. (1965). A method for discriminating between failure density functions used in reliability predictions. *Technometrics*, **7**, 1–10.

Patel, J.K. (1975). Bounds on moments of linear functions of order statistics from Weibull and other restricted families, *Journal of the American Statistical Association*, **70**, 670–673.

Patel, J.K. and Read, C.B. (1975). Bounds on conditional moments of Weibull and other monotone failure rate families, *Journal of the American Statistical Association*, **70**, 238–244.

Patra, K. and Dey, D.K. (1999). A multivariate mixture of Weibull distributions in reliability modeling. *Statistics and Probability Letters*, **45**, 225–235.

Pearson, K. (1965). *Tables of the Incomplete Gamma-Function*, Cambridge University Press, Cambridge.

Pearson, E.S. and Hartley, H.O. (1972). *Biometrika Tables for Statisticians*, Vol. II, Cambridge University Press, London.

Perlstein, D. and Welch, R.L.W. (1993). A Bayesian approach to the analysis of burn-in of mixed populations, *Proceedings 1993 Annual Reliability and Maintainability Symposium*, 417–421.

Percy, D.F. and Kobbacy, K.A.H. (1996). Preventive maintenance modelling—A Bayesian perspective, *Journal of Quality and Maintenance Engineering*, **2**, 15–24.

Percy, D.F., Bouamra, O., and Kobbacy, K.A.H. (1998). Bayesian analysis of fixed interval preventive maintenance models, *IMA Journal of Mathematics Applied in Business and Industry*, **9**, 157–175.

Peto, R. and Lee, P.N. (1973). Weibull distributions for continuous carcinogenesis experiments, *Biometrics*, **29**, 457–470.

Pham, T. and Almhana, J. (1995). The generalized Gamma-distribution—its hazard rate and stress-strength model, *IEEE Transactions on Reliability*, 44, 392–297.

Pham, H. and Phan, H.K. (1992). Mean life of k-to-1-out-of-n systems with non-identical Weibull distribution components, *Microelectronics and Reliability*, **32**, 845–850.

Phani, K.K. (1987). A new modified Weibull distribution function, *Communications of the American Ceramic Society*, **70**, 182–184.

Pierskalla, W.P. and Voelker, J.A. (1976). A survey of maintenance models: The control and surveillance of deteriorating systems, *Naval Research Logistics Quarterly*, **23**, 353–388.

Pike, M.C. (1966). A method of analysis of a certain class of experiments in carcinogenesis, *Biometrics*, **22**, 142–161.

Pintelton, L.M. and Gelders, L. (1992). Maintenance management decision making, *European Journal of Operations Research*, **58**, 301–317.

Pohl, E.A. and Dietrich, D.L. (1995). Environmental stress screening strategies for complex systems: A 3-level mixed distribution model, *Microelectronics and Reliability*, **35**, 637–656.

Polatoglu, H. and Sahin, I. (1998). Probability distributions of cost, revenue and profit over a warranty cycle. *European Journal of Operational Research*, **108**, 170–183.

Prentice, R.L., Williams, B.J., and Peterson, A.V. (1981). On the regression analysis of multivariate failure time data, *Biometrika*, **68**, 373–379.

Proschan, F. (1963). Theoretical explanation of observed decreasing failure rate, *Technometrics*, **5**, 375–383.

Qiao, H.Z. and. Tsokos, C.P. (1994). Parameter estimation of the Weibull probability distribution, *Mathematics & Computers in Simulation*, **37**, 47–55.

Qiao, H.Z. and Tsokos, C.P. (1995). Estimation of the three parameter Weibull probability distribution. *Mathematics & Computers in Simulation*, **39**, 173–185.

Quereshi, F.S. and Sheikh, A.K. (1997). Probabilistic characterization of adhesive wear in metals, *IEEE Transactions on Reliability*, **46**, 38–44.

Quesenberry, C.P. and Kent, J. (1982). Selecting among probability distributions used in reliability, *Technometrics*, **24**, 59–65.

Rad, P. and Olson. R. (1974). Tunneling machine research: Size distribution of rock fragments produced by rolling disk cutters, Bureau of Mines Report No. BUMINES-RI-78-82, Washington D.C.

Radcliffe, S.J. and Parry, A.A. (1979). Dispersion of life of bonded mos//2 solid lubricant coatings, *Wear*, **56**, 203–212.

Rademaker, A.W. and Antle, C.E. (1975). Sample size for selecting the better of two Weibull populations, *IEEE Transactions on Reliability*, **24**, 17–20.

Rahim, M.A. (1993). Economic design of $\bar{X}$ charts assuming Weibull in-control times, *Journal of Quality Technology*, **25**, 296–305.

Rao, B.R. and Talwalker, S. (1989). Bounds on life expectancy for the Rayleigh and Weibull distributions, *Mathematical Biosciences*, **96**, 95–115.

Rao, A.V., Kantam, R.R.L., and Narasimham, V.L. (1991). Linear estimation of location and scale parameters in the generalized gamma distribution. *Communications in Statistics—Series A: Theory and Methods*, **20**, 3823–3848.

Rao, A.V., Rao, A.V.D., and Narasimham, V.L. (1994). Asymptotically optimal grouping for maximum likelihood estimation of Weibull parameters, *Communications in Statistics—Series B: Simulation and Computation*, **23**, 1077–1096.

Reilly, P.M. (1976). The numerical computation of posterior distribution in Bayesian statistical inference, *Applied Statistics*, **25**, 201–209.

Rider, P.R. (1961). Estimating parameters of mixed Poisson, binomial, and Weibull distributions by the method of moments, *Bulletin International Statistical Institute*, **39**, 225–232.

Rigdon, S. (1989). Testing goodness-of-fit for the power law process, *Communications in Statistics—Series A: Theory and Methods*, **A-18**, 4665–4676.

Rigdon, S.E. and Basu, A.P. (1988). Estimating the intensity function of a Weibull Poisson process at the current time: Failure truncated case, *Journal of Statistical Computation and Simulation*, **30**, 17–38.

Rigdon S.E. and Basu, A.P. (1989). The power law process: A model for the reliability of repairable system, *Journal of Quality Technology*, **21**, 251–260.

Rigdon, S.E. and Basu, A.P. (1999). Estimating the intensity function of a power law process at the current time: Time truncated case, *Communications in Statistics—Series B: Computation and Simulation*, **B-19**, 1079–1104.

Rigdon, S.E. and Basu, A.P. (2000), *Statistical Methods for the Reliability of Repairable Systems*, Wiley, New York.

Rosen, P. and Rammler. B. (1933). The laws governing the fineness of powdered coal, *Journal of the Institute of Fuels*, **6**, 246–249.

Ross, R. (1996). Bias and standard deviation due to Weibull parameter estimation for small data sets, *IEEE Transactions on Dielectrics and Electrical Insulation*, **3**, 28–42.

Ross, S.M. (1970). *Applied Probability Models*, Prentice-Hall, New York.

Ross, S.M. (1980). *Stochastic Processes*, Wiley, New York.

Ross, S.M. (1983). *Introduction to Probability Models*, Wiley, New York.

Ray, D. and Gupta, R.P. (1992). Classifications of discrete lives, *Microelectronics and Reliability*, **32**, 1459–1473.

Roy, D. (1994). Classification of life distributions in multivariate models, *IEEE Transactions on Reliability*, **43**, 224–229.

Roy, D. and Gupta, R.P. (1992). Classifications of discrete lives, *Microelectronics and Reliability*, **32**, 1459–1473.

Roy, D. and Mukherjee, S.P. (1986). A note on characterization of the Weibull the distribution, *Sankhya*, **48**, 250–253.

Roy, D. and Mukherjee, S.P. (1988). Multivariate extensions of univariate life distributions, *Journal of Multivariate Analysis*, **67**, 72–79.

Salvia, A.A. (1996). Some results on discrete mean residual life, *IEEE Transactions on Reliability*, **35**, 359–361.

Sahin, I. and Potaloglu, H. (1998). *Quality, Warranty and Preventive Maintenance*, Kluwer Pub, Boston.

Salem, S.A. (1988). Bayesian estimation based on mixed Weibull and multicensored samples. *The Journal of Natural Sciences and Mathematics*, **28**, 127–140.

Sarkar, S.K. (1987). A continuous bivariate exponential distribution, *Journal of the American Statistical Association*, **82**, 667–675.

Scarf, P.S. (1997). On the application of mathematical models to maintenance, *European Journal of Operations Research*, **63**, 493–506.

Schifani, R. and Candela R. (1999). A new algorithm for mixed Weibull analysis of partial discharge amplitude distributions, *IEEE Transactions on Dielectrics and Electrical Insulation*, **6**(2), 242–249.

Schilling, E.G. (1982). *Acceptance Sampling in Quality Control*, Marcel Dekker, New York.

Schmid, U. (1997). Percentile estimators for the three-parameter Weibull distribution for use when all parameters are unknown, *Communications in Statistics—Series A: Theory and Methods*, **26**, 765–785.

Schneider, H. (1989). Failure-censored variable-sampling plans for log-normal and Weibull distributions, *Technometrics*, **31**, 199–206.

Schneider, H. and Weissfield, L.A. (1989). Interval estimation based on censored samples from the Weibull distribution, *Journal of Quality Technology*, **21**, 179–186.

Scholz, F.W. (1990). Characterization of the Weibull distribution, *Computational Statistics & Data Analysis*, **10**, 289–292.

Seguro, J.V. and Lambert, T.W. (2000). Modern estimation of the parameters of the Weibull wind speed distribution for wind energy analysis, *Journal of Wind Engineering and Industrial Aerodynamics*, **85**, 75–84.

Seki, T. and Yokoyama, S. (1993). Simple and robust estimation of the Weibull parameters, *Microelectronics and Reliability*, **33**, 45–52.

Seki, T. and Yokoyama, S. (1996). Robust parameter-estimation using the bootstrap method for the 2-parameter Weibull distribution, *IEEE Transactions on Reliability*, **45**, 34–51.

Seo, S.-K. and Yum, B.-J. (1991). Accelerated life test plans under intermittent inspection and Type-I censoring: The case of Weibull failure distribution, *Naval Research Logistics*, **38**, 1–22.

Seshadri, V. (1968). A characterisation of the normal and Weibull distribution, *Canadian Mathematical Bulletin*, **12**, 257–260.

Shaked, M. and Shanthikumar, J.G. (1994). *Stochastic Orders and Their Applications*, Academic, San Diego.

Shalaby, O.A. (1993). The Bayes risk for the parameters of doubly truncated Weibull distribution. *Microelectronics and Reliability*, **33**, 2189–2192.

Shalaby, O.A. and Elyousef, E.L. (1993). Bayesian analysis of the parameters of a doubly truncated Weibull distribution. *Microelectronics and Reliability*, **33**, 1199–1211.

Shapiro, S.S. and Brain, C.W. (1987). W-test for the Weibull distribution. *Communications in Statistics—Series B: Simulation and Computation*, **16**, 209–219.

Sharif, H. and Islam, M. (1980). Weibull distribution as a general model for forecasting technological change, *Technology Forecasting and Social Change*, **18**, 247–256.

Sheikh, A.K., Boah, J.K., and Hansen, D.A. (1990). Statistical modelling of pitting corrosion and pipeline reliability, *Corrosion*, **46**, 190–196.

Sherif, Y.S. and Smith, M.L. (1976). Optimal maintenance models for systems subject to failure—A review, *Naval Logistics Research Quarterly*, **23**, 47–74.

Sherif, Y. S. and Smith, M. L. (1976), Optimal maintenance models for systems subject to failure – A review, *Naval Logistics Research Quarterly* **23**:47–74.

Shimizu, R. and Davies, L. (1981). General characterization theorems for the Weibull and the stable distributions, *Sankhya*, **43**, 282–310.

Shimokawa, T. and Liao, M. (1999). Goodness-of-fit tests for type-I extreme-value and 2-parameter Weibull distributions, *IEEE Transactions on Reliability*, **48**, 79–86.

Shooman, M.L. (1968). *Probabilistic Reliability: An Engineering Approach*, McGraw-Hill, New York.

Shyur, H.J., Elsayed, E.A., and Luxhoj, J.T. (1999). A general model for accelerated life testing with time-dependent covariates, *Naval Research Logistics*, **46**, 303–321.

Sichart, K.V. and Vollertsen, R.-P. (1991). Bimodal lifetime distributions of dielectrics for integrated circuits, *Quality and Reliability Engineering International*, **7**, 299–305.

Sinha, S.K. and Sloan , J.A. (1988). Bayes estimation of the parameters and reliability function of the three-parameter Weibull distribution, *IEEE Transactions on Reliability*, **37**, 364–369.

Singh, N.K. and Bhattacharya, S.K. (1993). Bayesian analysis of the mixture of exponential failure distributions, *Microelectronics and Reliability*, **33**, 1113–1117.

Singpurwalla, N.D. (1988). An interactive PC-based procedure for reliability assessment incorporating expert opinion and survival data, *Journal of the American Statistical Association*, **83**, 43–51.

Singpurwalla, N.D. and Song, M.S. (1988). Reliability analysis using Weibull lifetime data and expert opinion, *IEEE Transactions on Reliability*, **37**, 340–347.

Sinha, S.K. (1987). Bayesian estimation of the parameters and reliability function of a mixture of Weibull life distributions, *Journal of Statistical Planning and Inference*, **16**, 377–387.

Sinha, S.K. and Kale, B.K (1980). *Life Testing and Reliability Estimation*, Wiley Eastern, New Delhi.

Sirvanci, M. (1986). Comparison of two Weibull distributions under random censoring. *Communications in Statistics—Series A: Theory and Methods*, **15**, 1819–1836.

Sirvanci, M. and Yang, G. (1984). Estimation of the Weibull parameters under type I censoring, *Journal of the American Statistical Association*, **79**, 183–187.

Slymen, D.J. and Lachenbruch, P.A. (1984). Survival distributions arising from two families and generated by transformations, *Communications in Statistics—Series A: Theory and Methods*, **13**, 1179–1201.

Smith, R.M. and Bain, L.J. (1975). An exponential power life-testing distribution. *Communications in Statistics—Theory and Methods*, **4**, 469–481.

Smith, F. and Hoeppner, D.W. (1990). Use of the four parameter Weibull function for fitting fatigue and compliance calibration data, *Engineering Fracture Mechanics*, **36**, 173–178.

Smith, W.L. and Leadbetter, M.R. (1963). On the renewal function for the Weibull distribution, *Technometrics*, **5**, 393–396.

Smith, R.L. and Naylor, J.C. (1987). A comparison of maximum likelihood estimators and Bayesian estimators for the three-parameter Weibull distribution, *Applied Statistics*, **36**, 359–369.

Soland, R.M. (1968). Bayesian analysis of the Weibull process with unknown scale parameter and its application to acceptance sampling, *IEEE Transactions on Reliability*, **17**, 84–90.

Soland, R.M. (1969). Bayesian analysis of the Weibull process with unknown scale and shape parameters, *IEEE Transactions on Reliability*, **18**, 181–184.

Soman, K.P. and Misra, K.B. (1992). A least square estimation of 3 parameters of a Weibull distribution, *Microelectronics and Reliability*, **32**, 202–305.

Spearman, M.L. (1989). A simple approximation for IFR Weibull renewal functions, *Microelectronics and Reliability*, **29**, 73–80.

Spivey, L.B. and Gross, A.J. (1991). Concomitant information in competing risk analysis, *Biometrical Journal*, **33**, 419–427.

Srivastava, J. (1989). Advances in the statistical theory of comparison of lifetimes of machines under the generalized Weibull distribution, *Communications in Statistics—Series A: Theory and Methods*, **18**, 1031–1045.

Srivastava, T.N. (1974). Life tests with periodic change in the scale parameter of a Weibull distribution, *IEEE Transactions on Reliability*, **23**, 115–118.

Stacy, E.W. (1962). A generalization of the gamma distribution. *Annals of the Mathematical Statistics*, **33**, 1187–1192.

Stacy, E.W. and Mihram. G.A. (1965). Parameter estimation for a generalised gamma distribution, *Technometrics*, **7**, 349–358.

Stein, W.E. and Dattero, R. (1984). A new discrete Weibull distribution, *IEEE Transactions on Reliability*, **33**, 196–197.

Stephens, M.A. (1974). EDF statistics for goodness-of-fit and some comparisons, *Journal of the American Statistical Association*, **69**, 730–737.

Stone, D.C. and Rosen, G.C. (1984). Some graphical techniques for estimating Weibull confidence intervals, *IEEE Transactions on Reliability*, **R-33**, 362–369.

Stuart, A. and Ord, J.K. (1991). *Kendall's Advanced Theory of Statistics*, Vol. 2, 5th ed., Oxford University Press, New York.

Tadikamalla, P.R. (1978). Application of the Weibull distribution in inventory control, *Journal of Operational Research Society*, **29**, 77–83.

Tadikmalla, P.R. (1980). Age replacement policies for Weibull failure rates, *IEEE Transactions on Reliability*, **29**, 88–90.

Talkner, P. and Weber, R.O. (2000). Power spectrum and detrended fluctuation analysis: Application to daily temperatures, *Physical Review E*, **62**, 150–160.

Tang, L.C., Sun, Y.S., Goh, T.N., and Ong, H.L. (1996). Analysis of accelerated-life-test data: A new approach, *IEEE Transactions on Reliability*, **45**, 69–74.

Thiagarajan, T.R. and Harris, C.M. (1979). Statistical tests for exponential service from M/G/1 waiting-time data. *Naval Research Logistics Quarterly*, **26**, 511–520.

Thomas, L.C. (1986). A survey of maintenance and replacement models for maintainability and reliability of multi-item systems, *Reliability Engineering*, **16**, 297–309.

Tillman, F.A., Hwang, C.L., and Kuo, W. (1980). *Optimization of System Reliability*, Marcel Dekker, New York.

Titterington, M., Smith, A.F.M., and Makov, U.E. (1985). *Statistical Analysis of Finite Mixture Distribution*, Wiley, New York.

Townson, P. (2002). Load-Maintenance Interaction: Modelling and Optimisation, Doctoral Thesis, The University of Queensland, Brisbane, Australia.

Townson, P., Murthy, D.N.P., and Gurgenci, H. (2002). Optimisation of design load levels for dragline buckets, in *Case Studies in Reliability and Maintenance*, W.R. Blischke and D.N.P. Murthy (eds.), Wiley, New York.

Tseng, S.T. (1998). Selecting more reliable Weibull populations than a control, *Communications in Statistics—Series A: Theory and Methods*, **17**, 169–181.

Tseng, S.T. and Chang, D.S. (1989). Selecting the most reliable design under Type-II censoring, *Reliability Engineering and System Safety*, **25**, 147–156.

Tseng, S.T. and Wu, H.J. (1990). Selecting, under Type II censoring, Weibull populations that are more reliable, *IEEE Transactions on Reliability*, **39**, 193–198.

Tsionas, E.G. (2000). Posterior analysis, prediction and reliability in three-parameter Weibull distributions, *Communications in Statistics—Series A: Theory and Methods*, **29**, 1435–1449.

Usher, J.S. (1996). Weibull component reliability-prediction in the presence of masked data, *IEEE Transactions on Reliability*, **45**, 229–232.

Usher, J.S., Alexander, S.M., and Thompson, J.D. (1991). Predicting the reliability of new products at IBM, *Proceedings Annual Reliability and Maintainability Symposium*, 208–213.

Valdez-Flores, C. and Feldman, R.M. (1989). A survey of preventive maintenance models for stochastically deteriorating single-unit systems, *Naval Research Logistics*, **36**, 419–446.

Vannoy, E.H. (1990). Improving "MIL-HDBK-217 Type" models for mechanical reliability prediction, *Proceedings of the Annual Symposium on Reliability and Maintainability*, 341–345.

Voda, V.G. (1989). New models in durability tool-testing: psuedo-Weibull distribution, *Kybernetika*, **25**, 209–215.

Wang, Y.H. (1976). A functional equation and its application to the characterization of the Weibull and stable distributions, *Journal of Applied Probability*, **13**, 385–391.

Wang, F.K. and Keats, J.B. (1995). Improved percentile estimation for the 2-parameter Weibull distribution, *Microelectronics and Reliability*, **35**, 883–892.

Wang, W. and Keceicioglu, D.B. (2000). Fitting the Weibull log-linear model to accelerated life-test data, *IEEE Transactions on Reliability*, **49**, 217–223.

Wang, Y., Chan, Y.C., Gui, Z.L., Webb, D.P., and Li, L.T. (1997). Application of Weibull distribution analysis to the dielectric failure of multilayer ceramic capacitors, *Material Science and Engineering B*, **47**, 197–203.

Watkins, A.J. (1994). Review: Likelihood method for fitting Weibull log-linear models to accelerated life-test data, *IEEE Transactions on Reliability*, **43**, 361–365,

Watkins, A.J. (1996). On maximum likelihood estimation for the two parameter Weibull distribution, *Microelectronics and Reliability*, **36**, 595–603.

Weibull, W. (1939). A statistical theory of the strength of material, *Ingeniors Vetenskapa Acadamiens Handligar*, **151**, 1–45.

Weibull, W. (1951). A statistical distribution function of wide applicability, *Journal of Applied Mechanics*, **18**, 293–296.

Weibull, W. (1977). *References on Weibull Distribution*, FTL A Report, Forsvarets Teletekniska Laboratorium, Stockholm.

White, J.S. (1964a). Weibull renewal analysis, Proc. of the Aerospace Reliability and Maintenance Conference, 29 June–1 July 1964, pp. 639–657.

White, J.S. (1964b). Least square unbiased censored linear estimation for the log-Weibull (extreme value) distribution, *Journal of the Industrial Mathematics*, **33**, 169–177.

White, J.S. (1969). The moments of log-Weibull order statistics, *Technometrics* **11**, 373–386.

Wilk, M.B. and Gnanadesikan, R. (1968). Probability plotting methods for the analysis of data, *Biometrika*, **55**, 1–17.

Wingo, D.R. (1987). Computing maximum likelihood parameter estimates of the generalized Gamma distribution by numerical root isolation. *IEEE Transactions on Reliability*, **36**, 586–590.

Wingo, D.R. (1998a). Methods for fitting the right truncated Weibull distribution to life test and survival data, *Biometrical Journal*, **30**, 545–551.

Wingo, D.R. (1998b). Parameter estimation for a doubly-truncated Weibull distribution, *Microelectronics and Reliability*, **38**, 613–617.

Witherell, C.E. (1994). *Mechanical Failure Avoidance*, McGraw-Hill, New York.

Wong, J.Y. (1993). Simultaneously estimating the 3 parameters of the generalized gamma distribution. *Microelectronics and Reliability*, **33**, 2225–2232.

Woodward, W.A. and Gunst, R.F. (1987). Using mixtures of Weibull distributions to estimate mixing proportions, *Computational Statistics & Data Analysis*, **5**, 163–176.

Wu, Z. and Sun, X. (1998). Statistical investigation for the behaviour of short fatigue cracks in medium carbon steel, *Jixie Qiangdu/Journal of Mechanical Strength*, **20**, 295–299.

Wyckoff, J. and Engelhardt, M. (1980). Cramér–Rao lower bounds for estimators based on censored data, *Communications in Statistics—Series A: Theory and Methods*, **9**, 1385–1399.

Xie, M. (1989). On the solution of renewal-type integral equations, *Communications in Statistics—Series B: Simulation and Computation*, **18**, 281–293.

Xie, M. and Lai, C.D. (1996). Reliability analysis using an additive Weibull model with bathtub-shaped failure rate function, *Reliability Engineering and System Safety*, **52**, 87–93.

Xie, M., Preuss, W., and Cui, L.R. (2001). Error analysis of some integration procedures for renewal equation and convolution integrals, *Journal of Statistical Computation and Simulation*, **73**, 59–70.

Xie, M., Gaudoin, O., and Bracquemond, C. (2002a). Redefining failure rate function for discrete distributions, *International Journal of Reliability, Quality and Safety Engineering*, **7**, 17–25.

Xie, M., Goh, T.N. and Tang, Y. (2002b). A modified Weibull extension with bathtub-shaped failure rate function, *Reliability Engineering and Systems Safety*, **76**, 279–285.

Xu, S. and Barr, S. (1995). Probabilistic model of fracture in concrete and size effects on fracture toughness, *Magazine of Concrete Research*, **47**, 311–320.

Yamada, S. and Osaki, S. (1983). Reliability growth models for hardware and software systems based on non-homogeneous Poisson processes: A survey, *Microelectronics and Reliability*, **23**, 91–112

Yamada, S., Histitani, J., and Osaki, S. (1993). Software reliability growth with a Weibull test effort, *IEEE Transactions on Reliability*, **42**, 100–105.

Yan, L., English, J.R., and Landers, T.L. (1995). Modeling latent and patent failures of electronic products, *Microelectronics and Reliability*, **35**, 1501–1510.

Yang, G. and Jin, L. (1994). Best compromise test plans for Weibull distributions with different censoring times, *Quality and Reliability Engineering International*, **10**, 411–415.

Yang, S.C. and Nachlas, J.A. (2001). Bivariate reliability and availability modeling, *IEEE Transactions on Reliability*, **50**, 26–35.

Yang, L., English, J.R., and Landers, T.L. (1995). Modeling latent and patent failures of electronic products, *Microelectronics and Reliability*, **35**, 1501–1510.

Yang, Z.L., See, S.P., and Xie, M. (2002). An investigation of transformation-based prediction interval for the Weibull median life. *Metrika*, **56**, 19–29.

Yazhou, J., Wang, M., and Zhixin, J. (1995). Probability distribution of machining center failures, *Reliability Engineering and System Safety*, **50**, 121–125.

Yeh, R.H. and Lo, H.C. (1998). Quality control for products under free repair warranty, *International Journal of Quality Management*, **4**, 265–275.

Yeh, R.H. and Lo, H.C. (2001). Optimal preventive-maintenance warranty policy for repairable products. *European Journal of Operational Research*, **134**, 59–69.

Yeh, R.H., Ho, W.T., and Tseng, S.T. (2000). Optimal preventive-maintenance warranty policy for repairable products. *European Journal of Operational Research*, **129**, 575–582.

Yehia, A.Y. (1993). On a characterization by residual moment, *Microelectronics and Reliability*, **33**, 1127–1132.

Yuan, J. and Shih, S. (1991). Mixed hazard models and their applications, *Reliability and System Safety*, **33**, 115–129.

Yuen, H.K. and Tse, S.K. (1996). Parameter estimation for Weibull distributed lifetimes under progressive censoring with random removals, *Journal of Statistical Computation and Simulation*, **55**, 57–71.

Zacks, S. (1984). Estimating the shift to wear-out systems having exponential-Weibull life distributions, *Operations Research*, **32**, 741–749.

Zacks, S. (1992). *Introduction to Reliability Analysis: Probability Models and Statistical Methods*, Springer, New York.

Zahedi, H. (1985). Some new classes of multivariate survival distribution functions, *Journal of Statistical Planning and Inference*, **11**, 171–188.

Zanakis, S.H. (1979). A simulation study of some simple estimators for the three-parameter Weibull distribution, *Journal of Statistical Computation and Simulation*, **9**, 101–116.

Zanakis, S.H. and Kyparisis, J. (1986). A review of maximum likelihood estimation methods for the three parameter Weibull distribution, *Journal of Statistical Computation and Simulation*, **25**, 53–73.

Zhao, J. and Zhang, X. (1997). Parameter estimation of reliability test data through optimization, *Binggong Xuebao/Acta Armamentarii*, **18**, 134–138.

Zheng, G. (2001). A characterization of the factorization of hazard function by the Fisher information under Type II censoring with application to the Weibull family, *Statistics and Probability Letters*, **52**, 249–253.

Zuo, M.J., Jiang, R., and C.M. Yam, (1999). Approaches for reliability modeling of continuous state devices, *IEEE Transactions on Reliability*, **48**, 9–18.

Zuo, M.J., Liu, B., and Murthy, D.N.P. (2000). Replacement-repair policy for multi-state deteriorating products under warranty. *European Journal of Operational Research*, **123**, 519–530.

Index

Accelerated
  failure, 32
  life model, 31,223
  life testing, 32,228,328
Acceptance sampling, 333,345
Additive
  mixture Weibull, 28,160
  Weibull, 29
Age policy, 340
Age-based maintenance, 339
Age replacement policy, 339
Aircraft windshield, 296
Alternating renewal process, 38, 264,278
Anderson-Darling (A-D) test, 91
Arrhenius Weibull model, 32,226
Asymptotic unbiasedness, 60
Auxiliary variable, 223

Baseline
  hazard function, 229
  intensity function, 273
Batch production, 333
Baye's estimate, 269
Bayesian
  approach, 235,236
  estimation, 84
  estimator, 64
  method, 71,172,188,268
  models, 20
Bayes' theorem, 64
Beta
  Function, 234
  integrated model, 135
Bimodal-mixed Weibull, 28,160
Bivariate exponential, 35,36
Bivariate
  extension, 252
  hazard function, 249
  model, 34,250
  Weibull, 20,36,250
Black box
  approach, 306,313
  model, 283
Block policy, 340
Bootstrap, 296
Burn-in, 335,346
Burr distribution, 233

Catastrophic failure, 305
Cramer-Rao Inequality, 61
Cramer-von Mises Test, 92
Censored data, 59,66,67
Censoring
  scheme, 328
  Type I, 59,63
  Type II, 59,63
Central moment, 47
Chi-square test, 90
Classification of
  multivariate distributions, 249
  shapes of density function, 46
  shapes of hazard function, 46
  WPP plots, 285
Cleaning web failures, 291
Clock-based maintenance, 339

Weibull Models, by D.N.P. Murthy, Min Xie, and Renyan Jiang.
ISBN 0-471-36092-9

Coefficient of
  excess, 49
  skewness, 47,75,107
  variation, 49,123
Cold standby, 329
Competing risk, 19,29
  model, 182,224
Component level
  maintenance, 339
  modeling, 308,312
Compound
  model, 182,233
  Weibull law, 234
  Weibull model, 33,233
Composite model, 30,208
Complementary
  risk model, 30,197
  Weibull distribution, 23,114
Complete
  data, 59,66,67,95
  failure, 305
Condition-based replacement, 339
Conditional density, 248
Confidence
  coefficient, 65
  interval, 65,268
Consistency, 60
Continuous
  mixture, 233
  mixture model, 33
  production, 333
Control charts, 334
Convolution, 311
  operator, 274
Counting process, 261,262
Covariance, 255
Covariates, 223,225
Cumulative
  damage, 305
  hazard function
  shock damage model, 311

Data dependent model, 283
Data
  analysis, 8
  complete, 59
  collection, 7
  censored, 59
  group, 59
  types, 58
Delayed renewal process, 277
Density function, 162,176,184,190,192,198, 203,210–216
  Parametric study, 50
  shapes, 46
Decision models, 341
Design
  limit testing, 328
  process, 325
Design-out maintenance, 339
Development
  process, 327
  tests, 327
Discrete
  failure rate function, 239
  Weibull models, 20,33,238
Dispersion, 49
Distribution function, 198,203,230, 231,248
Double Weibull distribution, 108
Doubly truncated Weibull, 27
Dragline optimization, 318

Efficiency, 60
Empirical
  distribution function, 86
  model, 283
  modeling, 7
Environment testing, 328
Estimation, 66
  interval, 60
  hybrid methods, 81
  methods, 61
  point, 60
Estimators, 61
  Bayesian, 64
  linear, 72
  maximum likelihood, 63
  moment, 61
  percentile, 62
  properties, 60
Euler constant, 52
Event truncation, 266,267
Excess life, 276,278
Expectation, 47
Explanatory variables, 225,273
Exponential probability plot, 110
Exponentiated
  Weibull, 25
  Weibull distribution, 127,225
Extended
  failure, 305
  generalized gamma, 27
  Weibull, 24
  Weibull distribution, 26,125,139
External shock, 310
Extreme value distribution, 23,111
Eyring model, 226

Failure
  gradual, 305
  mechanisms, 305
  models, 304,305
  overstress, 308
  wear-out, 310
Finite
  mixture model, 232
  Weibull mixture, 28
First failure, 307
Fisher information matrix, 73
Fisher-Trippet type III distribution, 107
Five-parameter Weibull, 27,146
Flatness, 49
Four-parameter
  generalized gamma, 143
  Weibull, 27,145
Fractile, 49
FRW policy, 337

Gamma
  distribution, 26
  function, 68
General
  bivariate model, 252
  curve fitting method, 171,188
Generalization, 19,23,24,35
Generalized
  Burr distribution, 233
  competing risk model, 183,192
  gamma model, 26
  Weibull, 24
  Weibull distribution, 31,225
  Weibull family, 138
Goodness-of-fit tests, 89,96,99,175,270,272
Graphical
  approach, 190.205,230,266,270
  parameter estimation, 194
  methods, 65,74,86,167,177,186
Gray box approach, 312
Grouped data, 64,66,67,266,268

Hazard
  function, 163,176,184,185,190,193,199, 204,210–216,231,240
  plot, 66,88,96
Histogram, 87
Hot standby, 329
Hybrid
  method, 81,173,189
  mixtures, 160
  sectional model, 208
  Weibull competing risk model, 191
  Weibull mixture, 179
Hypothesis tests, 273

Item
  conformance, 332
  nonconformance, 332
Incomplete gamma function, 27
Information matrix, 73
Informative prior distribution, 269
Inspection, 334
Intensity function, 272
  models, 20
Intermittent failures, 305
Interval
  censoring, 59
  estimates, 268
  estimation, 65,73
Inverse
  Gaussian distribution, 115
  transformation, 106
  Weibull competing risk model, 190
  Weibull distribution, 114,115,124,183
  Weibull model, 23,28,29,30
  Weibull multiplicative model, 203
  Weibull probability paper, 117
  Weibull transform, 116,117
  Weibull transformation, 190
IWPP plots, 117,176,190,204

Jack knife approach, 296
Joint
  density function, 38,64,119,255
  distribution, 255
  survivor function, 253

k-out-of-n system, 317
Kolmogorov-Smirnov (K-S) test, 91
Kurtosis, 49,52,107

Large data set, 290
Least-squares method, 66
Left truncated, 27
Leptokurtic, 49
Likelihood
  Function, 61,63,70,78,118,153,232,244, 267,273
  ratio test, 273
Linear
  estimators, 72
  transformation, 105
Log-likelihood function, 255

Log-ratio transformation, 270–272
Log transformation, 96,105
Log Weibull
  model, 32
  transformation, 111
Lot sizing, 335
Lower confidence limits, 74

Maintenance, 339,347
Manufacturing, 332,343
Marginal distributions, 250–252
Maximum-likelihood
  estimates, 79,255
  estimation, 73,118,145,232
  estimator, 63
  method, 171,219
Mean, 47
  residual life, 244
Median, 49,53,54
Medium data test, 289
Method
  based on WPP, 216
  least square, 66,137,150
  maximum likelihood, 68,78,188,195,227,231,241,267, 268,273
  moments, 67,75,171,241
  percentiles, 68,77
Minimal repair, 307
Minimum variance unbiased estimators, 60
Mixed-mode Weibull, 28,160
Mixture, 19
  distribution, 159,333
  model, 28,159
  with negative weights, 161
Mode, 53,54,115
Model
  analysis, 161,257,258,265,274
  black-box, 3
  data dependent, 3
  discrimination, 93
  parameters, 47,77,290
  physical based, 3
  properties, 107
  selection, 8,9,11,58,85,270,285,287, 289,290
  validation, 58,85,94,270,287,288, 290,322
Modeling
  empirical, 3
  failures, 306
  product failures, 303
  theory based, 3
Modified
  maximum-likelihood estimators, 80
  moment method, 76
  renewal process, 38,264
  Weibull, 25,27,28,121,
  Weibull distribution, 134,148
Modification, 19,23,34
Modulated
  gamma process, 38
  power law process, 38,263,272
Moment, 47,51,52,54,161,232,235,240, 244,275
  estimator, 61
  generating function, 49,51,109, 112,142
More generalized Weibull family, 139
MTTF, 232
Multimodal Weibull, 28
Multimode Weibull distribution, 160
Multiple-step stress accelerated life testing, 228
Multiplicative, 19
  models, 29,30,197,229
  Weibull model, 198
Multi risk model, 182
Multivariate
  distribution, 37
  exponential, 34
  extension of Weibull distribution, 256
  mixtures of Weibull distribution, 258
  model, 36,247,256
  Weibull, 20,247
  Weibull model, 34
More generalized Weibull, 26

Nonconforming items, 332
Non-informative prior, 269
Non-repairable component, 307
Non-Weibull distributions, 29
Null hypothesis, 89

Optimization method, 201
Order statistics, 49,53,112,234
Ordinary
  renewal process, 38,264
  Weibull renewal process, 274
Output inspection and testing, 334
Overstress
  failure, 308
  mechanisms, 305

Parallel structure, 317,318
Parameter Estimation, 177,194,205,216, 232,235,241,244,251,255,258,266, 273,277,285–288,290
Parametric regression models, 225
Partial failure, 305
Partition points, 30
Percentile, 49,53,54
Percentile estimator, 62,77
Performance degradation, 305
PH model, 229,231
Photocopier failures, 314
Piecewise model, 30,208
Platykurtic, 49
Plot
  Inverse Weibull, 116
  P-P, 87
  Probability, 87
  Q-Q, 87
  TTT, 53
  Transformed, 86
  Weibull, 87
Point estimates, 267
Post sale, 336,346
Posterior
  density function, 64
  distribution, 65,269
  mean, 269
Power law, 20,34
  intensity model, 39
  process, 37,263,265
  process model, 273
  transformation, 22,109
  Weibull renewal process, 261, 264,278
Power transformation, 36,96,105
Power Weibull model, 32,226
P-P plot, 87
Prior, 65,233
  density function, 64
Probability mass function, 238,239,243
Probability plot, 87
Probability-weighted moment method, 76
Product
  life cycle, 12
  reliability, 324
  structure, 306
Production
  batch, 333
  continuous, 333
Proportional
  hazard models, 20,32,223,229
  intensity model, 264,273
PRW policy, 338
Pseudo-Weibull, 24
  distribution, 122

Quality
  control, 343
  improvement, 335
  variations, 332
Q-Q plot, 87
Quantile, 49,53
  function, 25,26,127,138

Random
  parameter, 33,232
  censoring, 59,64
Rasch-Weibull process, 37
Ratio
  power transformation, 270–272
  type test, 97
Rayleigh distribution, 21
Reciprocal Weibull distribution, 114
Reduced log Weibull distribution, 112,113
Redundancy, 329
  Cold, 329
  Hot, 329
  Warm, 330
Reflected Weibull distribution, 106
Regression, 66
Regression models, 20
Reliability
  allocation, 326,341
  assessment, 327,342
  growth, 330
  improvement, 329,343
  prediction, 327,341
Renewal
  density function, 275
  process, 38
  process models, 20
  reward theorem, 277
  type equation, 276
Repair
  good as new, 307
  minimal, 307
Repairable component, 307
Right
  truncated, 27
  type I censoring, 59,63
  type II censoring, 59,63

Sample
  mean, 67
  variance, 67
Sectional, 19
  model, 30,208,225
  Weibull model, 228
Series
  structure, 317,318
  system model, 182
Shock induced stress, 309
Skewness, 52,75,107
Slymen-Lachenbruch distribution, 148
Small data set, 287
Stacy-Mihram model, 124
Standard deviation, 47
Statistical method, 227
Step-stress model, 229
Stochastic point process, 20,37,261
Stress-strength model, 308
Structure function, 316
Sub-populations, 28,29,182,183
Subset selection, 345
Sudden failure, 305
Supplementary variables, 32,225
Survivor function, 248
System level
  maintenance, 340
  modeling, 313,316

Taxonomy, 10,18
Testing, 342
  accelerated life, 32,228,328
  design limit, 328
  environmental, 328
  involving progressive censoring, 328
  to failure, 327
Tests
  Anderson-Darling, 91
  Chi-Square, 90
  Cramer-von Mises, 92
  Kolmogov-Smirnov, 91
Threefold Weibull mixture model, 166
Three-parameter generalized gamma, 140
Throttle failures, 293
Time truncation, 266,267
Time-varying parameters, 19,31,223
Transformation, 18,21,22,23,35,66,253
  Ratio-Power, 271
  Log-Ratio271
  Weibull, 28,50,53., 66
Transforms, 34
Truncated Weibull distribution, 146
Truncation, 27
TTT plot, 50,53,95
Two fold Weibull mixture model, 161, 163–165
Type I extreme value distribution, 135
Type II asymptotic distribution, 114

unbiasedness, 60
Uniform
  density function, 234
  prior, 234,235
Upper confidence limit, 73,74
Usage-based replacement, 339

Variance, 47,52
Varying
  parameters, 19,30
  stress model, 228

Warm standby, 330
Warranties, 336,346
Warranty
  cost, 337
  policies, 336
Wear-out
  failures, 310
  mechanisms, 305
Weibull
  AFT model, 225–227
  competing risk model, 183,226,227
Weibull distribution
  1-parameter, 21
  2- Parameter, 10
  3-Parameter, 9
  4-Parameter, 27
  5-Parameter, 27
  Additive-mixed, 28
  Bivariate, 34
  Competing risk, 29
  Complementary, 23
  Compound, 33
  Discrete, 33
  Double, 22
  Extended, 24,26
  Exponentiated, 25
  Generalized, 25
  Inverse, 23
  Log, 23
  Mixture, 28
  Modified, 27

Multi-modal, 28
Multivariate, 34
Multplicative, 30
Pseudo, 24
Reciprocal, 23
Reflected, 22
Sectional, 30
Truncated, 27
extension, 151
Gnedenko distribution, 81
PH model, 230
intensity
function, 37
model, 261,263
mixture, 159,226,227
model, 160,192,292,295
plot, 87
Poisson process, 37
probability plot, 95,99
process, 37
random parameters, 33
renewal
model, 39
process, 261,264
sectional model, 208
transformation, 28,50,53,66
Well
mixed, 169,177
separated, 169,177
White box approach, 308,316
WPP plot, 53,55,116,164,185,193,199, 210–216

WILEY SERIES IN PROBABILITY AND STATISTICS

ESTABLISHED BY WALTER A. SHEWHART AND SAMUEL S. WILKS

The ***Wiley Series in Probability and Statistics*** is well established and authoritative. It covers many topics of current research interest in both pure and applied statistics and probability theory. Written by leading statisticians and institutions, the titles span both state-of-the-art developments in the field and classical methods.

Reflecting the wide range of current research in statistics, the series encompasses applied, methodological and theoretical statistics, ranging from applications and new techniques made possible by advances in computerized practice to rigorous treatment of theoretical approaches.

This series provides essential and invaluable reading for all statisticians, whether in academia, industry, government, or research.

ABRAHAM and LEDOLTER · Statistical Methods for Forecasting
AGRESTI · Analysis of Ordinal Categorical Data
AGRESTI · An Introduction to Categorical Data Analysis
AGRESTI · Categorical Data Analysis, *Second Edition*
ALTMAN, GILL, and McDONALD · Numerical Issues in Statistical Computing for the Social Scientist
AMARATUNGA and CABRERA · Exploration and Analysis of DNA Microarray and Protein Array Data
ANDĚL · Mathematics of Chance
ANDERSON · An Introduction to Multivariate Statistical Analysis, *Third Edition*
*ANDERSON · The Statistical Analysis of Time Series
ANDERSON, AUQUIER, HAUCK, OAKES, VANDAELE, and WEISBERG · Statistical Methods for Comparative Studies
ANDERSON and LOYNES · The Teaching of Practical Statistics
ARMITAGE and DAVID (editors) · Advances in Biometry
ARNOLD, BALAKRISHNAN, and NAGARAJA · Records
*ARTHANARI and DODGE · Mathematical Programming in Statistics
*BAILEY · The Elements of Stochastic Processes with Applications to the Natural Sciences
BALAKRISHNAN and KOUTRAS · Runs and Scans with Applications
BARNETT · Comparative Statistical Inference, *Third Edition*
BARNETT and LEWIS · Outliers in Statistical Data, *Third Edition*
BARTOSZYNSKI and NIEWIADOMSKA-BUGAJ · Probability and Statistical Inference
BASILEVSKY · Statistical Factor Analysis and Related Methods: Theory and Applications
BASU and RIGDON · Statistical Methods for the Reliability of Repairable Systems
BATES and WATTS · Nonlinear Regression Analysis and Its Applications
BECHHOFER, SANTNER, and GOLDSMAN · Design and Analysis of Experiments for Statistical Selection, Screening, and Multiple Comparisons
BELSLEY · Conditioning Diagnostics: Collinearity and Weak Data in Regression

*Now available in a lower priced paperback edition in the Wiley Classics Library.

BELSLEY, KUH, and WELSCH · Regression Diagnostics: Identifying Influential Data and Sources of Collinearity
BENDAT and PIERSOL · Random Data: Analysis and Measurement Procedures, *Third Edition*
BERRY, CHALONER, and GEWEKE · Bayesian Analysis in Statistics and Econometrics: Essays in Honor of Arnold Zellner
BERNARDO and SMITH · Bayesian Theory
BHAT and MILLER · Elements of Applied Stochastic Processes, *Third Edition*
BHATTACHARYA and JOHNSON · Statistical Concepts and Methods
BHATTACHARYA and WAYMIRE · Stochastic Processes with Applications
BILLINGSLEY · Convergence of Probability Measures, *Second Edition*
BILLINGSLEY · Probability and Measure, *Third Edition*
BIRKES and DODGE · Alternative Methods of Regression
BLISCHKE AND MURTHY (editors) · Case Studies in Reliability and Maintenance
BLISCHKE AND MURTHY · Reliability: Modeling, Prediction, and Optimization
BLOOMFIELD · Fourier Analysis of Time Series: An Introduction, *Second Edition*
BOLLEN · Structural Equations with Latent Variables
BOROVKOV · Ergodicity and Stability of Stochastic Processes
BOULEAU · Numerical Methods for Stochastic Processes
BOX · Bayesian Inference in Statistical Analysis
BOX · R. A. Fisher, the Life of a Scientist
BOX and DRAPER · Empirical Model-Building and Response Surfaces
*BOX and DRAPER · Evolutionary Operation: A Statistical Method for Process Improvement
BOX, HUNTER, and HUNTER · Statistics for Experimenters: An Introduction to Design, Data Analysis, and Model Building
BOX and LUCEÑO · Statistical Control by Monitoring and Feedback Adjustment
BRANDIMARTE · Numerical Methods in Finance: A MATLAB-Based Introduction
BROWN and HOLLANDER · Statistics: A Biomedical Introduction
BRUNNER, DOMHOF, and LANGER · Nonparametric Analysis of Longitudinal Data in Factorial Experiments
BUCKLEW · Large Deviation Techniques in Decision, Simulation, and Estimation
CAIROLI and DALANG · Sequential Stochastic Optimization
CHAN · Time Series: Applications to Finance
CHATTERJEE and HADI · Sensitivity Analysis in Linear Regression
CHATTERJEE and PRICE · Regression Analysis by Example, *Third Edition*
CHERNICK · Bootstrap Methods: A Practitioner's Guide
CHERNICK and FRIIS · Introductory Biostatistics for the Health Sciences
CHILÈS and DELFINER · Geostatistics: Modeling Spatial Uncertainty
CHOW and LIU · Design and Analysis of Clinical Trials: Concepts and Methodologies, *Second Edition*
CLARKE and DISNEY · Probability and Random Processes: A First Course with Applications, *Second Edition*
*COCHRAN and COX · Experimental Designs, *Second Edition*
CONGDON · Applied Bayesian Modelling
CONGDON · Bayesian Statistical Modelling
CONOVER · Practical Nonparametric Statistics, *Second Edition*
COOK · Regression Graphics
COOK and WEISBERG · Applied Regression Including Computing and Graphics
COOK and WEISBERG · An Introduction to Regression Graphics
CORNELL · Experiments with Mixtures, Designs, Models, and the Analysis of Mixture Data, *Third Edition*
COVER and THOMAS · Elements of Information Theory

*Now available in a lower priced paperback edition in the Wiley Classics Library.

COX · A Handbook of Introductory Statistical Methods
*COX · Planning of Experiments
CRESSIE · Statistics for Spatial Data, *Revised Edition*
CSÖRGŐ and HORVÁTH · Limit Theorems in Change Point Analysis
DANIEL · Applications of Statistics to Industrial Experimentation
DANIEL · Biostatistics: A Foundation for Analysis in the Health Sciences, *Sixth Edition*
*DANIEL · Fitting Equations to Data: Computer Analysis of Multifactor Data, *Second Edition*
DASU and JOHNSON · Exploratory Data Mining and Data Cleaning
DAVID and NAGARAJA · Order Statistics, *Third Edition*
*DEGROOT, FIENBERG, and KADANE · Statistics and the Law
DEL CASTILLO · Statistical Process Adjustment for Quality Control
DENISON, HOLMES, MALLICK and SMITH · Bayesian Methods for Nonlinear Classification and Regression
DETTE and STUDDEN · The Theory of Canonical Moments with Applications in Statistics, Probability, and Analysis
DEY and MUKERJEE · Fractional Factorial Plans
DILLON and GOLDSTEIN · Multivariate Analysis: Methods and Applications
DODGE · Alternative Methods of Regression
*DODGE and ROMIG · Sampling Inspection Tables, *Second Edition*
*DOOB · Stochastic Processes
DOWDY, WEARDEN, and CHILKO · Statistics for Research, *Third Edition*
DRAPER and SMITH · Applied Regression Analysis, *Third Edition*
DRYDEN and MARDIA · Statistical Shape Analysis
DUDEWICZ and MISHRA · Modern Mathematical Statistics
DUNN and CLARK · Applied Statistics: Analysis of Variance and Regression, *Second Edition*
DUNN and CLARK · Basic Statistics: A Primer for the Biomedical Sciences, *Third Edition*
DUPUIS and ELLIS · A Weak Convergence Approach to the Theory of Large Deviations
*ELANDT-JOHNSON and JOHNSON · Survival Models and Data Analysis
ENDERS · Applied Econometric Time Series
ETHIER and KURTZ · Markov Processes: Characterization and Convergence
EVANS, HASTINGS, and PEACOCK · Statistical Distributions, *Third Edition*
FELLER · An Introduction to Probability Theory and Its Applications, Volume I, *Third Edition,* Revised; Volume II, *Second Edition*
FISHER and VAN BELLE · Biostatistics: A Methodology for the Health Sciences
*FLEISS · The Design and Analysis of Clinical Experiments
FLEISS · Statistical Methods for Rates and Proportions, *Third Edition*
FLEMING and HARRINGTON · Counting Processes and Survival Analysis
FULLER · Introduction to Statistical Time Series, *Second Edition*
FULLER · Measurement Error Models
GALLANT · Nonlinear Statistical Models
GIESBRECHT and GUMPERTZ · Planning, Construction, and Statistical Analysis of Comparative Experiments
GIFI · Nonlinear Multivariate Analysis
GHOSH, MUKHOPADHYAY, and SEN · Sequential Estimation
GIFI · Nonlinear Multivariate Analysis
GLASSERMAN and YAO · Monotone Structure in Discrete-Event Systems
GNANADESIKAN · Methods for Statistical Data Analysis of Multivariate Observations, *Second Edition*
GOLDSTEIN and LEWIS · Assessment: Problems, Development, and Statistical Issues
GREENWOOD and NIKULIN · A Guide to Chi-Squared Testing

*Now available in a lower priced paperback edition in the Wiley Classics Library.

GROSS and HARRIS · Fundamentals of Queueing Theory, *Third Edition*
*HAHN and SHAPIRO · Statistical Models in Engineering
HAHN and MEEKER · Statistical Intervals: A Guide for Practitioners
HALD · A History of Probability and Statistics and their Applications Before 1750
HALD · A History of Mathematical Statistics from 1750 to 1930
HAMPEL · Robust Statistics: The Approach Based on Influence Functions
HANNAN and DEISTLER · The Statistical Theory of Linear Systems
HEIBERGER · Computation for the Analysis of Designed Experiments
HEDAYAT and SINHA · Design and Inference in Finite Population Sampling
HELLER · MACSYMA for Statisticians
HINKELMAN and KEMPTHORNE: · Design and Analysis of Experiments, Volume 1: Introduction to Experimental Design
HOAGLIN, MOSTELLER, and TUKEY · Exploratory Approach to Analysis of Variance
HOAGLIN, MOSTELLER, and TUKEY · Exploring Data Tables, Trends and Shapes
*HOAGLIN, MOSTELLER, and TUKEY · Understanding Robust and Exploratory Data Analysis
HOCHBERG and TAMHANE · Multiple Comparison Procedures
HOCKING · Methods and Applications of Linear Models: Regression and the Analysis of Variance, *Second Edition*
HOEL · Introduction to Mathematical Statistics, *Fifth Edition*
HOGG and KLUGMAN · Loss Distributions
HOLLANDER and WOLFE · Nonparametric Statistical Methods, *Second Edition*
HOSMER and LEMESHOW · Applied Logistic Regression, *Second Edition*
HOSMER and LEMESHOW · Applied Survival Analysis: Regression Modeling of Time to Event Data
HØYLAND and RAUSAND · System Reliability Theory: Models and Statistical Methods
HUBER · Robust Statistics
HUBERTY · Applied Discriminant Analysis
HUNT and KENNEDY · Financial Derivatives in Theory and Practice
HUSKOVA, BERAN, and DUPAC · Collected Works of Jaroslav Hajek—with Commentary
IMAN and CONOVER · A Modern Approach to Statistics
JACKSON · A User's Guide to Principle Components
JOHN · Statistical Methods in Engineering and Quality Assurance
JOHNSON · Multivariate Statistical Simulation
JOHNSON and BALAKRISHNAN · Advances in the Theory and Practice of Statistics: A Volume in Honor of Samuel Kotz
JUDGE, GRIFFITHS, HILL, LÜTKEPOHL, and LEE · The Theory and Practice of Econometrics, *Second Edition*
JOHNSON and KOTZ · Distributions in Statistics
JOHNSON and KOTZ (editors) · Leading Personalities in Statistical Sciences: From the Seventeenth Century to the Present
JOHNSON, KOTZ, and BALAKRISHNAN · Continuous Univariate Distributions, Volume 1, *Second Edition*
JOHNSON, KOTZ, and BALAKRISHNAN · Continuous Univariate Distributions, Volume 2, *Second Edition*
JOHNSON, KOTZ, and BALAKRISHNAN · Discrete Multivariate Distributions
JOHNSON, KOTZ, and KEMP · Univariate Discrete Distributions, *Second Edition*
JUREČKOVÁ and SEN · Robust Statistical Procedures: Aymptotics and Interrelations
JUREK and MASON · Operator-Limit Distributions in Probability Theory
KADANE · Bayesian Methods and Ethics in a Clinical Trial Design
KADANE AND SCHUM · A Probabilistic Analysis of the Sacco and Vanzetti Evidence

*Now available in a lower priced paperback edition in the Wiley Classics Library.

KALBFLEISCH and PRENTICE · The Statistical Analysis of Failure Time Data, *Second Edition*
KASS and VOS · Geometrical Foundations of Asymptotic Inference
KAUFMAN and ROUSSEEUW · Finding Groups in Data: An Introduction to Cluster Analysis
KEDEM and FOKIANOS · Regression Models for Time Series Analysis
KENDALL, BARDEN, CARNE, and LE · Shape and Shape Theory
KHURI · Advanced Calculus with Applications in Statistics, *Second Edition*
KHURI, MATHEW, and SINHA · Statistical Tests for Mixed Linear Models
KLEIBER and KOTZ · Statistical Size Distributions in Economics and Actuarial Sciences
KLUGMAN, PANJER, and WILLMOT · Loss Models: From Data to Decisions
KLUGMAN, PANJER, and WILLMOT · Solutions Manual to Accompany Loss Models: From Data to Decisions
KOTZ, BALAKRISHNAN, and JOHNSON · Continuous Multivariate Distributions, Volume 1, *Second Edition*
KOTZ and JOHNSON (editors) · Encyclopedia of Statistical Sciences: Volumes 1 to 9 with Index
KOTZ and JOHNSON (editors) · Encyclopedia of Statistical Sciences: Supplement Volume
KOTZ, READ, and BANKS (editors) · Encyclopedia of Statistical Sciences: Update Volume 1
KOTZ, READ, and BANKS (editors) · Encyclopedia of Statistical Sciences: Update Volume 2
KOVALENKO, KUZNETZOV, and PEGG · Mathematical Theory of Reliability of Time-Dependent Systems with Practical Applications
LACHIN · Biostatistical Methods: The Assessment of Relative Risks
LAD · Operational Subjective Statistical Methods: A Mathematical, Philosophical, and Historical Introduction
LAMPERTI · Probability: A Survey of the Mathematical Theory, *Second Edition*
LANGE, RYAN, BILLARD, BRILLINGER, CONQUEST, and GREENHOUSE · Case Studies in Biometry
LARSON · Introduction to Probability Theory and Statistical Inference, *Third Edition*
LAWLESS · Statistical Models and Methods for Lifetime Data, *Second Edition*
LAWSON · Statistical Methods in Spatial Epidemiology
LE · Applied Categorical Data Analysis
LE · Applied Survival Analysis
LEE and WANG · Statistical Methods for Survival Data Analysis, *Third Edition*
LEPAGE and BILLARD · Exploring the Limits of Bootstrap
LEYLAND and GOLDSTEIN (editors) · Multilevel Modelling of Health Statistics
LIAO · Statistical Group Comparison
LINDVALL · Lectures on the Coupling Method
LINHART and ZUCCHINI · Model Selection
LITTLE and RUBIN · Statistical Analysis with Missing Data, *Second Edition*
LLOYD · The Statistical Analysis of Categorical Data
MAGNUS and NEUDECKER · Matrix Differential Calculus with Applications in Statistics and Econometrics, *Revised Edition*
MALLER and ZHOU · Survival Analysis with Long Term Survivors
MALLOWS · Design, Data, and Analysis by Some Friends of Cuthbert Daniel
MANN, SCHAFER, and SINGPURWALLA · Methods for Statistical Analysis of Reliability and Life Data
MANTON, WOODBURY, and TOLLEY · Statistical Applications Using Fuzzy Sets
MARDIA and JUPP · Directional Statistics
MASON, GUNST, and HESS · Statistical Design and Analysis of Experiments with Applications to Engineering and Science, *Second Edition*

*Now available in a lower priced paperback edition in the Wiley Classics Library.

McCULLOCH and SEARLE · Generalized, Linear, and Mixed Models
McFADDEN · Management of Data in Clinical Trials
McLACHLAN · Discriminant Analysis and Statistical Pattern Recognition
McLACHLAN and KRISHNAN · The EM Algorithm and Extensions
McLACHLAN and PEEL · Finite Mixture Models
McNEIL · Epidemiological Research Methods
MEEKER and ESCOBAR · Statistical Methods for Reliability Data
MEERSCHAERT and SCHEFFLER · Limit Distributions for Sums of Independent Random Vectors: Heavy Tails in Theory and Practice
*MILLER · Survival Analysis, *Second Edition*
MONTGOMERY, PECK, and VINING · Introduction to Linear Regression Analysis, *Third Edition*
MORGENTHALER and TUKEY · Configural Polysampling: A Route to Practical Robustness
MUIRHEAD · Aspects of Multivariate Statistical Theory
MULLER and STOYAN · Comparison Methods for Stochastic Models and Risks
MURRAY · X-STAT 2.0 Statistical Experimentation, Design Data Analysis, and Nonlinear Optimization
MURTHY, XIE, and JIANG · Weibull Models
MYERS and MONTGOMERY · Response Surface Methodology: Process and Product Optimization Using Designed Experiments, *Second Edition*
MYERS, MONTGOMERY, and VINING · Generalized Linear Models. With Applications in Engineering and the Sciences
NELSON · Accelerated Testing, Statistical Models, Test Plans, and Data Analyses
NELSON · Applied Life Data Analysis
NEWMAN · Biostatistical Methods in Epidemiology
OCHI · Applied Probability and Stochastic Processes in Engineering and Physical Sciences
OKABE, BOOTS, SUGIHARA, and CHIU · Spatial Tesselations: Concepts and Applications of Voronoi Diagrams, *Second Edition*
OLIVER and SMITH · Influence Diagrams, Belief Nets and Decision Analysis
PALTA · Quantitative Methods in Population Health: Extensions of Ordinary Regressions
PANKRATZ · Forecasting with Dynamic Regression Models
PANKRATZ · Forecasting with Univariate Box-Jenkins Models: Concepts and Cases
*PARZEN · Modern Probability Theory and Its Applications
PEÑA, TIAO, and TSAY · A Course in Time Series Analysis
PIANTADOSI · Clinical Trials: A Methodologic Perspective
PORT · Theoretical Probability for Applications
POURAHMADI · Foundations of Time Series Analysis and Prediction Theory
PRESS · Bayesian Statistics: Principles, Models, and Applications
PRESS · Subjective and Objective Bayesian Statistics, *Second Edition*
PRESS and TANUR · The Subjectivity of Scientists and the Bayesian Approach
PUKELSHEIM · Optimal Experimental Design
PURI, VILAPLANA, and WERTZ · New Perspectives in Theoretical and Applied Statistics
PUTERMAN · Markov Decision Processes: Discrete Stochastic Dynamic Programming
*RAO · Linear Statistical Inference and Its Applications, *Second Edition*
RENCHER · Linear Models in Statistics
RENCHER · Methods of Multivariate Analysis, *Second Edition*
RENCHER · Multivariate Statistical Inference with Applications
RIPLEY · Spatial Statistics
RIPLEY · Stochastic Simulation
ROBINSON · Practical Strategies for Experimenting
ROHATGI and SALEH · An Introduction to Probability and Statistics, *Second Edition*

*Now available in a lower priced paperback edition in the Wiley Classics Library.

ROLSKI, SCHMIDLI, SCHMIDT, and TEUGELS · Stochastic Processes for Insurance and Finance
ROSENBERGER and LACHIN · Randomization in Clinical Trials: Theory and Practice
ROSS · Introduction to Probability and Statistics for Engineers and Scientists
ROUSSEEUW and LEROY · Robust Regression and Outlier Detection
RUBIN · Multiple Imputation for Nonresponse in Surveys
RUBINSTEIN · Simulation and the Monte Carlo Method
RUBINSTEIN and MELAMED · Modern Simulation and Modeling
RYAN · Modern Regression Methods
RYAN · Statistical Methods for Quality Improvement, *Second Edition*
SALTELLI, CHAN, and SCOTT (editors) · Sensitivity Analysis
*SCHEFFE · The Analysis of Variance
SCHIMEK · Smoothing and Regression: Approaches, Computation, and Application
SCHOTT · Matrix Analysis for Statistics
SCHOUTENS · Levy Processes in Finance: Pricing Financial Derivatives
SCHUSS · Theory and Applications of Stochastic Differential Equations
SCOTT · Multivariate Density Estimation: Theory, Practice, and Visualization
*SEARLE · Linear Models
SEARLE · Linear Models for Unbalanced Data
SEARLE · Matrix Algebra Useful for Statistics
SEARLE, CASELLA, and McCULLOCH · Variance Components
SEARLE and WILLETT · Matrix Algebra for Applied Economics
SEBER and LEE · Linear Regression Analysis, *Second Edition*
SEBER · Multivariate Observations
SEBER and WILD · Nonlinear Regression
SENNOTT · Stochastic Dynamic Programming and the Control of Queueing Systems
*SERFLING · Approximation Theorems of Mathematical Statistics
SHAFER and VOVK · Probability and Finance: It's Only a Game!
SMALL and McLEISH · Hilbert Space Methods in Probability and Statistical Inference
SRIVASTAVA · Methods of Multivariate Statistics
STAPLETON · Linear Statistical Models
STAUDTE and SHEATHER · Robust Estimation and Testing
STOYAN, KENDALL, and MECKE · Stochastic Geometry and Its Applications, *Second Edition*
STOYAN and STOYAN · Fractals, Random Shapes and Point Fields: Methods of Geometrical Statistics
STYAN · The Collected Papers of T. W. Anderson: 1943–1985
SUTTON, ABRAMS, JONES, SHELDON, and SONG · Methods for Meta-Analysis in Medical Research
TANAKA · Time Series Analysis: Nonstationary and Noninvertible Distribution Theory
THOMPSON · Empirical Model Building
THOMPSON · Sampling, *Second Edition*
THOMPSON · Simulation: A Modeler's Approach
THOMPSON and SEBER · Adaptive Sampling
THOMPSON, WILLIAMS, and FINDLAY · Models for Investors in Real World Markets
TIAO, BISGAARD, HILL, PEÑA, and STIGLER (editors) · Box on Quality and Discovery: with Design, Control, and Robustness
TIERNEY · LISP-STAT: An Object-Oriented Environment for Statistical Computing and Dynamic Graphics
TSAY · Analysis of Financial Time Series
UPTON and FINGLETON · Spatial Data Analysis by Example, Volume II: Categorical and Directional Data
VAN BELLE · Statistical Rules of Thumb
VESTRUP · The Theory of Measures and Integration
VIDAKOVIC · Statistical Modeling by Wavelets

*Now available in a lower priced paperback edition in the Wiley Classics Library.

WEISBERG · Applied Linear Regression, *Second Edition*
WELSH · Aspects of Statistical Inference
WESTFALL and YOUNG · Resampling-Based Multiple Testing: Examples and Methods for *p*-Value Adjustment
WHITTAKER · Graphical Models in Applied Multivariate Statistics
WINKER · Optimization Heuristics in Economics: Applications of Threshold Accepting
WONNACOTT and WONNACOTT · Econometrics, *Second Edition*
WOODING · Planning Pharmaceutical Clinical Trials: Basic Statistical Principles
WOOLSON and CLARKE · Statistical Methods for the Analysis of Biomedical Data, *Second Edition*
WU and HAMADA · Experiments: Planning, Analysis, and Parameter Design Optimization
YANG · The Construction Theory of Denumerable Markov Processes
*ZELLNER · An Introduction to Bayesian Inference in Econometrics
ZHOU, OBUCHOWSKI, and McCLISH · Statistical Methods in Diagnostic Medicine

*Now available in a lower priced paperback edition in the Wiley Classics Library.